How the World's Religions Are Responding to Climate Change

A growing chorus of voices has suggested that the world's religions may become critical actors as the climate crisis unfolds, particularly in light of international paralysis on the issue. In recent years, many faiths have begun to address climate change and its consequences for human societies, especially the world's poor. This is the first volume to use social science to examine how religions are helping to address one of the most significant and far-reaching challenges of our time.

While there is a growing literature in theology and ethics about climate change and religion, little research has been previously published about the ways in which religious institutions, groups and individuals are responding to the problem of climate change. Seventeen research-driven chapters are written by sociologists, anthropologists, geographers and other social scientists. This book explores what effects religions are having, what barriers they are running into or creating, and what this means for the global struggle to address climate change.

Robin Globus Veldman is a Ph.D. candidate in the Department of Religion at the University of Florida, USA.

Andrew Szasz is Professor of Environmental Studies at the University of California, Santa Cruz, USA.

Randolph Haluza-DeLay is an associate professor of sociology at The King's University College in Alberta, Canada.

Routledge Advances in Climate Change Research

Local Climate Change and Society
Edited by M. A. Mohamed Salih

Water and Climate Change in Africa
Challenges and community initiatives in Durban, Maputo and Nairobi
Edited by Patricia E. Perkins

Post-2020 Climate Change Regime Formation
Edited by Suh-Yong Chung

How the World's Religions Are Responding to Climate Change
Social scientific investigations
Edited by Robin Globus Veldman, Andrew Szasz and Randolph Haluza-DeLay

How the World's Religions Are Responding to Climate Change

Social scientific investigations

Robin Globus Veldman,
Andrew Szasz and
Randolph Haluza-DeLay

Routledge
Taylor & Francis Group
LONDON AND NEW YORK

First published 2014
by Routledge
2 Park Square, Milton Park, Abingdon, Oxon OX14 4RN

Simultaneously published in the USA and Canada
by Routledge
711 Third Avenue, New York, NY 10017

Routledge is an imprint of the Taylor & Francis Group, an informa business

British Library Cataloguing in Publication Data
A catalogue record for this book is available from the British Library

Library of Congress Cataloging-in-Publication Data
Veldman, Robin Globus.
How the world's religions are responding to climate change : social scientific investigations / Robin Globus Veldman, Andrew Szasz and Randolph Haluza-DeLay.—1 [edition].
pages cm
1. Ecology—Religious aspects. 2. Ecotheology. 3. Climatic changes. I. Title.
BL65.E36V45 2013
201'.77—dc23
2013012410

ISBN: 978–0–415–64034–3 (hbk)
ISBN: 978–0–203–08279–9 (ebk)

Typeset in Times New Roman
by Swales & Willis Ltd, Exeter, Devon

Printed and bound in the United States of America by Publishers Graphics, LLC on sustainably sourced paper.

Contents

Illustrations

Figures

Tables

Contributors

Michael Agliardo, SJ, is an assistant professor of sociology at Loyola University Chicago. One focus of his research is religion in public life and religious environmentalism.

Ulil Amri, MA, works as a research anthropologist at the Research Center for Regional Resources, Indonesian Institute of Sciences in Jakarta, where he studies social-environmental movements, natural resource management, and the anthropology of development.

Samuel Awuah-Nyamekye is a senior lecturer in the Department of Religion and Human Values at the University of Cape Coast in Ghana, and completing a Ph.D. at the University of Leeds, U.K.

Matt Branch is a Ph.D. candidate in geography and human dimensions of natural resources and environment at Pennsylvania State University, where his research examines environmental policy in Bhutan.

Carolyn Peach Brown, Ph.D., is Assistant Professor and Director of Environmental Studies at UPEI. Prior to her doctoral studies she worked for ten years in community development in Africa.

Douglas R. Brown, Ph.D., is Director of Agriculture and Food Security at World Vision International and responsible for the promotion of sustainable, productive and resilient agricultural livelihoods.

Sadia Chérif holds a Ph.D. in sociology from the University of Bouaké, where he lectures and researches the design and implementation of effective interventions that promote sustainable agriculture development.

Georgina Drew, Ph.D., is a lecturer in Anthropology and Development Studies at the University of Adelaide in Adelaide, Australia. Her work examines religion, gender, development and environment in several locations in South Asia.

Joy H. Greenberg has a Ph.D. in mythological studies from Pacifica Graduate Institute and an M.F.A. in creative writing from California State University, Chico.

Randolph Haluza-DeLay is an associate professor of sociology at The King's University College in Alberta, Canada.

Julie Hermesse is a postdoctoral researcher at the Fonds National de la Recherche Scientifique, Belgium, and a member of the Laboratory for Prospective Anthropology of the University of Louvain.

Mike Hulme is Professor of Climate Change and Culture in the Department of Geography at King's College London, U.K., and was the Founding Director of the Tyndall Centre for Climate Change Research.

Noor Johnson is a postdoctoral fellow at Brown University's Environmental Change Initiative. Her research focuses on environmental change, bureaucracy, and local institutions in the Arctic.

Futaba Kazama is a professor at the International Research Center for River Basin Environment at the University of Yamanashi in Japan, interested in studying relationships between natural water systems and social life.

Guillermo Kerber holds a Ph.D. in sciences of religion (UMESP, Brazil). Presently, he coordinates the work of the World Council of Churches on climate change. A citizen of Uruguay, he is based in Geneva, Switzerland.

Mishka Lysack is an associate professor in social work, University of Calgary, focusing on renewable energy, the green economy, environmental advocacy, and faith and ethics.

Sujata Manandhar is a doctoral student studying integrated river basin management at the University of Yamanashi, Japan. She is interested in climate change adaptation in water and agriculture.

Maria Nita, Ph.D., is a Religious Studies and Philosophy lecturer at Weston College, UK. Maria's academic interests include interdisciplinary approaches to identity, heritage and communication.

Vishnu Prasad Pandey is a postdoctoral researcher at the International Research Center for River Basin Environment, University of Yamanashi in Japan, and studies water resources management, including climate change vulnerability and adaptation.

James Pender, M.Sc. in resource management, Edinburgh University, is academically qualified and experienced in both nature conservation and social development, with specializations in climate change adaptation, ornithology and "creation care."

Andreana Reale is currently exploring ministry in the Baptist Church, and is passionate about church being a home for rich and poor alike. This work was done while at the Centre for Risk and Community Safety at RMIT University, Melbourne.

Paul Sarfo-Mensah holds a Ph.D. in Natural Resources Management and is a senior research fellow/natural resources management expert in the Bureau of Integrated Rural Development at Kwame Nkrumah University of Science and Technology in Kumasi, Ghana.

Dietrich Schmidt-Vogt is a geographer and professor at the Kunming Institute of Botany of the Chinese Academy of Sciences, specializing in the landscape ecology of Asia.

Christopher Shore, M.B.A., is Director, Natural Environment and Climate Issues at World Vision International, and responsible for natural resource management and climate change mitigation and adaptation.

Andrew Szasz is Professor of Environmental Studies at the University of California, Santa Cruz.

Patricia K. Townsend is affiliated with the University at Buffalo Department of Anthropology and is a member of the board of New York Interfaith Power and Light.

Robin Globus Veldman is a Ph.D. candidate in the Department of Religion at the University of Florida. She researches American conservative Christians' views of climate change.

Bernard Daley Zaleha, J.D., is an environmental litigator and a Ph.D. student at the University of California, Santa Cruz, who studies religion's impact on environmental behavior.

Foreword

Mike Hulme

The rhythms of the sky have long been companions of human thought and ritual. From the frigid north to the torrid tropics they have induced wonder and fear, whilst also offering comfort and assurance. Alongside intense yet predictable diurnal and seasonal weather cycles comes the unreliable and unfeeling performance of climate from year to year and from generation to generation. A mere dry season is not a drought; and winter is not an ice age. It is little wonder then that human anxieties and hopes—and hence many of our spiritual longings—have been bound up with the skies. Religions have found many ways to make sense of these cruel fates, acknowledging our dependence on powers beyond our control and giving thanks for mercies and blessings received. Climate and religion have a long history of interdependency.

And so it is somewhat surprising that there has been such a willful exclusion of religions, and the religious, in the forging of modernist accounts of climate change and its multiple causes. I am thinking here especially of the United Nations expert body—the Intergovernmental Panel on Climate Change (IPCC)—which formed in 1988 (although the erasure in Western science of religious sentiment from accounts of climate and what it means to human beings has a longer history). The IPCC has managed successfully to prise the idea of climate and its capricious behavior away from its deep cultural anchors.

In recent years this "purification" (*pace* Latour) of climate has begun to change. As the scientized account of climate change and human agency has spread around the world it has run up against deeper narratives and stronger resistance than many scientists—and politicians and campaigners—might have thought. Climate change turns out to be not just one thing, a thing defined and simulated inside Earth System models. The meanings of climate change are multiple, and the ways in which human beings express, represent, engage with and resist climate change are too numerous to be controlled by a single dominating perspective, namely science. And many of these reactions have their roots in ancient religious doctrines, traditions and practices.

The scholarly community, dominated as it has been by the secularized cultures of Western Europe and some Anglophone countries, has been slow to awaken to this reality. Europe is becoming a post-secular society, in which faith-based reasoning is to be re-accommodated, even while much of the rest of the world never rejected it. Climate change therefore needs to be studied not just by meteorologists, ecologists and economists, but by sociologists, anthropologists and philosophers, and by theologians. The IPCC has offered a one-dimensional account of climate change. As many commentators have remarked, cultural innovation and effective climate policy need to tap into intrinsic, deeply held values and motives. One such commentator, the Alliance of Religions and Conservation, observed in 2007:

> The emphasis on consumption, economics and policy usually fails to engage people at any deep level because it does not address the narrative, the mythological, the metaphorical or the existence of memories of past disasters and the way out. The faiths are the holders of these areas and without them, policies will have very few real roots.[1]

It is in this context that the editors of this new volume—*How the World's Religions Are Responding to Climate Change: Social Scientific Investigations*—have done a great service. As social scientists they have taken these two things—climate change and religion—seriously and through this collection of studies brought into view the many different ways in which they interact in different settings. This collection of empirical studies encompasses Bangladesh, Bhutan, Canada, Côte d'Ivoire, Ghana, Guatemala, India, Indonesia, Nepal, Solomon Islands, the U.K. and the U.S.A.

Why then does religion matter for understanding climate change? The various authors emphasize the importance of religions in giving substance and power to social norms, social capital, lifestyle patterns, creation care and human dignity. And then there is simply the brute reality of the political power of religious institutions in the world! Arresting climate change is not just beyond science—it is also beyond the state. Religious movements and institutions, as has been found also with businesses, cities and NGOs, have the mobilizing power to enlist and de-list multitudes of citizens in causes of . . . of what? Well, whatever causes are invigorated by the idea of anthropogenic climate change; as this volume shows, there is no single authorized response to climate change.

This diversity is well captured in the contributions to this book, which embrace Buddhism, several variants of Christianity, Hinduism and Islam, alongside traditional indigenous beliefs and practices. Captured in these case studies are the religious filters that sift through scientific evidence and rhetoric to yield resonant interpretations and perceptions of climate change both consistent and not consistent with the scientific narratives. Captured here are the religious social movements that inspire transforming behaviors and political activism, or that provide barriers to such action. Captured here are the different institutional forms of religious practices as they engage and support the environmental poor. And captured here too are some of the multiple forms of public reasoning which yield epistemological and ethical underpinnings for all the above. Far from inspiring a replicable or universal response, the world's religions are engaging the idea of climate change for a diversity of reasons and in a diversity of ways.

Science is never quite enough, either in its promise of solid and reliable knowledge or in its efforts to animate social movements and resolve political contestation. Reading climate change and human agency through the eyes of the world's religions offers new insights and new inspirations about what it means to be human in an age of climate change. And this volume offers both.

Norwich, U.K.
November 2012

Note

1 Alliance of Religions and Conservation. 2007. "UN and ARC Launch Programme with Faiths on Climate Change." December 7. Available at http://www.arcworld.org/news.asp?pageID=207.

Acknowledgments

Chapter 1 has been revised and is reprinted with permission from Veldman, Robin Globus, Andrew Szasz and Randolph Haluza-DeLay. "Introduction: Climate Change and Religion – A Review of Existing Research." *Journal for the Study of Religion, Nature and Culture* 6.3, 255–275. © Equinox Publishing Ltd 2012

Chapter 2 is reprinted with permission from Drew, Georgina. "A Retreating Goddess? Conflicting Perceptions of Ecological Change near the Gangotri-Gaumukh Glacier." *Journal for the Study of Religion, Nature and Culture* 6.3, 344–362. © Equinox Publishing Ltd 2012

Chapter 16 is reprinted with permission from Johnson, Noor. "'Healing the Land' in the Canadian Arctic: Evangelism, Knowledge, and Environmental Change." *Journal for the Study of Religion, Nature and Culture* 6.3, 300-318. © Equinox Publishing Ltd 2012

Introduction

1 Social science, religions, and climate change

Robin Globus Veldman, Andrew Szasz and Randolph Haluza-DeLay

A growing chorus of voices has suggested that the world's religions may, individually and collectively, become critical actors as the climate crisis unfolds. Religions affect societies at every level, from the individual to the transnational, as worldviews and as institutions, as conservators of traditions and as resources for change. The social roles of religions and spiritualities vary tremendously among the peoples of the world, but what they say and do about climate change—whether they encourage attention to the issue or discourage it; whether they help their adherents recognize and cope with the challenge or persuade them to ignore or deny it—could decisively impact how societies all over the world respond to it. This volume explores the extent to which the world's religions are living up to their potential, what effects they are having, what barriers they are running into (or creating), and what this means for the global struggle to address climate change.

To say that a great deal has been written about climate change as of the first decade of the twenty-first century would be an understatement. In books, journal articles, policy planning documents, and news stories over the last quarter-century, a chorus of voices has spoken of the urgent challenge we face. Concentrations of carbon dioxide and other greenhouse gases (GHGs) in the atmosphere continue to grow, and climate scientists are nearly unanimous in warning of even greater, potentially catastrophic, consequences if nothing—or not enough—is done soon (e.g., Barnosky et al. 2012; Hansen 2012. Durack, Wijffels and Matear 2012 and Oppenheimer et al. 2007 also suggest that risks may be greater than the Intergovernmental Panel on Climate Change (IPCC) reports suggest).[1] At the time of writing, impacts have already been observed around the world in events such as heat waves, persistent drought, flooding, receding glaciers, and the bleaching of coral reefs as the oceans warm. Climate change has begun to affect human societies as well. Reduced food production, increased disaster vulnerability, and forced migration are just a few of the changes with which societies around the world are now being forced to cope. And, in a cruel irony of industrialization, some of the harshest effects are striking (or predicted to strike) regions whose inhabitants have contributed least to the problem. Amidst such conditions, religions find themselves amongst the many institutions in societies around the world that must become part of the response to climate change, as people work to accommodate a warmer and less predictable world.

What is "religion"?

Defining religion is notoriously difficult. Not all things recognized as "religion" have divine beings or supernatural aspects, nor is non-material or non-evidentiary faith an adequate defining characteristic.[2] Most definitions encompass a number of integrated dimensions, such as sacred narratives or texts, ethics, ritual, doctrine, collective and personal practices, and social institutions. Despite the common tendency among English speakers to equate the term with belief, then, religion is expressed in both practice and material culture. As for its role in society, early social scientists focused on religion as having a *social* function, such as providing shared values, producing social cohesion, ritualizing life stages, or legitimating social order. Among other effects, this approach conceptualized different faiths as functional alternatives, despite clearly different societal positions, especially in complex multi-faith and multi-cultural societies. Other scholars have conceptualized religion as a system of meaning that can provide answers to central questions about behavior, social order, and purpose. From this perspective, religious cosmologies make substantive epistemic and ontological claims about the nature of reality and the sources of knowledge, such as whether the world is "created" or purposeful or has aspects that are not verifiable by science alone. These meaning systems are not only ideational, but may also be embodied through social practices such as prayer, ritual, good works, and fidelity to the church, mosque, temple, or tribe. Another important characteristic of religions when they manifest themselves as institutions, congregations, temples, mosques, and faith-based organizations is that they constitute collective actors. Religions can therefore be understood as operating at multiple scales: individual and collective; local, regional, and transnational. All of these dimensions potentially affect how a religion engages with climate change. For example, the worldview it provides could shape how the individual or group interprets changes in the weather, and the resources to which it provides access could influence members' ability to adapt.

Importantly, religion and religions have different roles and effects in different societies. This variability cannot be stressed enough. One need only think about the many types of Islam and their variable roles at play in Malaysia, Pakistan, northern Mali, the United States, or France. Moreover, religions vary greatly in terms of their internal organization—not all faiths have clear structures or chains of authority or organization. Doctrinality also varies—not all faiths have clearly defined beliefs, or some faiths often defined as "religions" (such as Taoism) do not have recognizable conceptions of divine beings or cosmologies. Finally, there are widely differing degrees of rationality involved in religion. It is worth noting that how a religion engages with climate change in one region or country or even temple does not necessarily predict how it operates in other locales. For example, both Pope John Paul II and Pope Benedict XVI have repeatedly advocated for climate justice, yet Watson and Kochore (2012) describe how in northern Kenya the Catholic Church has contributed to increased sedentarization, which they argue reduces local people's ability to adapt to climate change. All this variability means that "religion" must be studied in context, especially in regard to the phenomena associated with global climate change.

Religions' potential to motivate concern and activism

Observers have suggested a number of reasons why the world's religions, once engaged, could be an important part of the societal response to climate change (see especially Gardner 2003; Gerten and Bergmann 2012; Kearns and Keller 2007; Posas 2007; Reder 2012; Tucker and Grim 2001; Wolf and Gjerris 2009).

First, religions may be able to encourage a response to climate change via their influence on believers' worldviews or cosmologies. These narratives provide meaning and purpose, and explain the individual's and group's place in the world, thereby creating the context for ethical deliberation. Such narratives also establish what is sacred, which in turn is often set aside to be preserved, respected and protected. Considerable research shows that religions shape adherents' perceptions of and behavior toward the natural environment (Sponsel 2005), so the idea that this influence extends to climate change seems plausible.

Second, religious leaders are able to engage a broad audience, many of whom accept and respect their moral authority and leadership. When one considers the global presence of religious groups, the number of people who could be engaged is impressively large: some estimate that up to 84 percent of the world population identifies with one of the major world religions.[3] Organized religions' influence may have declined in some parts of the world, such as Western Europe, but elsewhere it continues to have profound influence. As a pervasive and powerful force in the lives of the majority of the world's people, religions are, at least in theory, well positioned to mobilize millions of people on the issue of climate change.[4]

Third, religions have significant institutional and economic resources at their disposal. They run schools and social agencies and are, collectively, the third largest category of investors in the world.[5] In some countries they also wield considerable political power. At the institutional level this influence enables them to reach a broad audience not only through their own networks but also via interfaith coalitions such as the Parliament of the World's Religions, ecumenical groups such as the World Council of Churches, or faith-based transnational organizations such as World Vision. The international recognition and respect of leaders such as the Dalai Lama and the Pope are also symbolic resources that can be deployed to some effect on issues of their choosing.

Finally, religions have the potential to provide connectivity (e.g., in the form of social capital) that fosters the achievement of collective goals (e.g., Smidt 2003; Swart 2006). Church affiliation is one of the most common forms of association, reaching more people in some countries (including the United States) than political affiliation or other voluntary associations (Smidt 2003). Local faith communities are often among the first to respond in disaster situations (Wisner 2010). Worldwide, faith groups are hubs for the distribution of goods, services, and even emotional support to those who are at risk; this plays a role both locally and internationally via missionary and charity outreach activities (Clarke 2006). Such fostering of trust and strengthening of social ties positions religious association as potentially effective agents in response to climate change.

Impediments to religious engagement

While there are good reasons to hypothesize that religions can play a key role in the global response to climate change, it is also important to note that there are substantial obstacles to such action. In a study on environmental education in North American churches, Haluza-DeLay (2008) identifies four classes of barriers to environmental engagement that apply equally to climate change. *Paradigmatic* barriers are theological beliefs or worldviews that disable environmental concern. This includes having otherworldly goals such as salvation, rather than this-worldly ones such as the maintenance of the biosphere. Accordingly, leaders may encourage adherents to believe that the weather or climate is not directly influenced by people, or that a divine force is in control of the cosmos and that humankind is not so powerful as God or nature (Donner 2011; see also in this volume Chapter 2 by Drew, Chapter 8 by Hermesse, Chapter 9 by Chérif and Greenberg, Chapter 10 by Sarfo-Mensah and Awuah-Nyamekye, and Chapter 16 by Johnson).

Barriers of *applicability* include debate about the appropriate level of attention to give environmental concerns—or whether to give them any attention at all. Often, environmental concerns compete with other foci, like hunger, poverty, housing, or evangelism. Lack of *social critique* can be a barrier as well. Not recognizing the societal and cultural factors that affect the human–earth relationship can prevent religious groups from acting at all, or from developing an effective campaign that recognizes and addresses the roots of the problem. Lack of social critique is sometimes coupled with an overemphasis on individual or moralistic solutions (see Chapter 11 by Lysack in this volume). Wilkinson (2010) suggests that this is one reason why American evangelicals have been hesitant to engage (see also Chapter 14 by Zaleha and Szasz in this volume). Finally, there can be lack of *conviction*, because of attachment to one's current lifestyle, or lack of knowledge or motivation. These obstacles, especially the latter two, are not unique to religious adherents, but can take unique form within religious communities. For example, Chapter 12 by Michael Agliardo describes how some Catholic bishops in the U.S. are hesitant to take action on climate change because they worry that tackling such a controversial issue would squander their political capital.

While religions have great potential, religious discourses and actors face a daunting challenge. No matter how inspiring their creeds and persuasive their leaders, these may not be sufficient to counter well-entrenched habits and social structures that generate greenhouse gases, or the skepticism and resistance that climate skeptics have encouraged (Brulle, Carmichael and Jenkins 2012; Lahsen 2008; McCright and Dunlap 2000). There are also definite limits to their power in global civil society. Indeed, religions' institutional, economic, and political resources may be inadequate to overcome those invested in maintaining the carbon-based economy. Lastly, it is worth considering that while some religions are mobilizing in response to climate change—either to promote mitigation or to facilitate adaptation—others are directly or indirectly obstructive. Both helpful and obstructive responses may be actively present within the same institutional religions (as with other controversial issues, such as women's rights, scientific knowledge, tradition versus innovation, adaptation to new cultural phenomena, and so on), sparking conflicts that reduce their adherents' ability to adapt to climate change. A simple, unidirectional relationship between religions and climate change should therefore not be presumed.

What religions are doing about climate change

In spite of such impediments, religious groups have been commenting on climate change since the late 1980s, with activity accelerating in recent years. Many Christian denominations have issued unequivocal, powerful statements that declare climate change to be an urgent moral matter. Such statements generally argue that God wants people to care for and be good stewards of Creation. In some cases, the statements also contend that, because climate change is likely to hurt the world's poor first and worst, addressing it is an issue of social justice. For example, in the Roman Catholic Church, both Pope John Paul II and Pope Benedict XVI have advocated for "climate justice." Benedict did so repeatedly and with increasing passion in the years leading up to his retirement. Under the leadership of Ecumenical Patriarch Bartholomew, the Eastern Orthodox Church has also been quite active. Sometimes dubbed the "green patriarch," Patriarch Bartholomew has long been an outspoken advocate for tackling environmental degradation, including climate change (Makrides 2008).

In the United States, the national bishops conference of the Roman Catholic Church, all the "mainline" Protestant denominations, and other more liberal churches have all issued statements and developed educational programs regarding climate change. More surprisingly, some fairly conservative religions, such as the Seventh-Day Adventist Church, have also issued climate change statements, and one sees climate initiatives in even the conservative faiths, like the 2006 Evangelical Climate Initiative and the 2008 Southern Baptist Declaration on the Environment and Climate Change (see Kearns 2011; and Chapter 14 by Zaleha and Szasz, this volume). We should note, though, that such grand statements may not mean much if they do not trigger real actions. Numerous examples of particular actions are available (Veldman, Szasz and Haluza-DeLay 2012), but it appears to be generally true that, to date, denominational statements and education programs have had difficulty penetrating to the grassroots.

Some religious–secular partnerships are also in evidence. For example, Christian development organizations, like World Vision in Canada, have become members of non-religious coalitions like the Climate Action Network. At the "Many Heavens, One Earth" celebration convened by Alliance of Religions and Conservation (ARC), the leaders of nine world religions announced commitments to environmental action, to be funded by the United Nations Development Program (UNDP) in partnership with secular bodies such as the World Wildlife Fund and the World Bank.[6] ARC has also partnered with the UNDP to help religious leaders of 11 major faiths develop seven-year plans of action to address climate change. In 2000, ARC further developed a "Climate Change Partnership Initiative" that resulted in 15 countries officially agreeing to work with religions on climate change.[7]

Interfaith and ecumenical coalitions have been important sites for religious engagement with climate change. For example, before the Conference of the Parties (COP) meeting in 2011, 130 African faith leaders "of many creeds" released a joint statement declaring: "Only a profoundly renewed moral vision is likely to end 20 years of effective impasse in the climate negotiations."[8] Groups like the World Council of Churches (WCC) have increasingly focused on climate change as a paramount issue (see Chapter 18 by Kerber, this volume).

Websites that list official religious statements on climate change (such as the Yale Forum on Religion and Ecology), web searches, and searches for books from popular presses suggest that Christian faiths have been more active vis-à-vis the climate crisis than have other religious traditions. The apparently greater engagement of Christian groups could have several explanations. In countries such as the U.K. and the U.S., Christian organizations receive a disproportionate share of support from donors and governments, allowing them to become more influential relative to other faiths in the international development scene. Also, many Christian organizations are headquartered in developed countries, where they may have more wealthy, educated constituents who are informed and concerned about issues like climate change. Christian denominations may also be more visible because they are more hierarchically organized than some other religious traditions, such as Hinduism and Islam, which do not have centralized authorities that represent the faith globally (Clarke 2006). It is also true that there are greater resources available for scholarship in the Global North, where the publishing culture also facilitates its visibility. Finally, we are keenly aware that our observations may be skewed toward the North American and Anglophone context as a result of our (the co-editors') physical locations and scholarly networks.

That said, a number of non-Christian religious traditions have released statements on climate change, although it is not clear what the impact of such work has been, nor how representative these statements are of the broader tradition's concerns.[9] Many of these statements can be accessed at the Forum on Religion and Ecology's webpage on climate change.[10] Millais (2006) also contains statements about climate change from 16 religious groups (seven of which are not Christian) that are active in Australia. The Islamic Foundation for Ecology and Environmental Sciences (IFEES) in Britain has worked on climate change, and the South Woodford Islamic center has built a carbon-neutral place of worship (Gilliat-Ray and Bryant 2011). The faith-based relief and development organizations Islamic Relief Worldwide and Muslim Aid have also incorporated climate change into their programming.[11] Also within Islam, as part of the ARC–UNDP partnership mentioned above, the Grand Mufti of Egypt, an influential Islamic leader, pledged to a seven-year action plan that included constructing green mosques and greening pilgrimage cities. At the same meeting, American and Israeli Jews pledged to reduce their carbon footprint by limiting meat consumption, and Shintos in Japan and Taoists in China also pledged to convert thousands of temples to green energy ("World Religious Leaders Promise Climate Action" 2009). Finally, in the U.K., the non-profit Hindu Sevika Samiti participated in a campaign initiated in 2012 known as "Climate Week" by promoting lifestyle changes that would reduce participants' carbon footprints, as well as by holding talks about Hindu Dharma and the environment.[12] Numerous similar examples can be found scattered across the world. Again, we highlight uncertainty about the cumulative effects of religious engagement. Further research is needed to establish whether the examples given above are indicative of a growing trend or of relatively thin engagement on the part of world religions.

Types of scholarship on religion and climate change

To date much of the literature on religion and climate change is *theological* and *prescriptive*.[13] Such writing speaks from within a particular religious worldview, and argues that adherents should be concerned about climate change by appealing to religious texts and traditions. Comparatively little of the existing literature on the subject can be characterized as empirical or social scientific. Consequently, there are many significant gaps in the existing literature. Although writing on religion and climate is growing, it deserves more attention from social scientists.

That said, there are several patterns in the existing literature on religions and climate change. The first pattern is that the largest body of scholarship that does exist is about one tradition (Christianity) and one country (the United States). The bulk of that scholarship consists of work that links theologies and ethics of particular faith traditions to the climate crisis or global environmental change (e.g., McFague 2008; Northcott 2007; Primavesi 2009; Skrimshire 2010; Xia and Schönfeld 2011). A second and somewhat overlapping body of scholarship asserts on a more general level that religious participation will be productive in the fight to mobilize the world to combat climate change (Gardner 2003; Posas 2007; Schipper 2010; Tucker and Grim 2001; Wolf and Gjerris 2009). A third body of literature is directed toward communicating the issue of climate change to adherents of a particular faith. This group includes numerous works directed toward Christian audiences (e.g., Atkinson 2008; Hayhoe and Farley 2009; Merritt 2010), as well as a smattering of works directed toward those of other faiths (such as Abdul-Matin and Ellison 2010; Stanley, Loy and Dorje 2009). Fourth, a number of polls and surveys relating religious beliefs and climate change attitudes have been published, many of which focus on the attitudes of American evangelicals (e.g., Barna Group 2008; Pew 2006, 2009; Public Religion Research Institute 2011).

The emergence of a fifth body of literature is to be welcomed. Along with the work you have in your hands, other empirical social scientific investigations are beginning to appear. Gerten and Bergmann (2012) have included some empirical social scientific contributions about religion and climate change in their edited volume. Both Leduc (2010) and Wilkinson (2012) are thoroughly anthropological studies of the intersection of religion and climate change, as are several contributions in the edited collection by Crate and Nuttall (2009). There are sprinklings of articles in specialized scholarly journals (e.g., Roncoli, Ingram and Kirshen 2002; Salick, Byg and Bauer 2012). A collection of five research papers (two of which are included in this volume) also formed a special issue of the *Journal for the Study of Religion, Nature and Culture* (Veldman, Szasz and Haluza-DeLay 2012).

In spite of this recent uptick in social scientific studies (for recent overviews, see Veldman, Szasz and Haluza-DeLay 2012; Haluza-DeLay 2012; Hulme 2009; Kearns 2011, 2012), no comprehensive review of global religious responses to climate change exists at the time of this writing. Because of the small size of the literature and the substantial gaps in it, whatever observations we offer here should be regarded as tentative. Much more research will have to be done on the effectiveness of religious activism on climate change, the effectiveness of high-level institutional statements and coalition organizing, and the degree to which the grassroots is responding to initiatives undertaken by religious leadership.

This book: origins, organization, content

This book began as a call for papers for a special issue of the journal noted above. Because we wanted the analyses to be solidly social scientific, drawing on the best available data and analysis, we explicitly stated that we were not seeking proposals that adopted an ethical or theological approach (surprisingly, about a third of proposals adopted this approach anyway). Instead, we sought to assemble evidence of what religions were actually doing, advocating, and promising. We also wanted to range across *all* the world's religions—from the ones often labeled major or global religions to the ones labeled local, folk, minor, or indigenous. Though some religions have more adherents or international recognition, we felt that a more inclusive model would furnish greater insight into the nature and effectiveness of religious activity. Finally, we also hoped to present a geographically diverse picture.

Though we received nearly one hundred submissions from around the world, they were somewhat uneven. We received many submissions focusing on North America, but most of the Asian landmass, including both Russia and China, most of Western Europe, all of Eastern Europe, Scandinavia and Latin America were missing. The relative abundance of submitted research about various religions was also uneven. Relative to their demographic importance, Islam, Hinduism, Buddhism, and even Roman Catholicism were underrepresented. The contributions were also heavily skewed toward Protestant Christian faiths.

There are several possible causes for the gaps in our collection. The gaps may represent actual differences in how much or how little various religions have engaged (or not) with the issue of climate change. If some religions, in some societies, have not yet addressed the climate change issue, it is unlikely that any social scientist would have decided to conduct research on this missing engagement (given the notorious difficulty of studying non-events). In other places religions may have started to engage, but that engagement has yet to attract academic interest. Finally it is also possible that our efforts to solicit contributions from scholars in other parts of the world simply failed to reach some who are, indeed, already studying and writing about the ways that, in their societies, faith communities are engaging with the climate issue.

While the unevenness points out where more research is needed, the case studies we have assembled nevertheless begin to demonstrate the complex and nuanced ways in which climate change is affecting people around the world, and some of the roles that religions—as faiths, values, practices, and institutions—are playing in this interaction.

How the book is organized

Climate change is a global phenomenon with divergent regional or local manifestations. So too are religions. The chapters in this volume exhibit considerable variety in terms of geography, methodology, and religious tradition. The religious practices, beliefs, and institutions discussed range from being scientifically and politically engaged to being in complete denial of climate change. The scale of the religious phenomena ranges from the individual, to local community, to congregation or temple, to institutions, to national policy, and to global and transhistorical faith traditions.

Geographically, the contributions begin in the Global South—from Asia (with contributions from Nepal, Indonesia, Bangladesh, India, and Bhutan) across the Indian Ocean to Africa (with contributions from Ivory Coast and Ghana), across the Atlantic Ocean to Central America (Guatemala) and partway across the Pacific to Solomon Islands. In the Global North, five contributions from North America span all the way to the Inuit in the Arctic, while a contribution from the United Kingdom rounds out the geographical diversity. The final two chapters escape the container of the nation-state to examine an international faith organization and transnational organizing.

The decision to organize the book according to the location of the case studies in the Global North or South is deliberately guided by a postcolonial, global justice sensibility. The historical industrialization and contemporary patterns of consumption and social organization of Global North countries (including geographically Southern countries such as Australia that adopted the industrial model early) have driven anthropogenic climate change. By contrast, countries from the Global South have played a lesser historical role in emissions. In exchange, not only have they been disadvantaged by colonization and extraction of their raw resources for the world economic system, but many of them also will bear the brunt of climatic changes. Examples include Bangladesh, the low-lying islands of the Pacific, and marginal grazing land in sub-Saharan Africa.

While we have chosen to organize the book in this way, the story of religion and climate change is not really about geography. Many of the chapters have an interfaith component, either in actual religious interaction (as in the case of Mishka Lysack's analysis of interfaith organizing on climate change across Canada, or Maria Nita's study of Muslims and Christian climate change activists in the United Kingdom) or in comparative analysis (like Sadia Chérif and Joy H. Greenberg's description of Christian and indigenous African faiths in the Ivory Coast or the comparison of indigenous, Christian and Muslim faiths in Ghana by Paul Sarfo-Mensah and Samuel Awuah-Nyamekye). A second important observation is that religions are located differently within each country. For instance, Ulil Amri's contribution on Muslim organizational efforts to educate about climate change in Indonesia occurs within a context where that religion dominates. On the other hand, James Pender's explication of the Christian efforts in Bangladesh occurs in a context where Christians constitute only about 2 percent of the population (Muslims are 88 percent; the rest are Hindu). Across the 18 chapters that follow, Christianity features in 12 chapters, Islam in three, indigenous faiths in five contributions, Buddhism in two and Hinduism in one. This count itself means little. That Christianity is so prominent is partially because of the effort that Christians have put into climate change, but also an artifact of many factors described already or yet to be explained. Yet all the variability in the ways that religious institutions and people engage with climate change can only be appreciated by reading all the chapters. In the final chapter we return to synthesize these studies in order to draw more favorably on their variety.

As the table of contents shows, the nation-state is still the central unit for analysis, even of border-crossing religions. Still, we should keep in mind that, while religious expressions take particular shapes in part because of their immediate environment, religious practice is affected by factors that scale beyond the local. Ideas are but one of these transnational factors, as Noor Johnson shows in her analysis of Inuit in the Canadian Arctic who are affected by the confluence of local ecology and culture, transnational notions of indigeneity, and an evangelical Christian movement that originated in the South Pacific. The chapters in this volume also demonstrate the error of a simplistically functionalist understanding of religion, for religions clearly do not play the same role in each society. On the contrary, each case study bears witness to the necessity of attending to differences in the organizational structure, resources, and societal position of different religious traditions.

A brief description of the sections and chapters

Religions and climate change in the Global South

While the narrative of secularization still has currency in the Northern regions of the globe, religion remains a tremendously vital force in the Global South. Missionary faiths such as Buddhism, Islam, and Christianity have spread widely. At the same time, many indigenous religions are still vibrant, sometimes hybridized with global religions. The nine chapters in this part show religions' ongoing vitality and their sometimes troubling and sometimes beneficial relationship to climate change.

Georgina Drew illustrates how cultural-religious practices intertwined with the sacred Ganga River in India may be affected by the decline of the river's source, the Gangotri-Gaumukh glacier. Her ethnographic research identifies how faith in the enduring nature of the Goddess Ganga influences interpretations of environmental change, independent of the discourses of climate science. It also shows how ideas about the river's impermanence in Hindu texts can sometimes deter conservation efforts.

Sujata Manandhar, Dietrich Schmidt-Vogt, Vishnu Prasad Pandey and Futaba Kazama explore the role of religion and indigenous knowledge in adapting farming practices to a changing climate in a mountain region of Nepal. Villagers have noted signs of environmental change greater than changes they have faced in the past. Yet indigenous knowledge based on both the Buddhist faith and the even older Bon religion does not fully enable villagers to adapt, and may in fact mislead them about appropriate responses. The authors conclude, however, that these traditions' flexibility may help villagers to adapt. Also in the Himalaya, Matt Branch shows that Bhutan's internationally famous development model, gross national happiness (GNH), draws heavily on Buddhism and the authority of Bhutan's Buddhist monarchy to construct its legitimacy. Using mixed methods, Branch examines the GNH's connections to climate change and its ability to shape climate change policy and to balance economic goals with maintaining traditional culture and a healthy environment.

Though Bangladesh is not a major emitter of greenhouse gases, it is likely to experience significant, perhaps catastrophic, impacts if climate change is not meaningfully addressed soon. It is in this context that James Pender describes the climate change work of the Church of Bangladesh. According to his account, the church's emphasis on social service to the poor has led it to take an active role vis-à-vis climate change. Ulil Amri's research unpacks the roles played by Muslim civil society organizations in Indonesia's effort to combat climate change. The two biggest and oldest religious organizations have introduced an eco-theology and related praxis of "eco-jihad" in the struggle against climate change. Arguing that these organizations are playing a pivotal role, Amri details how they are using their institutional resources both to spread the word about climate change and to encourage adaptation.

Also focusing on institutional capacity, Andreana Reale shows that the Christian church structures in Solomon Islands facilitate a response to the increasingly visible effects of climate change on those low-lying islands. Significantly, the churches have an institutional presence that the national government does not. This builds resilience in Solomon Islands and, she argues, should be valued as part of a holistic disaster risk reduction strategy.

Based on long-term ethnographical research in the Guatemalan Mam Altiplano, Julie Hermesse analyzes locally mobilized discourses on climate change. Prophetic representations from both Mayan spirituality and millenarist Christian groups diagnose climate change as a cause of meteorological disturbances and future catastrophes. However, interpretations of climate events as being of divine origin or having anthropogenic influences differ among the local people. On the other side of the Atlantic, in the mountains of the Ivory Coast, farmers have offered competing religious narratives to explain recent drought. Sadia Chérif and Joy H. Greenberg's phenomenological-qualitative approach shows that neither animist traditions nor monotheist Christianity has been effective in dealing with climate change.

Paul Sarfo-Mensah and Samuel Awuah-Nyamekye round out the section. Their research shows that, in communities in Ghana where the influence of indigenous African religion is strong, the level of environmental degradation is not as serious as in other communities. However, the spread of non-indigenous religions—particularly Christianity—is causing reverence for and fear of nature to decline. Other factors that negatively affect indigenous conservation include government policy, the market economy, overdependence on Western technology, population growth, and urbanization. The researchers conclude that indigenous ways are important resources for addressing climate change.

Religions and climate change in the Global North

Whereas most of the contributions in the section on the Global South explore how religions affect the perception of and adaptation to climate change, most of the chapters in this section observe religious practices intermingling with political engagement. Often this engagement has taken place on the institutional level.

Writing as both researcher and activist, Mishka Lysack describes how religions in Canada are responding to climate change through advocacy in the public square. After providing an overview of the obstacles and constraints that hinder faith communities from taking action, he explores the opportunities for these communities to provide leadership. He includes elements of an action plan that could guide faith traditions to deepen their environmental citizenship and engagement.

Moving to the United States, three chapters examine how different Christian denominations are engaging with climate change. Michael Agliardo's chapter explores the current U.S. Catholic response to climate change and its significance for ongoing Catholic mobilization. As he explains, though leaders at the highest levels of the Catholic Church have made bold statements urging action on climate change, whether and how climate change is taken up depend upon the initiative of local actors. Patricia K. Townsend elaborates the process of engagement by the Presbyterian Church (U.S.A.), a mainline Protestant church. PC(USA) became an early proponent of addressing climate change, and, compared to other issues the denomination debated, climate change was relatively uncontroversial. Yet while many PC(USA) congregations have attempted to become more carbon-neutral, it is unclear whether resolutions crafted at the national level have been able to change the practices or perspectives of congregants. Bernard Daley Zaleha and Andrew Szasz analyze climate controversy within the Christian Right. Though climate activists argue that even culturally conservative Christians should care about climate change, those in political leadership depict climate policy as something American Christians should shun.

From the United Kingdom, Maria Nita focuses on individual faith practice and identity work among Christian and Muslim activists in the climate and Transition Towns movements. Though experiencing identity conflicts and resistance to their participation, religious activists also found points of convergence between their respective faiths and the biocentric teachings and practices promoted in these sustainability movements.

Noor Johnson extends the Canadian analysis to an underdeveloped region of an overdeveloped country via an ethnographic examination of Healing the Land (HTL), a new evangelical Christian movement in an Inuit community in the Arctic. Many of the "miraculous" changes that HTL leaders point to as signs of divine blessing are the same ones that scientists consider to be indicators of environmental change. But in contrast to scientific narratives of global climate change, HTL offers a resonant vision of integrated action through which individuals and communities can address social, environmental, and spiritual issues simultaneously.

Transnational aspects of religion and climate change

That the HTL movement originated in Fiji but found purchase in the Arctic illustrates the fluidity of contemporary religious traditions in the context of globalization. Numerous other chapters in the two parts dealing with the Global South and the Global North also illustrate how religious involvements cross boundaries. Two chapters take some form of the transnationality as their foci.

Carolyn Peach Brown, Douglas R. Brown and Christopher Shore conducted a survey of staff at a large Christian international development organization in order to understand what contributed to different levels of knowledge about climate change and different perceptions of the risk posed by it. Faith-based organizations play an important role in confronting global challenges like climate change, so staff and decision-maker views about, for instance, its impact on child well-being will affect the extent to which it is included in development programming.

Guillermo Kerber illuminates the World Council of Churches' international advocacy for climate change and climate justice. Involving an increasingly interfaith constituency, the WCC has adopted the position that climate change, like other environmental issues, must not be considered in an isolated manner but in relationship to economic, social, and cultural matters, including justice and peace. The chapter concludes with lessons learned from the WCC's various initiatives.

Together these 17 studies begin to sketch the relationships between religions and climate change through the lenses and tools of the social sciences. In the final chapter we discuss broader implications of the preceding chapters. We hope and expect that the volume will inspire further research, discussion, and debate over potential and actual contributions by religions around the world to climate change advocacy, mitigation, and adaptation.

Acknowledgments

We first wish to thank the people whose experiences furnished the raw material for this volume, and our contributors for giving analytical form to those experiences. Our thanks also go to Helena Hurd, Louisa Earls, and Charlotte Russell at Routledge for their dedication to seeing this project through. Finally, we thank Bron Taylor and the *Journal for the Study of Religion, Nature and Culture* for putting the three of us in touch and for providing a forum to begin to explore these ideas.

Notes

1 Because the scientific research on climate change is so extensive and fast changing, it is not feasible to summarize it here. Readers are directed to the website of the Intergovernmental Panel on Climate Change (http://www.ipcc.ch) or to numerous

web-accessible studies published by the National Oceanic and Atmospheric Administration, the National Center for Atmospheric Research, and the Environmental Protection Agency. Other nations' governmental agencies, the insurance industry, and even the World Bank are also good sources.

2 Conceptual entanglements are illustrated in that "faiths" such as Marxism or secular humanism may meet many definitions of "religion," both sociologically and conceptually.

3 The 84 percent estimate comes from www.adherents.com.

4 In the past decade, social theorists have increasingly acknowledged the ongoing salience of religious perspectives in contemporary societies, even observing a "theological turn" within social theory (Eagleton 2009; Gane 2008; Habermas 2010). It is also commonly observed that the secularization of society as experienced by Europe may be an exceptional case, and that secularism should be considered as a phenomenon in its own right rather than a counterpole to a religiously oriented society (Cannell 2010).

5 Alliance of Religions and Conservation, http://www.arcworld.org/about_ARC.asp.

6 http://www.arcworld.org/news.asp?pageID=351 (accessed January 28, 2013).

7 http://www.arcworld.org/projects.asp?projectID=137 (accessed January 28, 2013).

8 http://safcei.org/african-faith-leaders-a-renewed-moral-vision-is-vital-to-progress-in-climate-talks/ (accessed January 28, 2013).

9 The Buddhist Declaration on Climate Change (http://www.ecobuddhism.org/bcp/all_content/buddhist_declaration/, accessed January 30, 2013), for example, appears to derive from the U.S. context rather than to represent the Buddhist community worldwide. When examining such statements, it is critical to investigate who is representing or interpreting the tradition.

10 http://fore.research.yale.edu/climate-change/statements-from-world-religions. Ohio Interfaith Power and Light also keeps a list, which was more up to date as this volume went to press: http://www.ohipl.org/resources/faith-basedearthkeeping/climate-change-resources/climatestatements.

11 http://www.islamic-relief.com/InDepth/2-12-environment.aspx;http://www.muslimaid.org/index.php/media-centre/763-muslim-aid-calls-for-aradical-change-in-lifestyle-to-combat-climate-change.

12 http://www.hssuk.org/samiti/. Climate Week was billed as "Britain's biggest climate change campaign" on the group's website, http://www.climateweek.com.

13 Our comments are based on a review of the English language literature only, admittedly a significant limitation.

References

Abdul-Matin, Ibrahim and Keith Ellison. 2010. *Green Deen: What Islam Teaches about Protecting the Planet*. New York: Berrett-Koehler Publishers.

Atkinson, David. 2008. *Renewing the Face of the Earth: A Theological and Pastoral Response to Climate Change*. Norwich, U.K.: Canterbury Press.

Barna Group. 2008. "Evangelicals Go 'Green' with Caution." September 22. Available at http://www.barna.org/barna-update/article/13-culture/23-evangelicals-go-qgreenq-with-caution.

Barnosky, Anthony D., Elizabeth A. Hadly, Jordi Bascompte, Eric L. Berlow, James H. Brown, Mikael Fortelius, Wayne M. Getz, John Harte, Alan Hastings, Pabo A. Marquet, Neo D. Martinez, Arne Mooers, Peter Roopnarine, Geerat Vermeij, John W. Williams, Rosemary Gillespie, Justin Kitzes, Charles Marshall, Nicholas Matzke, David P. Mindell, Eloy Revilla and Adam B. Smith. 2012. "Approaching a State Shift in Earth's Biosphere." *Nature* 486, 52–58.

Brulle, Robert, Jason Carmichael and J. Jenkins. 2012. "Shifting Public Opinion on Climate Change: An Empirical Assessment of Factors Influencing Concern over Climate Change in the U.S., 2002–2010." *Climatic Change* 114. 2, 169–188.

Cannell, Fenella. 2010. "The Anthropology of Secularism." *Annual Review of Anthropology* 39, 85–100.

Clarke, Gerard. 2006. "Faith Matters: Faith-Based Organisations, Civil Society and International Development." *Journal of International Development* 18.6, 835–848.

Crate, Susan A. and Mark Nuttall, eds. 2009. *Anthropology and Climate Change: From Encounters to Action*. Walnut Creek, CA: Left Coast Press.

Donner, Simon. 2011. "Making the Climate a Part of the Human World." *Bulletin of the American Meteorological Society* 92.10, 1297–1302.

Durack, Paul J., Susan E. Wijffels and Richard J. Matear. 2012. "Ocean Salinities Reveal Strong Global Water Cycle Intensification during 1950 to 2000." *Science* 336.6080, 455–458.

Eagleton, Terry. 2009. *Reason, Faith, and Revolution: Reflections on the God Debate*. New Haven, CT: Yale University Press.

Gane, Nicholas. 2008. "Religion, Theology and Culture: Introduction." *Theory, Culture and Society* 25.7–8, 119–123.

Gardner, Gary. 2003. "Engaging Religion in the Quest for a Sustainable World." In *State of the World 2003*, ed. L. Starke, pp. 152–175. New York: W.W. Norton & Co.

Gerten, Dieter and Sigurd Bergmann, eds. 2012. *Religion in Environmental and Climate Change: Suffering, Values, Lifestyles*. London: Continuum.

Gilliat-Ray, Sophie and Mark Bryant. 2011. "Are British Muslims 'Green'? An Overview of Environmental Activism among Muslims in Britain." *Journal for the Study of Religion, Nature and Culture* 5.3, 284–306.

Habermas, Jurgen 2010. *An Awareness of What Is Missing: Faith and Reason in a Post-Secular Age*. Cambridge, U.K.: Polity Press.

Haluza-DeLay, Randolph B. 2008. "Churches Engaging the Environment: An Autoethnography of Obstacles and Opportunities." *Human Ecology Review* 15, 71–81.

——. 2012. "Religion." In *Encyclopedia of Global Warming and Climate Change*, 2nd edn., ed. S.G. Philander, pp. 1176–1179. Thousand Oaks, CA: Sage.

Hansen, James. 2012. "Game Over for the Climate." *New York Times*, May 9. Available at http://www.nytimes.com/2012/05/10/opinion/game-over-for-the-climate.html?_r=1.

Hayhoe, Katharine and Andrew Farley. 2009. *A Climate for Change: Global Warming Facts for Faith-Based Decisions*. New York: FaithWords.

Hulme, Mike. 2009. *Why We Disagree about Climate Change*. Cambridge, U.K.: Cambridge University Press.

Kearns, Laurel. 2011. "The Role of Religions in Activism." In *The Oxford Handbook of Climate Change and Society*, ed. John S. Dryzek, Richard B. Norgaard and David Schlosberg, pp. 414–428. Oxford, U.K.: Oxford University Press.

——. 2012. "Religious Climate Activism in the United States." In *Religion in Environmental and Climate Change: Suffering, Values, Lifestyles*, ed. Dieter Gerten and Sigurd Bergmann, pp. 97–124. London: Continuum.

Kearns, Laurel and Catherine Keller. 2007. *Ecospirit: Religions and Philosophies for the Earth*. New York: Fordham University Press.

Lahsen, Myanna. 2008. "Experiences of Modernity in the Greenhouse: A Cultural Analysis of a Physicist 'Trio' Supporting the Backlash against Global Warming." *Global Environmental Change* 18, 204–219.

Leduc, Timothy B. 2010. *Climate, Culture, Change: Inuit and Western Dialogues with a Warming North*. Ottawa, ON: University of Ottawa Press.

Makrides, Vasilios. 2008. "Christianity—Greek Orthodox." In *The Encyclopedia of Religion and Nature*, ed. Bron R.Taylor, pp. 338–341. London: Thoemmes Continuum.

McCright, Aaron M. and Riley E. Dunlap. 2000. "Challenging Global Warming as a Social Problem: An Analysis of the Conservative Movement's Counter Claims," *Social Problems* 47.4, 499–522.

McFague, Sallie. 2008. *A New Climate for Theology: God, the World, and Global Warming*. Minneapolis, MN: Fortress Press.

Merritt, Jonathan. 2010. *Green like God: Unlocking the Divine Plan for Our Planet*. New York: FaithWords.

Millais, Corin, ed. 2006. *Common Belief: Australia's Faith Communities on Climate Change*. Sydney: Climate Institute.

Northcott, Michael S. 2007. *A Moral Climate: The Ethics of Climate Change*. London: Darton, Longman & Todd.

Oppenheimer, Michael, Brian C. O'Neill, Mort Webster and Shardul Agrawala. 2007. "The Limits of Consensus." *Science* 317, 1505–1506.

Pew. 2006. *Science in America: Religious Beliefs and Public Attitudes*. Washington, DC: Pew Research Center for the People and the Press and Pew Forum on Religion and Public Life. Available at http://www.pewforum.org/Science-and-Bioethics/Science-in-America-Religious-Belief-and-Public-Attitudes.aspx.

——. 2009. *Fewer Americans See Solid Evidence of Global Warming*. Washington, DC: Pew Research Center for the People and the Press and Pew Forum on Religion and Public Life. Available at http://people-press.org/report/556/global-warming.

Posas, Paul J. 2007. "Roles of Religion and Ethics in Addressing Climate Change." *Ethics in Science and Environmental Politics* 2007, 31–49.

Primavesi, Anne. 2009. *Gaia and Climate Change: A Theology of Gift Events*. Oxford, U.K.: Routledge.

Public Religion Research Institute and Religion News Survey. 2011. "Climate Change and Evolution in the 2012 Elections." Available at http://publicreligion.org/research/2011/09/climate-change-evolution-2012/.

Reder, Michael. 2012. "Religion in the Public Sphere: The Social Function of Religion in the Context of Climate and Development Policy." In *Religion in Environmental and Climate Change: Suffering, Values, Lifestyles,* ed. Dieter Gerten and Sigurd Bergmann, pp. 32–45. London: Continuum.

Roncoli, Cara, Keith Ingram and Paul Kirshen. 2002. "Reading the Rains: Local Knowledge and Rainfall Forecasting among Farmers of Burkina Faso." *Society and Natural Resources* 15, 411–430.

Salick, Jan, Anja Byg and Kenneth Bauer. 2012. "Contemporary Tibetan Cosmology of Climate Change." *Journal for the Study of Religion, Nature and Culture* 6.4, 447–476.

Schipper, E. Lisa F. 2010. "Religion as an Integral Part of Determining and Reducing Climate Change and Disaster Risk: An Agenda for Research." In *Der Klimawandel: Sozialwissenschaftliche Perspektiven*, ed. M. Voss, pp. 377–393. Wiesbaden, Germany: VS-Verlag.

Skrimshire, Stefan. 2010. *Future Ethics: Climate Change and the Apocalyptic Imagination*. London: Continuum.

Smidt, Corwin, ed. 2003. *Religion as Social Capital: Producing the Common Good*. Waco, TX: Baylor University Press.

Sponsel, Leslie E. 2005. "Anthropologists." In *The Encyclopedia of Religion and Nature*, ed. Bron R. Taylor, pp. 94–96. London: Thoemmes Continuum.

Stanley, John, David R. Loy and Gyurme Dorje, eds. 2009. *A Buddhist Response to the Climate Emergency*. Somerville, MA: Wisdom Publications.

Swart, Ignatius. 2006. "Churches as a Stock of Social Capital for Promoting Social Development in Western Cape Communities." *Journal of Religion in Africa* 36.3–4, 346–378.

Tucker, Mary Evelyn and John A. Grim. 2001. "Introduction: The Emerging Alliance of World Religions and Ecology." *Daedalus* 130.4, 1–22.

Veldman, Robin Globus, Andrew Szasz and Randolph Haluza-DeLay. 2012. "Climate Change and Religion: A Review of Existing Research." *Journal for the Study of Religion, Nature and Culture* 6.3, 255–275.

Watson, Elizabeth E. and Hassan Hussein Kochore. 2012. "Religion and Climate Change in Northern Kenya: New Moral Frameworks for New Environmental Challenges?" *Journal for the Study of Religion, Nature and Culture* 6.3, 319–343.

Wilkinson, Katharine K. 2010. "Climate's Salvation: Why and How American Evangelicals Are Engaging with Climate Change." *Environment* 52.2, 47–57.

——. 2012. *Between God and Green: How Evangelicals Are Cultivating a Middle Ground on Climate Change*. Oxford, U.K.: Oxford University Press.

Wisner, Ben. 2010. "Untapped Potential of the World's Religious Communities for Disaster Reduction in an Age of Accelerated Climate Change: An Epilogue and Prologue." *Religion* 40.2, 128–131.

Wolf, Jakob and Mickey Gjerris. 2009. "A Religious Perspective on Climate Change." *Studia Theologica* 63. 2, 119–139.

"World Religious Leaders Promise Climate Action." 2009. Public Radio International, Minneapolis, MN. Available at http://www.pri.org/stories/science/environment/world-religious-leaders-climate-change1710.html (accessed January 30, 2013).

Xia, Chen and Martin Schönfeld. 2011. "A Daoist Response to Climate Change." *Journal of Global Ethics* 7. 2, 195–203.

Part I

The Global South

2 A retreating Goddess?

Conflicting perceptions of ecological change near the Gangotri-Gaumukh glacier

Georgina Drew

Extensive ecological transformations are in progress around the globe, and awareness is growing over the degree to which human activities are responsible. In addition to the contributions of climate science, many public debates help to circulate varied understandings of the issues and the challenges ahead. The emerging emphasis on climate stewardship and justice from Christian, Buddhist, and Hindu practitioners, among others, shows the degree to which religious orientations influence the way that many perceive the proper responses to the stakes at hand. In some cases, however, cosmological orientations and religious teachings are a resource that people draw upon to make sense of ecological change in ways that do not necessarily support the established science or motivate participation in conservation efforts. This chapter focuses on the multiple points of discussion and contention over climate change phenomena as I observed them in the Indian Himalayas from 2008 to 2009. By exploring a particular case, that of how people understand and respond to the potentially threatened Ganges River, or "Ganga" as I will refer to it here, I examine somewhat contradictory discourses about the current moment of change.

To highlight the varied perceptions and epistemological tensions that inform people's responses to transformations along the Ganga, I focus on circulating perceptions in three arenas. First I address the influence of international climate science and some of the controversial predictions of glacial retreat. The second realm that I engage is a space of devotion to the Ganga at a location some 85 miles below the river's glacial source. I share perspectives about the Ganga's agency and resilience that I encountered when speaking to river devotees in places such as Manikarnika ghat in Uttarkashi. The third realm pertains to understandings of the Ganga's potential impermanence that are informed by Hindu texts and cosmologies.

My exploration of these arenas demonstrates the production of meanings about environmental change and the influence that distinct perceptions have on behavior. The investigation is relevant to a growing body of literature on the social causes and implications of climate change phenomena. This scholarship recognizes the inextricable linkages between humans and the environment evident in ecological anthropology, historical ecology, and political ecology, among others. The vast literature originating from these fields attends to the values that people ascribe to themselves and the environment along with the relationships that connect them.

Importantly, more attention is emerging within particular fields, such as anthropology, on the role that climate change can play in altering subjectivities, cosmological orientations, and cultural practices (Crate and Nuttall 2009; Orlove, Wiegandt and Luckman 2008). A focus for some scholars is the role that religion has, or could have, in helping to inform responses to climate change phenomena. This is a complex area of inquiry that highlights the degree to which religious views and practices draw from and influence the ecological contexts in which human societies are embedded. These connections are particularly important to examine along the Ganga. The river is considered divine by millions, is an integral part of Hindu sacred geography and cultural practice (Eck 1996), and is also an iconic symbol of religious reverence for nature. The physical, spiritual, and religious grace that the river bestows, however, could be threatened by the ecological transformations linked to climate change. To begin exploring these threats and the debates that the topic of the Ganga's potential disappearance inspires, I first turn to the international concern for the Himalayan glaciers that provide runoff for the river.

A glacial range and a river in peril? Context and controversy

Glacial melt is one of many potentially deleterious effects of the warming temperatures associated with climate change. In the scholarship and the media, numerous estimates have forewarned of vast hydraulic changes in places such as the Himalaya where the runoff from glaciers feeds rivers that help support some 1.5 billion people in South Asia and China. Often, the picture painted is one of dwindling resources combined with occasional floods as the glaciers lose a bulk of their mass in the hot summer months. In the feared scenarios, food security would likely be compromised in downstream regions. This is worrisome in countries like India that struggle to meet production targets under already extreme and erratic weather conditions tied to the annual monsoon cycle. While the long-term food and water concerns merit proactive measures, especially among those who are the most vulnerable socio-economically, there are still uncertainties about how the transformation of the Himalayan glaciers will progress and what the shifts will mean for the lived realities of people residing nearby.

A problem that climate scientists have struggled with is the paucity of long-term data on glacial movements in the Himalayas. Whereas the records for glaciers in areas like North America, Europe, and the Arctic can follow their movements over hundreds of years in some cases, consistent ground-based data on Himalayan glaciers extends back only a few decades. In South Asian countries like India, for instance, much of the early research relied on limited field investigations and aerial photography.

Among the glaciers studied, the Gangotri-Gaumukh glacier is one of the better documented. This ice mass feeds into the upper stretch of the Ganga in the Indian state of Uttarakhand. It is located near India's northwestern border with the Tibetan Autonomous Region of China. Other than a few sketches and photographs taken over the years since 1937, the regular monitoring of the glacier's snout only began in 1971 when the Geological Survey of India undertook the task (Dobhal 2009: 72). Satellite imagery and remote sensing were introduced later in the 1990s. Some of the data shows that the Gangotri-Gaumukh glacier retreated 1.25 miles from 1780 to 2001 (Naithani, Nainwal and Sati 2001: 94). In the 25 years before 2004 alone, the glacier retreated more than half a mile.[1]

Unfortunately, the ground-based data has been inadequate for the entire glacial range of which the Gangotri-Gaumukh glacier is a part. Scant funding and the difficulty of accessing high-altitude glaciers have contributed to the problem. The dearth of historical information, other than what can be derived from photos and satellite images, has made the movement of Himalayan glaciers in places like India subject to speculation. Because of this, postulations regarding the quantity and quality of ice loss have sometimes been based on a limited selection of glaciers, including the Gangotri-Gaumukh range, rather than the region's ice masses as a whole. And, based on the relatively stable movements of a few terminal points (snouts), some Indian glaciologists have even dismissed the fears over Himalayan glacial decline (Raina 2009).

At the other end of the spectrum are those who have predicted the imminent disappearance of the Himalayan glaciers. This perspective was fueled by a 2007 report affiliated with the Intergovernmental Panel on Climate Change (IPCC) that cited the year 2035 as a likely timeframe for Himalayan glaciers to vanish (Cruz et al. 2007: 493). That date was quoted in numerous environmental circles from 2007 up until the middle of 2009. Many publications drew on this prediction to paint a picture of coming doom. The use of such discourses was tempered in mid- to late 2009 when, after substantial criticism and calls to revoke the 2035 estimate, the IPCC conceded that the timeframe was based on reports of dubious scientific rigor. But, IPCC representatives cautioned, the Himalayan glaciers are still threatened by ecological shifts and warming temperatures. They and others called for an escalation of data collection on Himalayan glaciers and cautioned against complacency given the potentially high stakes.

Also at issue is the observation that the rate and scope of glacial retreat can vary on a case-by-case basis. A study by Immerzeel, Ludovicus and Bierkens (2010) asserts that the contribution of glacial water and snow melt into the Ganga may only be a fraction of its overall flow as a result of limited upstream precipitation, the relatively smaller size of the glaciers in that system, and the recharge contribution of monsoon-dominated downstream climates (ibid.: 1383). The Brahmaputra and Indus, by contrast, depend on glacial melt to a much greater extent. Given the geographic variation of melt rates and the divergent ways that river systems will be affected, such scholarship concludes that, while Asia's "water towers" are threatened by climate change, the effects on water availability and food security will differ by region and thus cannot be generalized (ibid.; Rees and Collins 2006). This reminder tempers some apocalyptic visions for the coming changes while encouraging us to think in terms of the differential impacts that will be felt along distinct waterways and the adaptation strategies that could be formed in response.

Beyond the need for more careful research, what the debates underscore is that our knowledge of climate change processes and outcomes can be a moving and uncertain terrain. In addition to differing opinions among scientists and ecologists, there are other perceptions that soften narratives of impending ecological collapse. In the Himalaya, there exist multiple and conflicting interpretations of environmental change and its significance, especially in relation to the longevity of the Ganga. The perspectives, which are influenced by everyday Hindu practice and regional cosmological orientations, indicate diverse connections to dynamic bodies of water and ice. These entities are infused with varied religious meanings and perceived to exert various levels of agency over the elements. In the discussion that follows, I highlight the perceptions of ecological change and glacial melt that residents of the Indian Himalaya provided to me from along the upper stretch of the Ganga.

Assessing knowledge of ecological change at Manikarnika ghat, Uttarkashi

In 2008 and 2009, I conducted ethnographic research in Garhwal, an ethnolinguistic region of Uttarakhand state, and in urban centers such as New Delhi to understand emerging religious-environmental conflicts over the use of the Ganga's Himalayan stretch. At the time, I sought to examine the points and counterpoints that circulated about the construction of new hydroelectric dams on the Ganga. A related and overlapping interest was the concern for the river's longevity as a result of factors such as anthropogenic climate change. The current discussion focuses on the latter issue and is based on the commentary provided to me by people living along the Ganga in Uttarkashi, a district capital of Garhwal.

Uttarkashi sits at an elevation of 3,800 feet. It is located roughly 230 miles northeast of New Delhi and approximately 85 miles downstream from the Gangotri-Gaumukh glacier. One of Uttarkashi's claims to notoriety is that it is considered to be one of the famous "Kashi" or abodes of Lord Shiva. "Uttar" means "upper" in Sanskrit, so the city's name translates to "Upper Kashi." In ancient texts, it is also referred to as Saumyakashi. The Skanda Purana (Nautiyal 1994), one of the many books associated with the sacred Puranas, elaborates at length on the importance of this particular Kashi. The prose of the Skanda Purana emphasizes Uttarkashi's significance as a *tirtha* —or sacred place of pilgrimage—in which many Hindu gods reside. It is also a strategic place for pilgrims and travelers to rest on their way up to the seasonal town of Gangotri, where the Ganga is said to have descended from the heavens at the bequest of a mortal, King Bhagirath. Gangotri is home to a renowned temple with an embodiment of the Ganga, and it is the gateway to a narrow trail leading 15 miles up to the glacier.

Given Uttarkashi's relative proximity to the Gangotri-Gaumukh glacier, I anticipated that I would encounter many of the familiar discussions about glacial melt and climate change that were already in international circulation at the time. Soon after I arrived, however, I was surprised to find that the estimates for glacial retreat and the scientific discourses on climate change were not yet topics about which many of the people I spoke with expressed extensive familiarity. My conversations with people at Manikarnika ghat were particularly indicative that not everyone was aware of climate change or that, at the very least, they did not perceive it or fear for its effects in the same way I was trained to.

Manikarnika ghat is one of the most important places to worship the Ganga in Uttarkashi. The ghat, or series of steps that descend to the riverbank, is situated at a point where the Ganga rounds a bend on the outer part of the city. It is a prominent site for people to gather on auspicious days in the Hindu calendar, pray to the Goddess Ganga, and cleanse themselves in the sacred waters. Its location adjacent to a main market also makes it a convenient stop that people make time for in their daily routines.

I visited Manikarnika ghat regularly throughout the period of my fieldwork. It was while sitting on the steps leading to the river that I was able to observe in detail the practices of reverence and devotion to the Ganga, which have their own rhythm and mountain flair compared to the more notorious places of worship in the Indian plains at Haridwar and Varanasi (Benares). For major celebrations, men, women, and children frequented the ghat along with a myriad of regional gods alight in palanquins decorated with red and gold fabrics and carried on long wooden poles atop the shoulders of men. Whatever the occasion, it was often the women—representing a range of socio-economic backgrounds—who stayed the longest, charged as they were with woven baskets full of spices, flowers, and foodstuffs that they ceremonially offered to the Goddess.

Upon reaching the waters, the women would place the palms of their hands together in a gesture of greeting and reverence. After offering a murmured prayer or Sanskrit mantra, they would then take out a bronze pot from the basket, which they dipped into the waters of the Ganga. Devotees held this water up to the Sun and its God, Surya, before pouring it back into the river's flow. They repeated this about three times and most often they would spin clockwise while they let the water descend back into the current that rushed around their feet. Devotees then proceeded to offer the other items in their baskets: a pinch of turmeric, a note of sandalwood, and a bit of the vermilion dye (*sindoor*) that married women place at or along the part in their hair. They would also sprinkle dry rice and milk sweets into the waters. When asked why these items were offered, amused women explained patiently to me, the white foreigner (*gori*), that the Goddess enjoys these tastes and adornments as much as any other woman.

In addition to observing practices of worship, Manikarnika ghat was a place where I witnessed several acts of Goddess possession. These moments of rapture were often very brief. The people experiencing possession included rural women visiting the city for the day and the seemingly middle-class women who lived in Uttarkashi. There was no telling who would be possessed at any given moment. Women would just approach the waters, perform their ritual practices, and then immerse themselves. When the Goddess came to them while bathing, they would suddenly scream and shake uncontrollably. If the current was particularly swift, nearby devotees would grab the possessed woman to keep her from falling deeper into the rushing waters as she writhed and flailed. Lasting between 30 and 90 seconds, these moments of possession would desist almost as quickly as they had begun. On a particularly auspicious day that celebrates the Ganga's "birthday" (Ganga Dussehra), the commemoration of her descent into earth from the heavens, I counted seven instances of Goddess possession in a half-hour period. The women that had been possessed weren't always sure who had entered their bodies but they most often explained that it was the family Goddess (*ghar ki devi*) that they worshiped at home and who protected them. In response to my questioning, one informant explained why these Goddesses would come to them at such moments: "The Goddess who watches over our home gets happy when we are near Ganga," she said, "and she comes to worship her through us." This comment exemplified the degree to which reverence for the Ganga forms part of human social life as well as the sociality of regional Gods and Goddesses.

Along with being a central location for worship of the Ganga in Uttarkashi, Manikarnika ghat was one of the locations with a high variability of flow. In the summertime, when rain was plentiful and the rate of upstream glacial melt was high, the river would swell. In the fall and winter, the water could be nearly absent from the ghat. Devotees would either have to bathe in a small trickle that flowed past the steps of Manikarnika or they would cross a dry embankment. It was because of the high variations in the river's flow past Manikarnika that I chose it as a location to ask people about their perceptions of change in the region.

In speaking with roughly 30 women at Manikarnika over a series of some ten visits to the ghat, I came to realize that very few of them were concerned about the changes they had seen in the riverbed. Not one of my informants at Manikarnika, for instance, expressed a fear that the Ganga would one day cease to flow. When I reminded them of the scant levels of water in the fall and winter months, people told me stories about the river's movements and capricious nature, including anecdotes that the Ganga's path used to cut through a different part of the city hundreds of years ago. In recent memory, they said, the Ganga used to flow near Manikarnika ghat no matter how little water there was. Over the last few decades, the river has shifted and now flows most of the year on the far side of the riverbed opposite to the ghat.

The general lack of concern for the Ganga's condition did not mean that the women I spoke with at Manikarnika ghat hadn't noticed and weren't worried about ecological shifts. Many, in fact, commented on the growing water shortages and unpredictable weather patterns. They noted that these phenomena inhibited the timely maturation of crops and endangered the agricultural cycle. Water scarcity—in terms of the water that was available via groundwater and springs—was a topic that many elaborated upon by sharing examples of water stress from villages near and far.

My conversations with people at Manikarnika, therefore, indicated that, although people had observed environmental change, they often perceived that those transformations would not lead to a significant decline of water in the Ganga. This meant that many people viewed the Ganga as an exceptional entity that was independent of the variations in the water cycle that occurred around it.

Because of this perception, these river devotees did not perceive that the Ganga's flow was threatened.

To further explore this apparent lack of a preoccupation for the Ganga's longevity, I tried to pose hypothetical questions about the river's loss. I asked people at Manikarnika, for instance, "Could you ever imagine a time when the Ganga would stop flowing?" Numerous people replied to this question with a gasp, a strong "No," or a caution that the Ganga's disappearance would signal the end of the world. The difficulty imagining life without the river demonstrated the monumental value that people placed on its existence.

One of the reasons why it was challenging to engage people on the subject of the Ganga's potential loss perhaps has to do with the particular significance of Manikarnika ghat. In the aforementioned Skanda Purana, for instance, Lord Shiva speaks to his devotees and tells them that he can be found residing in Uttarkashi during even the worst of times and that they should worship the Ganga at Manikarnika ghat to attain salvation. Shiva further asserts that the performance of sacred rituals will be more beneficial when done at Manikarnika than at other sites because the location will retain its holiness far into the cycle of physical and moral corruption that marks the Hindu epoch of Kaliyug, the current and most degenerate period in a four-part cycle of cosmological time.[2] For river devotees, such scriptures likely influence ideas about Uttarkashi's importance while providing reassurance that the Ganga's flow through the city at Manikarnika will continue to provide grace in times of decay.

In my initial visits to Manikarnika ghat, I would ask people what they knew about climate change. Many of the people I spoke with were unfamiliar with this term and its Hindi equivalent. In place of a common language, I would share my understanding of climate change and sometimes even the predictions of glacial retreat that I had read and which had filled me with fear for the future of Himalayan water resources.

The responses I received contrasted sharply to the urgency with which I thought about glacial melt. When I stated the estimates for retreat, for instance, they were often met with nonplussed expressions. Numerous people even rejected the estimates outright. Others looked dubious when presented with the predictions but deferred to information with a shrug of the shoulders or a seemingly unconvinced tone of concession. Some even suggested that I speak with the priests of various nearby Hindu temples because they assumed that these religious authorities would have a definitive understanding of the potential transformations in question.

When I inquired about the state of the upstream glaciers, some mentioned that they had seen changes from the previous form while on pilgrimage to the source or that they had heard that the ice mass was undergoing a transformation from friends and other pilgrims. From these sources of information, they knew that Gangotri-Gaumukh no longer looked like the mouth of a cow in accordance with the namesake for the glacier's snout, "Gaumukh."

There were several reasons cited for the changing ice formations upstream. Some respondents pointed out that pollution has grown in the world and that this is a cause of the changes to the regional ecology. Conceptualizations of pollution, however, were not limited to plastics, carbon emissions, and the brown clouds that cover many of the world's cities. They also included corrupt human nature or "internal pollution." This moral degeneracy was linked with the inevitable increase in sin (*paap*) that accompanies the aforementioned Kaliyug. Reinforcing this view, a few informants at Manikarnika ghat pointed to litter and sewage in the river to argue that people are "no longer good" and have lost respect for the sacred Ganga. Numerous river devotees reasoned, moreover, that the Goddess was upset at such sinful behavior and that the changing ice formations at Gaumukh were signs of her displeasure. That is to say, even though the Goddess is associated with the water that *flows* from the glacier and technically not the glacier itself, many informants indicated that she was in command of the glacial mass.

Since the Goddess Ganga is a powerful agent with a degree of control over her riverine condition and that of the glacier, many asserted that she could revive her physical manifestations if humans were to amend and improve their moral caliber. Stressing this, a fortune-teller who channeled one of the regional Goddesses proclaimed that the glaciers didn't have to go back, that they could also surge. She reasoned that the Goddess Ganga could turn the tide of retreat if people adhered to higher standards of conduct—a point to which her children sitting nearby readily agreed. This shows how some viewed the glacial recession to be potentially reversible given the context of action and response in which humans and nonhumans engage. The ways in which the condition of the ice mass was linked to behavior were reminiscent of studies in places as far away as Alaska, where Athapaskan and Tlingit natives understood glacial movements to be linked with human conduct in landscapes where glaciers can hear, smell, and react to actions of which they disapprove (Cruikshank 2005). These ways of knowing are not just narratives about "nature" but modes of sharing a multitude of historical, cultural, and social values that challenge positivism and scientific frames of reference (ibid.: 258).

Hindu interpretations of change: the influence of texts and cosmologies

By the summer of 2009, the number of public and private discourses circulating in Uttarkashi about "climate change" had increased. When overhearing or engaging in climate-related discussions, I found that Uttarkashi's teachers, business owners, and middle-class professionals were prominent among those who had quickly become versed in the topic. They were exposed to the issue via newspapers, through coverage of the topic disseminated on cable television, and in some instances from the children who attended schools with updated environmental studies curricula.

When I spoke with such people, I asked them to clarify what climate change entailed. These informants associated the term with a global rise in temperature, a lack of snowfall in Himalayan winters, and changing monsoon patterns. They were also aware of the global inequities associated with climate change phenomena. Media coverage of the international climate summit at Copenhagen was particularly influential. It educated people about the tensions between the "developed" countries emitting the bulk of the world's carbon and the "developing" countries that defended their rights to economic growth even if it meant increasing emission levels. With this came hostilities directed towards the biggest polluters (such as the U.S.A. and China), and concerns that the pace of India's rapid economic growth could stall because of dwindling resources, declining agricultural production, and the myriad other repercussions of a global ecology tipped out of balance.

Even as people began talking more about climate change, their concerns often stopped short when they were asked about the Ganga's future. Yes, some of my media-savvy informants conceded, the Gangotri-Gaumukh glacier is receding as a result of "global warming" and the continuity of the Ganga's flow could be in jeopardy. Yet, even when citing concerns for the glacier that feeds the Ganga and the alarming lack of snow and snow melt, many rejected the idea that the river—and especially its Goddess—could completely disappear from the earth. In some cases, informants pointed to the climate change controversies and the lack of scientific agreement on glacial melt to support this perspective. In other cases, informants argued that the Ganga is more than the sum of its surface flows and the glaciers that feed into them. This point was made to assert that climate change phenomena cannot entirely sever people's connections with the divine that resides in the sacred waters. In an interview, for instance, a schoolteacher drew from Hindu framings to say that the Goddess Ganga can never leave the earth because she simultaneously flows on three levels: in the heavens, in the riverbed, and beneath the surface of the earth. The subterraneous zone is known in Hindi as *patal*. The teacher cited the Saraswati River as an example of a river that has disappeared from the surface of the earth but that continues to flow underground. As she and others reminded me, Hindus still worship the Saraswati even though the river dried up many centuries ago. For her, this meant that the Ganga would continue to bless humans from *patal* as long as they pray and keep her in their thoughts. In this logic, the glaciers may melt and ultimately disappear but the Goddess that flowed from them will never recede from devotees' hearts. The idea that the Ganga will continue to live in the heavens, in the underworld, and within her devotees, some suggest, is evidence that the Ganga will always remain an important entity in Hindu practice regardless of what happens to its waters.

Another surprising revelation was that, although Uttarkashi was heavily populated with hundreds of Hindu religious leaders and their centers of spiritual retreat (*ashram*), hardly more than a dozen of them were involved in efforts to address the environmental and development challenges that impact the river's course. I was initially confused by the apparent lack of concern for the Ganga's future condition that I encountered when I spoke with Hindu sadhus, priests, and swamis in Uttarkashi. One of the Himalayan city's main attractions, after all, is the Ganga itself. In addition to its cultural importance, the river's contemplation is believed to increase spiritual focus and progress. I imagined spiritual aspirants, or *sadhak*, would be motivated to undertake environmental campaigns to safeguard their access to the Ganga. At least for the period of my research, I was proven wrong. What, I wondered, could explain what seemed to me to be an anomaly? Initially,

I imagined that perhaps some of the spiritual aspirants and religious leaders had adopted a philosophical stance, preferring to contemplate the illusory nature of reality as advocated in the Advaita Vedanta of religious leaders past such as Shankaracharya. This proved an inadequate explanation, however, because many of my informants' daily lives were intertwined with the Ganga's flow and they acknowledged the importance of respecting the material conditions of existence and of honoring lived experience.

Over time, I realized that scriptures that predict the *eventual* disappearance of the Ganga were influential. When I spoke with learned priests and swamis, they often explained their thinking on the river's inevitable decline by drawing from ancient texts such as the Puranas (which they often named generically without offering more specification) or perhaps the *Srimad Bhagavatum*. Such informants cited these sources, composed millennia ago, to confirm that the Ganga's destiny is to one day leave the earth. These respondents clarified, however, that the Ganga is predicted to disappear in the third or the fourth stage of Kaliyug. And, they asserted, we are only in the first phase of the epoch—a period which, depending on the calculation, can last up to 432,000 years.[3]

Hinduism and ecological sustainability: some limits

The complexities I have illustrated thus far indicate that religious reverence for nature does not always lead to the enactment of ecologically sound behaviors, especially when multiple circulating epistemologies are layered within regional policies and politics. This observation complicates some of the scholarship that posits that reverence for the natural world promotes sustainable ecologies. Sponsel (2001), for instance, argues that enactments of religious reverence for the nonhuman can foster more environmental stewardship, because "Religions are alternative ways of representing nature; affording it spiritual, moral, and cultural meanings and values; and defining the place of humans in nature, including how they should and shouldn't act toward nonhuman beings and other natural phenomena" (2001: 185). Based on the points that I have highlighted, his presentation of "spiritual ecology" is perhaps overly optimistic regarding the role that religion has in reorienting our ecological interactions. In an edited volume on religion in environmental and climate change, Vogt (2011) likewise asserts that the religions of the world have a unique role to play in fostering sustainability, because, among other reasons, "All religions identify themselves as embodying long-term thinking" (2011: 76). My discussion, however, has shown how long-term perspectives on the cycles of time can also deter proactive environmental behaviors and initiatives.

Commenting specifically on Hindu "traditions" there are some who argue that religious practices can help in the quest for sustainability by reminding us to think about our intrinsic physical and spiritual interconnectivity with nature (Gosling 2001). Shiva (1988), for instance, argues that Hindu reverence for nature is a counterweight to the conceptual separation between humans, ecologies, and the divine that post-enlightenment modernity has helped to create and which has fueled humankind's extensive practices of resource extraction and destruction.

Although communion with nature is prominent in many religious and spiritual practices, the case of ecological change on the Himalayan reaches of the Ganga shows that concern for a natural entity can be offset by a myriad of factors and perceptions that have distinct historical, cultural, and political influences. Supporting this observation, Nagaranjan (1998) asserts that it is problematic to think that ritual care for a place, object, or natural entity directly leads to "ecological behavior" and conservation. Critiquing the above-mentioned work of Vandana Shiva, she argues that reverence for nature might be a part of some Indian sociocultural systems but it does not necessarily lead to environmental protection, because, even though "non-Western religions" may have a reverence towards landscapes and they may contain innumerable "embedded ecologies," these beliefs do not necessarily lead to ecological practices that resemble conservationism in the sense that the West has come to know it (Nagaranjan 1998: 284).

The work of Alley (2000) concurs with elements of the above caution. Her writing draws from ethnographic fieldwork to explain the contradictions of pollution along the Ganga in Varanasi in the 1990s. Alley's detailed study demonstrated how faith in the Goddess's perpetuity, despite the degraded appearance of her riverine form, could quell some of the preoccupations for the Ganga's condition that environmental activists presented. She also showed how the focus on personal redemption in moments of reverence to the Ganga could limit concern for the river's toxicity and high fecal content. This means that "religious rituals cannot be confused with civic ethics" and we thus cannot assume that reverence for nature will inspire its protection in isolation from other concerns and influences (2000: 377).

This is not to say that religion, and particularly religiously inspired affect for entities like the Ganga, does not *at times* mobilize environmental efforts. Haberman's (2006) ethnography on the Yamuna documented the rise of religiously motivated environmental campaigns to save the river from pollution. My work with dam oppositions on the Ganga, detailed elsewhere (Drew 2011), showed that many of the movements were driven by a desire to ensure the continuous flow of a resource that people believe to be culturally and spiritually enriching and to which their identities as mountain residents and devout Hindus were linked. The exceptional significance that the Ganga held in everyday life inspired many to oppose its obstruction by hydroelectric projects. Although dams presented a short-term and highly visible threat to people's daily practices and climate change poses longer-term and perhaps harder-to-imagine impacts, a discourse of care for the river grew within these movements that eventually spilled over into some of the concerns of how the glaciers should be treated.

During the latter period of my research, for instance, some Uttarkashi residents became sensitized to glacial retreat while working in tandem with anti-dam activists from New Delhi and elsewhere. Incensed by the implications of the environmental reports they were briefed on, a few people from the dam opposition movements began campaigning for the glacier's conservation. One woman, who identifies as semi-literate, adamantly demanded that foot traffic from the temple town at Gangotri up to the glacial snout at Gaumukh be completely stopped, even though a cap was already in place that restricted the per-day number of visitors to 150. Her arguments against glacier tourism combined information learned from environmentalists and Hindu religious leaders with her own perceptions of the drivers of ecological change. She argued that, as more people travel to Gaumukh for adventure or to beg for spiritual salvation, their wastes and "impure" activities, such as meat and alcohol consumption, physically and ritually pollute the area around the fragile ice. The heat from the bodies approaching the glacier, she further asserted, warms the ice mass and increases the rate of its deterioration. The woman was especially angered by the disrespect that such activities showed the Goddess at her earthly source. For these reasons, she declared the tourism industry to be ecologically *and* religiously unsound, much to the ire of the residents and shop owners who survive on income from pilgrimage. Undeterred, she drew from notions of religious duty and right action, sometimes referred to as *dharma*, to argue that people should behave in ways that please the Gods and support the ecological contexts in which they reside.

Several other campaigns emerged during the same period (and have likely expanded), creating an assortment of occasionally overlapping efforts to clean the river, advocate for responsible tourism, promote a lighter ecological footprint along the Ganga's banks, and raise awareness about the long-term implications of climate change for everyday life in the region. The result was an increased process of exchange, the negotiation of diverse perspectives, and the emergence of new discourses to explain the stakes at hand.

Conclusions

By examining circulating discourses and epistemologies about ecological change and glacial melt near the Gaumukh-Gangotri range from which the Ganga flows, and scrutinizing the related scientific debates, regional interpretations, and Hindu religious framings, I have sought to illuminate the contexts in which people produce meanings about environmental transformations. I did so while highlighting the faith in the Ganga's perpetuity that a selection of people on Manikarnika ghat expressed despite observations of ecological change elsewhere. I also showed that many of my informants conceived of pollution and environmental degradation to be a signifier of our moral corruption. We have lost respect for the world and the Gods that inhabit it, they argued, and these Gods have become angry with us. To correct this, some offered suggestions that exceeded the call to merely cut back on carbon emissions. Perhaps, one woman asserted, if we overcame our moral shortcomings the glaciers would surge again.

With the exception of my informants at Manikarnika, residents living along the Ganga in Uttarkashi acknowledged the possibility that glaciers such as Gangotri-Gaumukh may be melting because of our global emissions and environmental hubris. At the same time, however, they added that these changes may not necessarily lead to the Ganga's total disappearance, because she will continue to exist in the heavens and in the underworld of *patal* through which she also flows. The comments of informants and religious figures also indicated how Hindu texts and cosmologies can support complacency or inaction, because they asserted that the Ganga's destiny (regardless of human activity) is to one day disappear from sight. Despite this, some of the people who perceived that human activities near the river's glacier were partly responsible for its deterioration began to draw from Hindu conceptions of right action to urge the protection of the ecological context in which the Goddess Ganga resides.

The points addressed reveal the different ways that people learn about and express concern for glaciers, their embedded ecosystems, and the divine entities with which they are associated. Although this discussion emphasized how certain perceptions and interpretations of ecological change inhibited proactive environmental efforts, it also indicates the growing awareness of human responsibilities vis-à-vis natural and sacred entities such as the Ganga. This is promising, as it signals the evolution and the potential hybridization of epistemologies about climate change phenomena that are beginning to compel people to action.

Notes

1 "Retreat of the Gangotri Glacier." Available at http://earthobservatory.nasa.gov/IOTD/view.php?id=4594 (accessed June 9, 2009).

2 The *yug* mark cycles of time in a progression from righteousness to degeneracy and back. The first, Satyayug, is a period of pure goodness. Tretayug, which follows, is characterized by a growing rate of sinfulness. In Dvaparayug, the world is supposedly half-bad and half-good. In Kaliyug, an age of corruption and turmoil, there is little goodness in the world. Each of these yugs has internal phases lasting thousands of years. When Kaliyug ends, the cycle starts over.

3 The *Srimad Bhagavatum* notes that Kaliyug may last 1,200 celestial years (Canto 3, Chapter 11, Text 19). Since one celestial year is equal to 360 human years, this equates to 432,000 years.

References

Alley, Kelly. 2000. "Separate Domains: Hinduism, Politics, and Environmental Pollution." In *Hinduism and Ecology: The Intersection of Earth, Sky, and Water*, ed. Christopher Key Chapple and Mary Evelyn Tucker, pp. 355–388. Cambridge, MA: Harvard University Press.

Crate, Susan A. and Mark Nuttall, eds. 2009. *Anthropology and Climate Change: From Encounters to Actions*. Walnut Creek, CA: Left Coast Press.

Cruikshank, Julie. 2005. *Do Glaciers Listen? Local Knowledge, Colonial Encounters, and Social Imagination*. Vancouver: University of British Columbia Press.

Cruz, R.V., H. Harasawa, M. Lal, S. Wu, Y. Anokhin, B. Punsalmaa, Y. Honda, M. Jafari, C. Li and N. Huu Ninh. 2007. "Asia." In *Climate Change 2007: Impacts, Adaptation and Vulnerability. Contribution of Working Group II to the Fourth Assessment Report of the Intergovernmental Panel on Climate Change,* ed. M.L. Parry, O.F. Canziani, J.P. Palutikof, P.J. van der Linden and C.E. Hanson, pp. 469–506. Cambridge, U.K.: Cambridge University Press.

Dobhal, D.P. 2009. "Climate Change and Glacier Retreat in the Indian Himalaya: An Overview." In *Climate Change at the Third Pole: The Impact of Climate Change Instability on Himalayan Ecosystems and Himalayan Communities*, ed. V. Shiva and V.K. Bhatt, pp. 67–76. New Delhi: Navdanya/Research Foundation for Science, Technology, and Ecology.

Drew, Georgina. 2011. "Ganga Is 'Disappearing': Women, Development, and Contentious Practice on the Ganges River." Ph.D. thesis, Department of Anthropology, University of North Carolina, Chapel Hill.

Eck, Diana L. 1996. "Ganga: The Goddess Ganges in Hindu Sacred Geography." In *Devi: Goddesses of India*, ed. John S. Hawley and Donna M. Wulff, pp. 137–153. Berkeley: University of California Press.

Gosling, David L. 2001. *Religion and Ecology in India and Southeast Asia*. New York: Verso.

Haberman, David L. 2006. *River of Love in an Age of Pollution: The Yamuna River of Northern India*. Berkeley: University of California Press.

Immerzeel, Walter W., P.H. Ludovicus and Marc F.P. Bierkens. 2010. "Climate Change Will Affect the Asian Water Towers." *Science* 328, 1382–1385.

Nagaranjan, Vijaya Rettakudi. 1998. "The Earth as Goddess Bhu Devi: Towards a Theory of Embedded Ecologies in Folk Hinduism." In *Purifying the Earthly Body of God: Religion and Ecology in Hindu India*, ed. Lance E. Nelson, pp. 269–296. Albany: State University of New York.

Naithani, A.K., H.C. Nainwal and K.K. Sati. 2001. "Geomorphological Evidences of Retreat of the Gangotri Glacier and Its Characteristics." *Current Science* 8. 1, 87–94.

Nautiyal, Shivanand, trans. 1994. *Skandapurānātargata—Kedārakhandam*. Allahabad: Hindi Sahitya Sammelan.

Orlove, Ben, Ellen Wiegandt and Brian H. Luckman, eds. 2008. *Darkening Peaks: Glacier Retreat, Science, and Society*. Berkeley: University of California Press.

Raina, V.K. 2009. "Himalayan Glaciers: A State-of-Art Review of Glacial Studies, Glacial Retreat and Climate Change." Government of India, Ministry of Environment and Forests. Available at http://moef.nic.in/downloads/public . . ./MoEF%20Discussion%20Paper%20_him.pdf (accessed February 7, 2010).

Rees, Gwyn H. and David N. Collins. 2006. "Regional Differences in Response of Flow in Glacial Fed Himalayan Rivers to Climatic Warming." *Hydrological Processes* 20, 2157–2169.

Shiva, Vandana. 1988. *Staying Alive: Women, Ecology, and Development*. London: Zed Books.

Sponsel, Leslie E. 2001. "Do Anthropologists Need Religion and Vice Versa? Adventures and Dangers in Spiritual Ecology." In *New Directions in Anthropology and Environment*, ed. Carole L. Crumley, with A. Elizabeth van Deventer and Joseph J. Fletcher, pp. 177–198. Walnut Creek, CA: AltaMira Press.

Vogt, Markus. 2011. "Climate Justice from a Christian Point of View: Challenges for a New Definition of Wealth." In *Religion in Environmental and Climate Change*, ed. Dieter Gerten and Sigurd Bergmann, pp. 85–106. New York: Continuum International.

3 Religion, indigenous knowledge and climate change in a mountain region

A case study of Thini village, Mustang, Nepal

Sujata Manandhar, Dietrich Schmidt-Vogt, Vishnu Prasad Pandey and Futaba Kazama

Introduction

Even though their contribution to climate change is quite small, climate change disproportionately affects developing countries (Huq et al. 2003). Mountainous regions in such countries are particularly susceptible. Even a small change in temperature can have dramatic impacts as ice and snow melt, which, in turn, affects water availability, biodiversity and agriculture (Tse-ring et al. 2010). These impacts are amplified when, as is often the case, these countries have little capacity to adapt. Nepal, the subject of this chapter, is such a nation. Climate change is taking a toll not only on the physical environment, but also on its social and economic wellbeing, and on the cultural integrity of its mountain communities (Adger et al. 2009).

Indigenous knowledge is a collective and adaptive type of knowledge that is passed on from generation to generation both orally and through practice. It is often described as being spatially and culturally context specific (Mistry 2009; Sillitoe 2000). In developing societies, religion is not only a medium for maintaining cultural values, but also a medium for preserving and passing on indigenous knowledge.

As such, religion can contribute to practical and ethical decision making concerning livelihoods and lifestyles, and to how human societies respond to challenges such as climate change (Wolf and Gjerris 2009).

That religious beliefs, traditions and indigenous practices are factors influencing people's response to climate change and variability has already been witnessed in mountain areas of Nepal (Manandhar et al. 2011; Regmi and Adhikari 2007). It is important, then, to understand how local people interpret and adjust to climate change and climate variability using indigenous knowledge based on religion (Sillitoe and Marzano 2009). In order to achieve this aim, it is necessary to document and evaluate religious beliefs, and indigenous knowledge and practices in relation to people's observations and interpretations of climate change. This chapter is an attempt to analyze the religious basis of land use practices in Thini village of Nepal, and to discern whether religion helps or hinders the capacity of Thini villagers to adapt to climatic vagaries.

Data collection and analysis

We used a variety of qualitative methods, observation, focus group discussions, and interviews with key informants. Qualitative methods create a concept of how the subjects under study—including their experiences, actions and interactions—relate to the context of study in different ways (Flick 2006). A focus group discussion with the elderly and elite community members in the village provided us with information on religion and indigenous cultural heritage, climate and environmental changes, and the villagers' responses to them. Key informants such as a village headperson and teachers were also interviewed to gather additional information. Triangulation, a fieldwork strategy consisting of the combination and cross-checking of various methods of data production (Holmelin 2010), was adopted to ensure the reliability of the information gathered in the field survey. By means of combining observations with interviews and focus group discussion, a better understanding of the various aspects of the issues was obtained. Additional information was collected from secondary sources such as reports, papers and books.

Description of the study area

The setting

Nepal is characterized by diverse eco-climatic variations, a result of an enormous range of altitude, from 60 to 8,848 meters above sea level. Nepal is divided into three distinctive altitudinal regions: Terai (lowlands), Hill (mid-hills) and Mountain (uplands), all within a distance of 150 kilometers of each other from south to north. Mustang is a Mountain district located in northwestern Nepal. Within Mustang district, the Thak Khola region extends from Ghasa village in the south to Jomsom (district headquarters of Mustang) in the north along the Kali Gandaki River. The region, occupied primarily by Thakali ethnic communities, has three distinct divisions: Thaksatsya, Panchgaun and Baragaun (meaning 700 Thakalis, five villages, and 12 villages, respectively) (Vinding 1998). Thini is one of the oldest original Thakali villages in the division of Panchgaun of lower Mustang. Thini village was selected as a study area because of its rich social and cultural heritage, and because of its accessibility, as it lies close to Jomsom, the district headquarters in the Thak Khola region of Mustang district (Figure 3.1).

The village includes four wards (ward numbers one, two, three and four) and belongs to the Jomsom Village Development Committee (VDC). VDC and ward are the fourth- and fifth-level units of altogether six units in the administrative hierarchy of Nepal. Thini is a large settlement of 84 households, with houses built wall to wall in a compact cluster. Geographically, Thini village lies in the trans-Himalayan rain shadow zone, which is the reason for its dry climate and desert-like landscape. The climate of Thini is, however, more humid than the climate in upper Mustang, adjacent to the north (NTNC 2008).

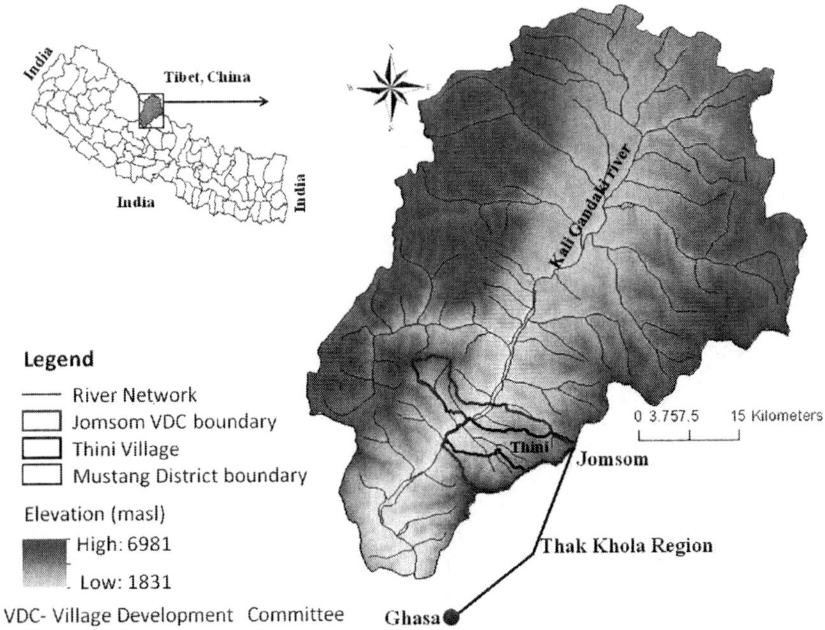

Figure 3.1 Map of the study area in western Nepal

Source: International Centre for Integrated Mountain Development (ICIMOD); NGIIP 2010; elevation from Jarvis et al. (2008).

Figure 3.2 Thini village with Mount Nilgiri in the backdrop

Source: Photo courtesy of Sujata Manandhar.

Farming system and livelihood options

Traditionally, the main livelihood activities of Thini residents consisted of agriculture, livestock and trade. In the past, some Thakalis of Thini were involved in the salt trade as brokers and traders. Presently they are engaged in different sectors such as teaching, government and non-government jobs and tourism, and some are employed abroad (NTNC 2008; Poudel 2008). For the majority of villagers, agriculture is still an important means of living. In the terraced fields, the main crops grown are cereals such as barley, naked barley, buckwheat and maize, as well as potatoes and other vegetables. Farming is mainly for subsistence. Apples and some other fruits are grown in orchards. Most of the cereals grown are consumed locally; some fruits and vegetables are sold at the local market, while most of the fresh apples and their derivatives (such as dried apples and apple brandy) are exported to cities. Recent development of roads has expanded the apple market and increased farmers' income.

Religion

Ethnically, the Thakalis of Thini village are categorized as Yhulkasompaimhi Thakalis; with respect to religion they are Buddhists practicing Tibetan Buddhism. Owing to their unique culture and history, the Thakalis of the Thak Khola have been studied extensively (e.g., Bista 1971; Fort 2000; Fuerer-Haimendorf 1975, 1978; Heide 1988; Messerschmidt 1982; Poudel 2008; Rai 1994; Vinding 1998). Thakali culture has a long history of religious transformations in response to political changes. From the seventh century to the present, religious preferences changed from Shamanic dhom, through Bon and Buddhism, to modern Hinduism and scientific atheism (Bista 1971; Fuerer-Haimendorf 1975; Messerschmidt 1982; Vinding 1998). For example, when their former trading partners in Tibet were no longer accessible following the closing of the border by the Chinese, the Thakalis of the Thak Khola began large-scale long-distance trade with Hindu populations, and when they did so they reoriented themselves from Buddhism towards Hinduism in order to accommodate the dominant status hierarchy of their new trading partners (Fuerer-Haimendorf 1978). This evidence reflects the cultural and economic flexibility of the Thakalis, and their ability to adjust to changing circumstances.

Before the introduction of Buddhism in the region, Thakalis believed in the Bon religion. There are two denominations, of Bon-Bhonnak (black) and Bhonggar (white). Adherents of the former sacrificed animals for religious purposes, while the latter did not (Poudel 2008). After the introduction of Buddhism, both Bon sects gradually turned towards Buddhism. According to Vinding (1998), Buddhism has existed in the Thak Khola region since the twelfth century. Most rituals and practices relating to birth, marriage, death and farming are regulated according to the Tibetan Buddhist astrological system.

Religion and environmental ethics

Buddhism is characterized by a non-anthropocentric view of the environment. Plants, animals and humans are perceived as components of an interrelated net of being (Clarke 1995; Cook 1989). Another characteristic element of the Buddhist worldview is that it stresses harmony with the environment, hence the importance accorded to regulating activities, especially land use practices, in accord with cosmic patterns, for example astrological cycles (Aubriot 2004). Both Buddhism and Hinduism deify nature, that is, they believe in the presence of supernatural beings in trees, stones and other natural objects (Ingles 1997). Such views explain, to some extent, the conservation ethic that is inherent in Buddhism and Hinduism, which is manifested in Buddhism by a restriction against killing animals. This includes using few or no pesticides.

As we have said, though, one can also see vestiges of earlier faiths, still. Animal sacrifice, for example, is alien to Buddhism, but animal sacrifice is still practiced in the predominantly Buddhist communities of Thini. Animal sacrifice in Thini is a remnant of earlier beliefs, in this case black Bon.

Religion and agricultural practices

The farming system in Thini is subsistence-oriented. Farming practices in the region are based on religious practices and beliefs. The religious basis of these practices is a composite of religions that the Thakalis have adhered to in the course of their history. The dominant component is Buddhism, but there are also vestiges of previous, animistic beliefs. Let us consider some of the ways that farming practices are shaped by religious beliefs.

First, farmers look to the Buddhist lama for guidance about when to plant. In Thini, traditional village committees (*Gaun Samiti*) exist side by side with official political institutions (Poudel 2008; Vinding 1998). The *Gaun Samiti*, which consists of male members of the Thini households, nominates two *Thuimis* (one headperson, or *Mukhiya*, and one vice-headperson), one *Bahidar* (village secretary) and four *Chhowas* (village messengers and caretakers of irrigation), representing one from each ward every year. It is the *Thuimis*' responsibility to arrange meetings and make final decisions regarding conflicts and all kinds of activities. Important dates, such as the date for buckwheat sowing, are determined by a Buddhist lama, the *Khaiwa*, who is also the village astrologer, expert in the Tibetan Buddhist astrological system. Therefore, the *Khaiwa*, in addition to being a religious authority, is also the keeper of traditional knowledge relating to farming in Thini. The post and related knowledge are usually inherited from father to son (Vinding 1998). The Tibetan Buddhist system of reckoning time is based upon a 12-year cycle where the years are named after animals. The cycle of 12 animals is combined with a cycle of five elements—fire (*me*), earth (*sa*), iron (*phay*), water (*chu*) and wood (*shin*)—to form a 60-year cycle. The relationship between any two of these elements is checked to find the right time to undertake a particular activity, for example on which date to sow buckwheat. This procedure is applied to buckwheat only, which is the opening crop in farmer's fields every agricultural year, followed by barley. The *Thuimis* obtain the sowing date from the lama, and the *Chhowas* disseminate this information to the villagers either by loud announcements made at selected spots or by visiting people. Farmers believe that if the planting process is carried out according to the lama's instruction weather and climate will favor them. They also believe that there will be no problem of pests and disease, and that the harvest will be good. However, the lama's declarations are not absolute. If, for some reason, farmers are not able to sow on the recommended date, they can decide to sow earlier; they are allowed to do so as long as they inform the *Mukhiya* and pay a penalty.

Thini villagers also seek spiritual help during times of drought. Groups of local women visit the nearby monastery and worship Lord Buddha through various religious rituals and prayers in the hope of rain. Villagers stated that it rained every time after such rituals were performed.

The villagers worship a big white stone, located in Thini, believing that that will protect crops from insects (and children from diseases). The stone is believed to be a symbol of the god. It has the contours of a yak and is thus named *Hyaley uma*, which means "yak" in the Thakali language. In past years, villagers have sacrificed a yak to appease *Hyaley uma* and believe that *Hyaley uma* has protected their crops and children from potential dangers. Since Buddhism prohibits killing living animals, villagers came to feel that sacrificing animals (in this case the yak) is wrong. The village chief initiated a new tradition in which instead of sacrificing an actual yak to *Hyaley uma* a dummy yak made from barley flour is "sacrificed" instead. This represents further evidence of the cultural flexibility of the Thakalis.

Figure 3.3 A big white stone in Thini worshipped as the god *Hyaley uma*

Source: Photo courtesy of Karna Bahadur Thakali.

Climate change impacts on agriculture

As the above examples show, until recently religion- and indigenous knowledge-based land use practices have, according to villagers' perceptions, proven adequate to help them cope with prevailing environmental conditions. Recent climatic changes have, however, emerged as a formidable challenge.

Nepal belongs to a group of countries which, as a result of their heavy reliance on natural resources for economic productivity and livelihoods, are going to be hardest hit by climate change and are, in fact, already feeling the impacts. On average, the temperature in Nepal is increasing at the rate of 0.06 °C per year. Temperatures are rising progressively faster at higher elevations (A.B. Shrestha and Aryal 2011). Rainfall is becoming more erratic and unpredictable throughout the country (Gurung and Bhandari 2008; A.B. Shrestha et al. 2000). A study by Manandhar et al. (2011) showed that, as rainfall and average annual maximum temperatures have increased in the last 30 years, snowfall in the Mustang region has decreased. These observations of increasing variability are clear indicators that climate change is already happening in Nepal, making dealing with climate change unavoidable.

Farmers in the Mustang region are keenly aware of these changes and they see both benefits and problems. Warmer temperatures, for example, are seen as beneficial for crop diversification, but, on the other hand, higher temperatures have also led to an increase in pests. Outbreaks of pests result in reduced harvests, especially in the case of Thini, where farmers refuse to use herbicides and pesticides for religious reasons.

Thini farmers have also complained that decreasing snowfall has reduced the levels of moisture in the ground, and that has severely affected the germination of barley seeds in the field and the regeneration of grass in pasturelands, leading to late and poor harvests of barley, and lower livestock productivity.

As these examples show, some changes in the climate benefit crops, but others constitute a danger. Heavy frost, which used to occur during the buckwheat season, has been less of a problem lately as a result of increases in temperature, and that has benefited buckwheat production. Negative influences are mainly due to recent irregularities in precipitation, such as rain, snowfall and hailstones, as experienced by villagers in Thini. Hailstorms have occurred when the buckwheat was in full bloom (around mid-July), causing damage to flowers, and leading to partial production loss.

Climate change undermines the relevance of tradition or religion for doing agriculture

Traditional beliefs are losing their ability to guide good agricultural practices. With hailstorms now coming later in the growing season, the lama's predictions of good weather conditions may prove to be misleading. Setting the date of buckwheat planting by consulting a lama (astrology) is becoming more irrelevant in the face of a changing climate. As temperatures have risen, pests have increased despite regular worship of the god *Hyaley uma*. Outbreaks of pests have been so devastating that the farmers were left with no choice but to abandon their Buddhist principles and use pesticides. These experiences and events reveal that religion-based indigenous knowledge no longer fully supports the capacity of Thini's villagers to adapt to climatic vagaries.

As we have already observed, rainfall patterns have changed. One of our interviewees, an old woman from Thini, told us that in the past rain would follow immediately after local women had performed religious rituals in the monastery, but that now they were no longer able to influence the weather in this way.

Summary and conclusion

A rapidly changing climate and more pronounced weather variability are expected to exacerbate stresses on ecosystems, resources and societies in the mountain regions of Nepal (K.L. Shrestha 2005). This study documents and provides insight into how land use based on religion and indigenous knowledge becomes increasingly problematic as climate changes in a mountain region of Nepal. Sowing buckwheat on the date instructed by a lama, subsistence farming without pesticide use, and worshipping both Buddha and *Hyaley uma* in hopes of protection from drought and pests are some of the indigenous practices that are based on Buddhism and beliefs that predate the advent of Buddhism.

Our research shows that religion-based indigenous knowledge and practices threaten to become inadequate in the face of current climatic variability, and may be further challenged by future climate change. At the same time, and compounding the problem, these traditional beliefs can also mislead people as they try to understand why these changes are happening. The old woman from Thini whom we quoted above bemoaning that religious rituals were no longer able to influence the weather also said that it has stopped raining at the appropriate time because the god is angry with the villagers.

To end on a more hopeful note, we also wish to observe that, although traditional religious beliefs are becoming less helpful as a guide to land use practices, there is enough flexibility built into the Thakalis' religious system for adjustments to changing circumstances to be possible in indigenous practices or even with respect to people's religious affiliation, as has happened in the past. Such flexibility, allowing for measures such as adapting the cropping calendar to their own weather observations or information from meteorological stations, while paying the penalty to the *Mukhiya* for not observing the dates determined by religious leaders, may help them to better deal with the vagaries of a changing climate. Future research should assist the people of this region as they endeavor to adjust to environmental change within their own cultural context, on the one hand by studying the validity of religion-based indigenous land use practices in the face of climate change and on the other hand by enhancing local capacity to monitor climate variability or change, and to adapt land use and livelihoods to their effects.

Acknowledgments

The authors would like to acknowledge the Japanese Government (Monbukagakusyo: MEXT), the GCOE Program of the University of Yamanashi for supporting this study, and Ms. Juliet Lu for her immense support in language editing and improving the chapter. We are also thankful to Ms. Lalita, Samita Thakali and the families of Thini, Jomsom for providing additional information and clarification on the studied topic.

References

Adger, W.N., S. Dessai, M. Goulden, M. Hulme, I. Lorenzoni, D.R. Nelson, L.O. Naess, J. Wolf and A. Wreford. 2009. "Are There Social Limits to Adaptation to Climate Change?" *Climatic Change* 93, 335–354.

Aubriot, O. 2004. "L'eau, miroir d'une société." *Irrigation paysanne au Népal central*. Paris: CNRS Editions.

Bista, D.B. 1971. "The Political Innovators of Upper Kali Gandaki." *Man* 6. 1, 52–60.

Clarke, G.E. 1995. "Thinking through Nature in Highland Nepal." In *Asian Perceptions of Nature: A Critical Approach*, ed. O. Bruun and A. Kalland, pp. 88–102. Richmond, Surrey, U.K.: Curzon Press.

Cook, F.H. 1989. "The Jewel Net of Indra." In *Nature in Asian Traditions of Thought: Essays in Environmental Philosophy*, ed. J.B. Callicot and R.T. Ames, pp. 213–229. Albany: State University of New York Press.

Flick, U. 2006. *An Introduction to Qualitative Research*. London: Sage.

Fort, M. 2000. "Glaciers and Mass Wasting Processes: Their Influence on the Shaping of the Kali Gandaki Valley (Higher Himalaya of Nepal)." *Quaternary International* 65/66, 101–119.

Fuerer-Haimendorf, C. von. 1975. *Himalayan Traders*. London: John Murray.

——. 1978. "Trans-Himalayan Traders in Transition." In *Himalayan Anthropology*, ed. James F. Fisher, pp. 339–358. The Hague: Mouton.

Gurung, G. and D. Bhandari. 2008. "An Integrated Approach to Climate Change Adaptation." *Leisa Magazine* 24. 4, 6–8.

Heide, S. von der. 1988. *The Thakalis of North-Western Nepal*. Kathmandu, Nepal: Ratna Pustak Bhandar.

Holmelin, N. 2010. "Farming Flexibility in Mustang, Nepal: Potentialities and Constraints under Conditions of Climate Change." Unpublished master's thesis, Department of Geography, University of Bergen, Norway.

Huq, S., A. Rahman, M. Konate, Y. Sokona and H. Reid. 2003. "Mainstreaming Adaptation to Climate Change in Least Developed Countries (LDCS)." Available at http://www.un.org/special-rep/ohrlls/ldc/LDCsreport.pdf (accessed April 16, 2010).

Ingles, A. 1997. "The Influence of Religious Beliefs and Rituals on Forest Conservation in Nepal." In *Nature Is Culture: Indigenous Knowledge and Socio-Cultural Aspects of Trees and Forests in Non-European Cultures*, ed. K. Seeland, pp. 57–66. London: Intermediate Technology Publications.

Jarvis, A., H.I. Reuter, A. Nelson and E. Guevara. 2008. "Hole-Filled Seamless SRTM Data V4." International Centre for Tropical Agriculture (CIAT). Available at http://srtm.csi.cgiar.org/SELECTION/inputCoord.asp.

Manandhar, S., D. Schmidt-Vogt, S.R. Perret and F. Kazama. 2011. "Adapting Cropping

Systems to Climate Change in Nepal: A Cross Regional Study of Farmers' Perception and Practices." *Regional Environmental Change* 11. 2, 335–348.

Messerschmidt, D.A. 1982. "The Thakali of Nepal: Historical Continuity and Socio-Cultural Change." *Ethno History* 29. 4, 265–280.

Mistry, J. 2009. "Indigenous Knowledges." In *International Encyclopedia of Human Geography*, ed. R. Kitchin and N. Thrift, pp. 371–376. Oxford, U.K.: Elsevier Science.

NGIIP (National Geographic Information Infrastructure Programme). 2010. "GIS Data of Nepal." Kathmandu, Nepal: Ministry of Land Reform and Management, Government of Nepal.

NTNC (Nepal Trust for Nature Conservation). 2008. *Sustainable Development Plan of Mustang, Kathmandu*. Jawalakhel, Lalitpur, Nepal: NTNC/GoN/UNEP.

Poudel, D.P. 2008. "Livelihood and Common-Pool Resources: A Study of Thini Village, Mustang, Trans-Himalayan Region of Nepal." Unpublished master's thesis, Department of Geography, University of Bergen, Norway.

Rai, R.K. 1994. *Along the Kali Gandaki: The Ancient Salt Route in Western Nepal*. Delhi: Book Faith India.

Regmi, B. and A. Adhikari. 2007. "Human Development Report 2007/2008." Available at http://hdr.undp.org/en/reports/global/hdr2007-8/papers/Regmi_Bimal%20and%20Adhikari_Anu_NEPAL.pdf (accessed April 20, 2011).

Shrestha, A.B. and R. Aryal. 2011. "Climate Change in Nepal and Its Impact on Himalayan Glaciers." *Regional Environmental Change* 11 Suppl. 1, S65–S77.

Shrestha, A.B., C.P. Wake, J.E. Dibb and P.A. Mayewski. 2000. "Precipitation Fluctuations in the Nepal Himalaya and Its Vicinity and Relationship with Some Large Scale Climatological Parameters." *International Journal of Climatology* 20, 317–327.

Shrestha, K.L. 2005. "Global Change Impact Assessment for Himalayan Mountain." *Global Environmental Research* 9. 1, 69–81.

Sillitoe, P. 2000. "Indigenous Knowledge, Science and the 'Poorest of the Poor,'" *Anthropology Today*, 16. 6, 3–7.

Sillitoe, P. and M. Marzano. 2009. "Future of Indigenous Knowledge Research in Development." *Futures* 41, 13–23.

Tse-ring, K., E. Sharma, N. Chettri and A. Shrestha, eds. 2010. *Climate Change Vulnerability of Mountain Ecosystems in the Eastern Himalayas*. Kathmandu, Nepal: ICIMOD.

Vinding, M. 1998. *The Thakali: A Himalayan Ethnography*. London: Serindia Publications.

Wolf, J. and M. Gjerris. 2009. "A Religious Perspective on Climate Change." *Studia Theologica: Nordic Journal of Theology* 63. 2, 119–139.

4 Climate change projects in the land of gross national happiness

Does religion play a role in environmental policy in Bhutan?

Matt Branch

> Throughout the centuries, the Bhutanese have treasured their natural environment and have looked upon it as the source of all life. This traditional reverence for nature has delivered us into the twentieth century with our environment still richly intact. We wish to continue living in harmony with nature and to pass on this rich heritage to our future generations.
>
> (His Majesty the Fourth King, Jigme Singye Wangchuck, in NEC 2002: 4)

Introduction

Bhutan, a small Buddhist country nestled in the Himalayas, is perhaps internationally best known for its development philosophy, gross national happiness (GNH), a holistic approach to governance that seeks to balance economics with personal well-being. A unique form of sustainable development, GNH has a strong commitment to environmental preservation. This is perhaps part of why Bhutan is considered to have such a healthy natural environment. Located in one of the world's biodiversity hotspots, the country is a net carbon sink, with an estimated 4.7 million metric tonnes of carbon sequestration occurring via its forests annually (National Biodiversity Centre 2011).

However, Bhutan is projected to suffer disproportionately from the effects of climate change, more than most other countries in the Global South. Climate change impacts are projected to be more intense in places with a steep topography, and Bhutan is amongst the steepest countries in the world, rising from less than 100 meters above sea level to peaks over 7,500 meters in a very short distance (Xu et al. 2009). In addition to increased severe weather events, the Himalayas' melting glaciers, sometimes referred to as the world's "third pole," may produce higher river flow fluctuations and glacial lake outburst floods (GLOFs). Bhutan has suffered three GLOFs since 1950, the most recent from the Lugge Tso glacial lake in 1994 which destroyed buildings, agricultural land, and livestock in the Pho Chu valley (Watanbe and Rothacher 1996). Increased cyclone activity has caused even more damage in the country via massive flooding and landslides, with the worst event occurring in 2009.

This chapter examines GNH and climate change to explore the relationship between religion and development in Bhutan. In particular I argue that GNH is both a product and a producer of a particular political culture, ostensibly rooted in Buddhism. For centuries Buddhism has played an important role in social and political life in Bhutan, and the roots of GNH are often attributed to the influence of Buddhist culture. Since its advent, GNH has both been infused with Buddhist meanings and also been a key aspect of the active reinterpretation of what Buddhism means. Additionally, I argue that although there is not much connecting climate change and GNH policy *thus far* there are some examples of how they are being discursively connected that demonstrate both the evolution of these linkages and why it is mutually beneficial for both donors and development agents in Bhutan.

To understand the intricate relationship between GNH and climate change in Bhutan, this chapter begins by examining the evolution of the relationship between religion and political life in Bhutan, from the formation of the country to its current manifestation today as GNH. The chapter then shifts to Bhutan's environmental policies, focusing on climate change mitigation and adaptation projects. These policies can be understood as an extension of this relationship between politics and religion, but with an important outside influence from international donors funding various environmental projects. The chapter continues by examining the goals and funding sources of climate change adaptation projects (which far outnumber mitigation projects in Bhutan) as a way to understand how religion and politics interact with these international forces. The chapter then discusses the relationship between GNH and climate change, both in policy documents and in interviewees' perceptions. The relationship between these two is still very much in nascent stages, and there is likely to be growth in this arena. The chapter concludes by summarizing the relationship between GNH, religion, and climate change policies in the country.

GNH and religion in Bhutan

There is a long-standing relationship between Buddhism and governance in Bhutan. As in Tibet, government of the pre-modern Bhutanese state relied on a principle known as the dual system (transliterated as *chos srid gnyis*). The term refers to the unity of religion (*chos*) and politics (*srid*) and implies both an institutional and an ideological interdependence between religious and worldly affairs (Ardussi 2004: 11; Kinga 2009: 10; Sinha 1968: 13). Under such a system, legitimacy of rule was also tied closely to religion. The state derived its right to rule from religious models of leadership, prophecy, karma, and social contract (Ardussi 2004). Although the understanding of the dual system has evolved over time as the country has shifted from a state ruled by religious figures to a monarchy and later to a democracy, the idea of a unified system of religion and politics continues to shape Bhutan. For example, the democratic constitution of Bhutan specifically identifies the dual system as one of the key roles of the monarchy (RGoB 2008: 2).

One interpretation of the dual system occurred under Bhutan's unifier and first political ruler, the Zhabdrung Ngawang Namgyal, who deliberately created social, political, and religious traditions to distinguish Bhutan from Tibet that continue to shape life in Bhutan today. A reincarnated lama, the Zhabdrung unified the state through his religious authority as much as his political and military strength. The state derived its right to rule from religious models of leadership, prophecy, and karma, and a type of social contract (Ardussi 2004). The Zhabdrung modified existing practices from religious institutions to create state-specific rituals like the *zhungdrel*, a procession used for formal state occasions, and *driglam namzha*, a system of courtly etiquette and conventions such as dress and architecture (Penjore 2011).

At the beginning of the twentieth century, Bhutan's political system radically changed when Ugyen Wangchuck consolidated enough power to have himself crowned king. Ugyen Wangchuck formalized his position through the signing of a contract (*genja*) by both secular and religious dignitaries. The contract positioned the monarchy above the monk body, placing the legitimacy of the throne on biological descent, and obtaining relative autonomy from religious institutions (Kinga 2009: 195–196). However, the ritual and symbolism of the monarchy remained rooted in religion. To this day Bhutan's king retains the privilege of wearing a yellow sash on formal occasions—an honor shared only with the head of the monk body, the Je Khenpo.

The concept of GNH was created by His Majesty the Fourth King, Jigme Singye Wangchuck, who coined the term in the 1970s in response to criticisms that Bhutan was not developing rapidly enough. Thus, GNH joined a growing wave of critiques and responses to the failings of GDP as a development metric. GNH was originally just a guiding principle for development and governance in Bhutan, with little international attention until 1998 when the prime minister of Bhutan, Jigme Y. Thinley, presented the concept of GNH at a UN meeting in Korea, which catapulted the concept to international prominence (Zurick 2006). The idea captured the attention of development workers, economists, and philosophers alike, and quickly became the most prominent idea connected to the country. International conferences ensued, as well as international funding, rapidly pushing the concept forward as an alternative development path.

Committed to putting the idea of GNH into practical operation, the government of Bhutan outlined four "pillars" of GNH. They are: economic growth and development, preservation and promotion of cultural heritage, preservation and sustainable use of the environment, and good governance (RGoB 2002). Around 2004, there was a lively debate over how GNH should be applied, as a guiding principle or as a development index with quantifiable measurements. The latter won out, and the Centre for Bhutan Studies was tasked with creating a national survey. The surveys are massive undertakings, with up to 290 questions connecting to 72 indicators in nine domains (these are: psychological well-being, health, time use, education, culture, good governance, ecology, community vitality, and living standards) (Ura 2008). Policy impacts from the survey are just beginning to be seen; for example, the first survey found that only a small portion of the country meditated on a daily basis, so daily meditation was introduced into the school system to encourage the practice.

Buddhism has also been an important part of the development of GNH. At the second GNH conference in 2004 the prime minister argued that GNH is just the latest incarnation of a long-standing Buddhist social contract wherein the happiness of the people is the concern of the state, which is why he claimed: "Traditional polity in Bhutan, drawing much on the Buddhist culture, was certainly guided towards GNH" (Thinley 2007). Likewise, abundant examples citing the connection between Buddhism and GNH exist in academic literature (Bates 2009; Mancall n.d.), news media, and policy papers (Ura 2009). Uddin, Taplin and Yu note that:

> GNH is rooted in the Buddhist philosophy and religion, which interprets nature as a living system rather than just a resource base to be exploited for material gain. In fact, the expression of GNH in Bhutan is essentially a summarization of the basic tenents [*sic*] of Vajrayana Buddhism, which encourages a culture of harmony and compassion [with nature].
>
> (Uddin, Taplin and Yu 2007: 2088)

While these papers largely cite Buddhism as a source of GNH, it is just as true that GNH has also encouraged particular interpretations of Buddhism.

In this regard, GNH has been part of the shift towards emphasizing the Buddhist values behind ecological policy and psychological aspects of Buddhism in Bhutan, incorporating both into its discourse. Whereas previously state Buddhism in Bhutan was part of a broad regulation of society or legitimation of the state, today GNH has shifted the focus of Buddhist governance toward the personal virtue of citizens. The emphasis on the individual, psychological aspects of Buddhism can be seen in Karma Ura's suggested education policy designed to promote GNH values. He places "meditation for mental training" and "prayers as an aspect of attitude training" at the center of his policy (Ura 2009: 26). Interpretations of GNH like this tend to focus on practices that were previously relatively uncommon, like meditation, or emphasize the psychological aspects of practices that were common but probably conducted for very different reasons, like prayers. One of the effects of this emphasis in the GNH discourse has been the increasing conflation of Buddhism and happiness.

Along with psychological well-being, GNH plays up the relationship between Buddhism and environmental protection. Although the bond between religion and landscape is not new, the particular formulation seen today is. Previously, saints were concerned not only with converting human beings to Buddhism, but also gods and demons that inhabit specific geographic locations. Similarly, some scholarship has pointed to the possible role of spirituality in limiting local natural resources use (Allison 2004). Under GNH, the association of Buddhism with geographically specific practices has given way to a construction of universal Buddhist morality (one certainly influenced by Western conceptions of Buddhism) that resists the materialism associated with modernization. Framing a series of interviews about GNH, Ross McDonald argues that, "In a Buddhist way of seeing, craving, a disregard of others and disrespect for nature lead inevitably toward conflict and sorrow" (2010: ix). Buddhist ideals of compassion and overcoming ignorance coincide, in this formulation, with environmentalist attitudes.

Of course, the shifts in the political culture that have coincided with and resulted from GNH do not mean older forms disappear. Frequently, the old and the new perspectives blend together. A recent petition started by a Bhutanese scholar at Oxford against creating a golf course in central Bhutan has argued that the golf course would threaten the lives of sentient beings and the sanctity of religious sites in the area, and the petition also opposes the materialism of golf with Bhutan's spirituality (G.K. Dorji 2011). GNH with its emphasis on conservation steps into and builds on this context, transforming in the process Buddhist involvement in development.

Each pillar of GNH has generated its own discursive field that allows for debate, action, and shared identity. This chapter will focus on the environmental aspects of GNH, specifically those relating to climate change. However, as already discussed, GNH is relatively new to political discourse in Bhutan. The next section will examine traditional practices and government policies that have helped the country become a net carbon sink.

Bhutan's environmental record

In several interviews, when asked about the relationship between climate change and GNH, several respondents suggested looking back to policies and traditional practices that existed before GNH was official policy or even a discussion. These respondents primarily pointed toward Bhutan's strong conservation history as being the primary reason for its current situation as a carbon sink. This section will first examine both the traditional practices and the early policies that led to the high percentage of forestation in the country, currently estimated at 70.4 percent (Pelden 2011).

Prior to 1961, when Bhutan shifted its policies toward a more Western form of development (such as coining currency, creating five-year plans, establishing a postal system, etc.), people's interactions with the environment were largely governed by pre-Buddhist animistic traditions. Forest spirits (*Sadag*), mountain deities (*Tsen*), and water spirits (*Lu*) populated the landscape, and people were not the sole owners of the land (L. Dorji 2003: 7). Each valley has its own local deities, both beneficent and malevolent, which were appeased according to local traditions (Zurcher and Choden 2004). Community enforcement of norms and behaviors regarding the treatment of these spirits limited and shaped the way people used the natural environment (Wangchuk 2001).

Sacred groves are a well-known institution in South Asia (Malhotra et al. 2001), but other local traditions also influenced people's behaviors. For example, *reedum* was/is a seasonal prohibition of forest use in Bhutan, most significantly limiting access to the forest during the summer monsoon, which is also when the landslide risk is the highest. While *reedum* is not as influential in people's actions as it used to be, it is beginning to regain prominence, especially with the recent introduction of community forestry in the country. Also important is the tradition of *sokshing* and *tsamdrog*, which are communally managed forests and pasturelands respectively. These commons promoted a sense of shared responsibility and management for the landscape; the government has formalized these two institutions, and they now require official registration. The last significant form of traditional landscape management is *tseri*, often referred to as slash-and-burn or long-term swidden agriculture. A type of shifting agriculture, *tseri* was the traditional form of most agriculture in the country, especially in the mid- and low altitudes. The 1969 Forest Policy Act outlawed *tseri* on public lands (where it was most commonly practiced), but recently there has been renewed interest in *tseri*, as research has demonstrated that it improves biodiversity in forests (during fallow periods) and crops would suffer less from the consequences of climate change.[1] Notwithstanding the importance of traditional practices, forests would not be as extensive as they are today if government policies had not also protected them.

Forest protection has been a goal of the Bhutanese monarchy since the first king was crowned in 1907. In 1911 the king sought assistance from the British to protect the southern forests from illegal logging, and in the 1920s several Bhutanese rangers were trained in India, initiating formal government management of the forests (RSPN 2010: 21). The Bhutan Forest Act of 1969 was a watershed act that formalized many interactions with the forests, and initiated a period of modernized forest management. It was the first act ever passed by the National Assembly upon its inauguration and nationalized all forests that were not under direct personal ownership (Social Forestry Division 2010: 8). Five years later came the first formal forest policy, which outlined Bhutan's goal of maintaining 60 percent of the total land cover under forest in perpetuity. This goal has since been included in the country's democratic constitution, which passed into law in 2008. Bhutan's commitment to 60 percent has also become an environmental marker and is often one of the first things mentioned when discussing environmental policy in the country. The most recent major act regarding forest policy in the country is the Bhutan Forest and Nature Conservation Act of 1995, which shifted away from modern forest management and recognized the importance of traditional forms of forest management (with the exception of *tseri*). The 1995 act also outlined rules for social and community forestry, a new development that has seen rapid success in recent years.

Success in meeting the 60 percent forest cover mandate rests in no small part on the country's high level of conservation areas and national parks. The first formal designation of a conservation area in Bhutan was in 1966 when what is now known as Royal Manas National Park was declared a wildlife sanctuary. Several other parks were created, but these were largely just parks on paper until the 1990s (NCD 2004: 4). In 1991, The Bhutan Trust Fund for Environmental Conservation (BTF), the world's first conservation trust fund, was established with an initial endowment of US$20 million funded by the UNDP, WWF, and several European countries, with the explicit purpose of staffing and legitimizing conservation areas in the country. BTF is still in operation, having more than doubled its endowment, and now supports a wide range of environmental projects in the country. In 1999 Bhutan instituted a series of biological corridors covering just less than 10 percent of the country, connecting all of the conservation areas (NCD 2004: 5). The latest addition to the conservation complex in Bhutan was in 2008, with the Wangchuck Centennial Park, which placed an additional 10 percent of the country under protection, bringing the total percentage just above 50 percent (Pelden 2009). This, of course, would not be possible if Bhutan did not subscribe to a people-in-parks model, which emphasizes buffer zones and seasonal grazing areas. While people within conservation areas are still able to collect firewood, commercial logging is not permitted. Thus far, the biggest monetary gain that has come out of Bhutan's 60 percent commitment has been through eco-tourism; however, the Royal Government of Bhutan is hoping to profit from the Reduced Emissions from Deforestation and Degradation (REDD) carbon trading scheme under the United Nations Framework Convention on Climate Change (UNFCCC). There is some skepticism about whether or not this will actually turn into an income-generating program for Bhutan, but research is underway to better quantify the ecosystem services that the forests are providing.

Bhutan's conservationist policies, matched with low population density, have resulted in the country having high forest cover and biodiversity. The high ratio of acres of forest per capita makes Bhutan one of a very small number of countries that are absorbing more carbon dioxide than is being produced within the state's borders. Although some would credit this to the insights of the monarchy, early conservation in the country had nothing to do with climate change. Additionally, Bhutan's steep geography and low population density have also strongly contributed to the country's current environmental condition. The policies described above have, however, laid the groundwork for international recognition of Bhutan as a "green state" and internally created a robust network of environmentalist professionals.

Bhutan's climate change policies and practices

Bhutan's status as a net carbon sink has meant that it has not had reason to focus on climate change mitigation projects. However, given its geography, it has had impetus for adaptation projects, and the majority of climate change projects in Bhutan reflect this need. This section will first look at the county's main mitigation project and potential future projects, and then focus primarily on adaptation projects that are currently underway in the country.

Bhutan, situated as it is entirely in the Himalayas, acts like a giant staircase, generating rapidly flowing rivers. Despite its small size, the country has four main river systems and an estimated 30,000 megawatts of hydropower potential (Garud and Gurung 2005: 10). As of 2011, Bhutan harnesses about 1,500 megawatts of this amount, accounting for over 50 percent of the government's funds (Meenawat and Sovacool 2011). Thus far all hydropower projects have been "run of the river" projects, which do not require a large reservoir, and are ostensibly more environmentally friendly. However, "run of the river" hydropower plants are more susceptible to seasonal variation, and there is a general concern that climate change will negatively impact the dependability of these power plants in Bhutan. Because of this, two of the proposed hydropower plants will be larger dams with significant reservoirs. Hydropower has largely had a negative reputation among environmentalists worldwide because of its major impacts on river ecosystems, but is finding renewed appreciation in the era of climate change. Hydropower emits drastically lower amounts of carbon dioxide than conventional fuel sources, and internationally some hydropower projects are finding funding through carbon trading funds such as the e7 Fund for Sustainable Energy Development, which has approved the Chendebji micro-hydroelectricity project, a 70-kilowatt project that offsets 524 metric tonnes of carbon dioxide every year. This project was largely financed by the Kansai Electric Power Company, a Japanese firm.

As a UN-designated least developed country, Bhutan has contributed little to the causes of climate change, and yet will likely suffer more than countries that have contributed the most. Its geography and dependence on the Himalayan glaciers will likely amplify the effects of climate change. Because of this, much of the activities that are directly related to climate change in the country revolves around adaptation, rather than mitigation. Bhutan was the first country to receive funds from the UNFCCC's Least Developed Country Fund (LDCF). Additionally, many people argue that they are already experiencing the effects of climate change, through a less predictable monsoon season, less snow in the winter, and stronger and more erratic weather.

Bhutan has thus far released several important documents regarding climate change. In its Initial National Communication to the UNFCCC, the government determined that the "six areas considered most vulnerable to climate change are: (i) forests and biodiversity, (ii) agriculture, (iii) water resources, (iv) glacial lake outbursts, (v) health, and (vi) landslides" (NEC 2000: 38). Bhutan was the first country to apply for LDCF funds, and was subsequently the first to receive funds from the LDCF, which was established to help countries prepare adaptation programs. Through the LDCF funding Bhutan produced the National Adaptation Programme of Action to Climate Change (NAPA) in 2006, which expanded on these areas, and prioritized nine projects of immediate importance:

1 Disaster Management Strategy (Pilot Implementation of Food Security and Emergency Medicine)
2 Artificial Lowering of Thorthormi Glacier Lake
3 Weather Forecasting System to Serve Farmers and Agriculture
4 Landslide Management & Flood Prevention (Pilot Schemes in Critical Areas)
5 Flood Protection of Downstream Industrial and Agricultural Area
6 Rainwater Harvesting
7 GLOF Hazard Zoning (Pilot Scheme—Chamkar Chu Basin)
8 Installation of Early Warning System on Pho Chu Basin
9 Promote Community-based Forest Fire Management and Prevention

(NEC 2006: vi)

International funding has supported two main projects in Bhutan related to climate change adaptation. The first is a collaboration between the Royal Government of Bhutan, Global Environment Facility (GEF), and UNDP, called "Reducing Climate Change-Induced Risks and Vulnerabilities from Glacial Lake Outburst Floods in the Punakha-Wangdi and Chamkhar Valleys." The project has a total budget of just over US$8 million, with funding coming from LDCF (51 percent), RGoB (32 percent), Austrian Development Agency (10 percent), UNDP (6 percent), and WWF (less than 1 percent) (NEC 2009: 31). This project has three main components (projects 2, 8, and 9 above), lowering the water levels of Thorthormi, a glacial lake that is highly likely to break its moraine dam in the next 15–20 years, the installation of an early warning system on the Pho Chu river (which is downstream of Thorthormi lake), and a community awareness and capacity building component in at-risk areas (Meenawat and Sovacool 2011: 528). Of the three components, the lowering of Thorthormi lake is by far the most technical and expensive. Thorthormi is located at very high altitudes and is inaccessible for most of the year. The goal is to lower the lake by 5 meters, but in the first year of work it had been lowered by only 67 centimeters (ibid.).

The second major climate change mitigation project is the Glacial Lake Outburst Flood Risk Reduction Project, which is primarily a research and public awareness project. The project cost is US$730,000 and it is funded through UNDP. Its primary goals are to create an inventory of glacial lakes and their risks, establish stronger emergency policies, and hold workshops. However, the UN is not the only funder of climate change adaptation projects in Bhutan. Denmark's DANIDA has funded more than US$875,000 in projects, primarily focusing on training and providing institutional support and infrastructure (NEC 2009: 55–63). There are numerous other smaller projects in the country that both government agencies and NGOs such as the Royal Society for the Protection of Nature (RSPN) oversee.

The National Environment Commission (NEC) is currently working on the country's second NAPA document, as well as a second national communication to the UNFCCC. This will undoubtedly bring in other sources of foreign funding for adaptation projects. This is important for the country, as domestic funding for climate change adaptation projects has to compete with the multiple other demands of a developing nation. The country's NAPA document is explicitly stated to be a tool to attract foreign assistance from the UN as well as other bilateral donors (NEC 2009: 31). However, this also brings up an important point regarding the religious foundations of climate change policy in Bhutan. If the funding is primarily coming from outside countries for climate change policy, then Bhutan likely has limited ability to influence the projects or ensure that they are in line with GNH (Nizhumizu 2008). For example, the following information was related in the methods section of the first NAPA document:

> The task force considered the application of Bhutan's (internationally popular) concept of "Gross National Happiness" (GNH) as a main criterion for prioritization of NAPA activities/projects, but decided that it would be preferable to break down the single "indicator" into several components that would

> be closer to internationally accepted criteria used to appraise project proposals. The task force then took the criteria recommended by the LDC Expert Group . . . as the starting point for the determination of suitable criteria.
>
> (NEC 2006: 30)

The decision to shift away from GNH as the main criterion for selecting development projects was likely a smart move when considering foreign funders for projects, but detracts from the ability of GNH as a guiding development principle to influence the country's direction. The next section will discuss the perceived relationship between GNH and climate change.

GNH's relationship to climate change

Although part of a longer history, GNH is still a new and evolving concept. It was not a prevalent discourse even in Thimphu until ten years ago, and many rural Bhutanese people still have not heard of the term, much less think about how the government's policies relate to GNH. In particular, GNH has been part of a reimagining of the meaning and role of Buddhism in society—the latest shift in the aforementioned dual system. GNH as good Buddhist rule is coming to be popularly understood among Bhutan's civil servants as facilitating an individual's ability to become happier and achieve enlightenment.

Environmental stewardship and protection have also been an emphasis of GNH itself. In addition to listing sustainable development and environmental conservation as two of the four pillars of GNH, Bhutan has proven to be open to trying novel policies in support of these four pillars. For example, the country banned the use of plastic bags to try to cut down on trash (with somewhat limited enforcement), enforces traditional architecture with the exception of metal roofing, which has dramatically cut down on timber consumption, and is currently pushing to be the first country to go 100 percent organic in agriculture by 2020. Bhutan could justifiably be called a "green state" (Eckersley 2004). Similarly, Buddhist concern over harming sentient beings was a contributing factor to government policy banning hunting and fishing in Bhutan.

Climate change also has a relatively short history in Bhutan's environmental policies, and thus far climate change policy, GNH, and Buddhism have been only loosely connected. Many informants, when asked about GNH connections with climate change, said that the reason why Bhutan has not had to implement explicit climate change mitigation policies is because government policies have always been in line with nature and preservation. One informant explained, "GNH was initially created in response to GDP, to consumerism and a sole focus on economy which are the causes of climate change. So we can think about GNH as an answer to climate change—not *the* answer, but one of them" (field notes, August 11, 2011).

Despite this idea that GNH is an answer to climate change, there have not been any policy documents that have explicitly dealt with GNH and its relationship to climate change; however, policy documents are now calling for the need for GNH to address climate change. For example, the "Strategizing Climate Change in Bhutan" white paper called for a rewriting of national documents to include climate change as part of the discourse (NEC 2009: 33). A draft document on biodiversity and climate change is even more explicit:

> While the environment is considered a crucial pillar of the national development philosophy of gross national happiness, the specific concerns of climate change and its impact on the environment in general and biodiversity in particular is lacking. Nevertheless, with the reality of climate change, it has become necessary for the country to address climate change concerns through the development of coherent policies and implementation of ecosystem based approaches.
>
> (National Biodiversity Centre 2011: 26)

One interviewee, when asked about how climate change relates to GNH, expressed the view that climate change would impact GNH, but that it should not be incorporated into the metric, rather that the government should keep it in mind when developing policies. Another said that climate change would clearly impact people's happiness, since it would be impacting their livelihoods.

Despite the short history of both GNH and climate change policies in Bhutan, we can see that the political climate in the country has facilitated interest in climate change mitigation projects, especially since funding for the projects has come from international donors. What is interesting about Bhutan is the evolution of discursive connections that can be seen developing between these two disparate concepts. As NGOs and government agencies pursue climate change funds, they are creating discursive connections between GNH and climate change. This creates a dual benefit: 1) GNH's reach is expanded into a new discursive field and brings in more funding opportunities from outside countries; and 2) funders of these policies are able to conceptualize their projects in a new way, that of facilitating happiness and not just mitigating climate change. Thus these projects are more appealing both for the international donors and for the local agencies. A perhaps purposeful or perhaps accidental byproduct is the creating of fields of interest for both GNH and climate change governance.

Conclusion

At first glance, GNH and religion seem to have little relevance for climate change policy in Bhutan. The problems of GLOFs and extreme weather events suggest strictly instrumental solutions that have more to do with mere survival than happiness or Buddhism. Indeed, the government has yet to create any official documents about how GNH relates to climate change, much of the funding for climate change projects is coming from international donors with international agendas, and domestic policies are largely being constructed because of these international discourses.

However, the relevance of GNH to climate change policy can be seen in the process of connecting GNH and climate change already underway. The Gross

National Happiness Commission (GNHC) and the NEC are working together to link climate change adaptation activities with other GNH goals such as poverty alleviation and gender equity (Meenawat and Sovacool 2011: 529). This is evidenced in two of the current projects that the NEC is linking to climate change projects, namely a rural biogas project and Bhutan's goal of 100 percent electrification by 2013. Similarly, GNH is brought up in discussions about climate change, and is being discursively connected to the issues that climate change is bringing about.

Bhutan's position as a net carbon sink demonstrates that the causes of climate change are largely coming from outside of Bhutan's borders; thus the government has little ability to mitigate climate change and must work to facilitate people's ability to adapt to a changing climate. Climate change has clear impacts on the environmental conservation pillar of GNH, but GNH's pillars are interrelated, and climate change policy is clearly impacting Bhutan's economic development and governance as well. There are general fears that climate change will have a negative impact on the country's economic development through the loss of reliable hydropower. Climate change is also expected to impact cultural preservation, as people's livelihoods will have to adapt to new systems and less reliable agriculture could increase urban migration. The RSPN is currently looking into traditional practices that might facilitate agricultural resilience, but this is likely only to slow, not reverse, the impacts of climate change. The last pillar, good governance, stands also to gain from climate change in the short term, mostly because of the increased revenue and government activities that are being sponsored by various international climate change programs (also benefiting the economy in the short term). Through expanding the government's activities and working on goodwill projects that largely do not come at the expense of Bhutanese taxpayers, goodwill towards the government will increase. One interviewee from the NEC noted that he had been tasked with writing a statement about the relationship between GNH and climate change. In this way, climate change projects are providing new territory for GNH to evolve in and become re-defined within a domestic discourse. Thus, GNH serves as a creative and legitimating cultural resource for policies regardless of their origin.

While climate change projects have to be approved by the GNHC, and are therefore ostensibly related to GNH, the projects themselves are funded by international organizations that bring along with them their own priorities and agendas. Thus climate change policies and projects in Bhutan are at best syncretistic with GNH, and, as seen in the example of the first NAPA document's task force, funders' priorities have taken precedence over GNH, and only later have government officials been tasked with connecting GNH to climate change policy. It is largely government officials and NGO employees who are creating links between GNH and climate change, and not monastic figures. Because GNH is tied to both religion and government, it is flexible enough to be discursively connected to climate change policy and projects, if only after the fact. Religion is being used discursively by some groups to advocate for or against climate change policies, but, at least in the example outlined in this chapter, a particular disposition towards climate change is not inherent in religious beliefs. Rather, interest in accepting or denying climate change, especially at the state level, relies much more on political economy. Studying the connection between GNH and climate change policy in Bhutan offers a number of important insights. Multiple players are influencing environmental policy in Bhutan, from the king down to everyday citizens. Likewise groups as diverse as the monk body, government officials in organizations like the NEC, and international donor organizations are still defining GNH. These disparate groups are not simply independent variables, but are responding to each other and influencing the shape of GNH. As one part of the current manifestation of the dual system, GNH continues to operate as a way to guide change in Bhutan, to respond to outside influence, and to place change into a meaningful context.

Note

1 Changes in groundwater have led to concerns that established cropland irrigation might not support crops; additionally one professional climate change specialist in the country claimed that the increased wind-storms of recent years are a result of climate change and are causing damage to corn and wheat crops. This damage would be much more limited in a traditional *tseri* system.

References

Allison, E. 2004. "Spiritually Motivated Natural Resource Protection in Eastern Bhutan." In *The Spider and the Piglet: Proceedings of the First International Seminar on Bhutan Studies*, ed. K. Ura and S. Kinga, pp. 529–563. Thimphu, Bhutan: Centre for Bhutan Studies.

Ardussi, J. 2004. "Formation of the State of Bhutan ('Brug gzhung) in the 17th Century and Its Tibetan Antecedents." *Journal of Bhutan Studies* 11, 10–32.

Bates, W. 2009. "Gross National Happiness." *Asian-Pacific Economic Literature* 23. 2, 1–16.

Dorji, G.K. 2011. "Signature Campaign against Course." *Kuensel*. Available at www.kuenselonline.com.

Dorji, L. 2003. "Assessing the Evolution, Status, and Future Implications of Forest Resources Management in the Inner Himalayas of the Kingdom of Bhutan." Ph.D. dissertation, Asian Institute of Technology, Pathumthani, Thailand.

Eckersley, R. 2004. *The Green State: Rethinking Democracy and Sovereignty*. Cambridge, MA: MIT Press.

Garud, S. and B. Gurung, eds. 2005. *Bhutan Energy Data Directory*. Thimphu, Bhutan: Department of Energy, Royal Government of Bhutan.

Kinga, S. 2009. *Polity, Kingship, and Democracy: A Biography of the Bhutanese State*. Thimphu, Bhutan: Bhutan Times.

Malhotra, K.C., Y. Gokhale, S. Chatterjee and S. Srivastava. 2001. *Cultural and Ecological Dimensions of Sacred Groves in India*. New Delhi and Bhopal, India: Indian National Science Academy and the Indira Gandhi Rashtriya Manav Sangrahalaya.

Mancall, M. n.d. "Gross National Happiness and Development: An Essay." In *Gross National Happiness and Development: Proceedings of the First International Seminar on Operationalization of Gross National Happiness*, ed. Karma Ura and Karma Galay, pp. 1–50. Thimphu, Bhutan: Center for Bhutan Studies. Available at www.un.int/wcm/

webdav/site/bhutan/shared/documents/GNH and Development 1st GNH conference.pdf.

McDonald, Ross. 2010. *Taking Happiness Seriously: Eleven Dialogues on Gross National Happiness*. Thimphu, Bhutan: Centre for Bhutan Studies.

Meenawat, H. and B. Sovacool. 2011. "Improving Adaptive Capacity and Resilience in Bhutan." *Mitigation and Adaptation Strategies for Global Change* 16. 5, 515–533.

National Biodiversity Centre. 2011. "Biodiversity Persistence and Climate Change in Bhutan." Draft version. Ministry of Agriculture, Royal Government of Bhutan, Thimphu, Bhutan.

NCD (Nature Conservation Division). 2004. *Bhutan Biological Conservation Complex (Living in Harmony with the Nature)*. Thimphu, Bhutan: Ministry of Agriculture, Royal Government of Bhutan. Available at Awsassets.panda.org/downloads/b2c2_20landscapeconservation_20plan.pdf.

NEC (National Environment Commission). 2000. *Initial National Communication under the UNFCCC*. Thimphu, Bhutan: Ministry of Agriculture, Royal Government of Bhutan.

——. 2002. *Bhutan: The Road from Rio*. Thimphu, Bhutan: Ministry of Agriculture, Royal Government of Bhutan. Available at http://www.nec.gov.bt/nec1/wp-content/uploads/2012/10/RoadfromRio.pdf.

——. 2006. *Bhutan National Adaptation Programme of Action to Climate Change*. Thimphu, Bhutan: Ministry of Agriculture, Royal Government of Bhutan. Available at http://www.preventionweb.net/english/professional/policies/v.php?id=8501.

——. 2009. *Strategizing Climate Change for Bhutan*. Thimphu, Bhutan: Royal Government of Bhutan. Available at www.rrcap.unep.org/nsds/uploadedfiles/file/bhutan.pdf.

Nizhumizu, M. 2008. *Portrait of a Leader: Through the Looking-Glass of His Majesty's Decrees*. Thimphu, Bhutan: Centre for Bhutan Studies.

Pelden, S. 2009. "More than Half the Country 'Protected.'" *Kuensel*. Available at www.kuenselonline.com.

——. 2011. "Forest Cover 70.4 Percent." *Kuensel*. Available at http://www.kuenselonline.com/2010/modules.php?name=News&file=article&sid=19686 (accessed July 31, 2011).

Penjore, D. 2011. "Rows of Auspicious Seats: The Role of bzhung gral phun tshog pa'I rten 'brel Ritual in the Founding of the First Bhutanese State in the 17th Century." Paper presented at the International Convention of Asian Scholars, Honolulu, HI.

RGoB (Royal Government of Bhutan). 2002. *Ninth Five Year Plan 2002–2007*. Available at http://www.gnhc.gov.bt/wp-content/uploads/2011/04/5yp09_main.pdf.

——. 2008. *The Constitution of the Kingdom of Bhutan*. Available at http://www.constitution.bt/.

RSPN (Royal Society for the Protection of Nature). 2010. *Bhutan's Natural Heritage: A Legacy of the Monarchs*. Thimphu, Bhutan: RSPN.

Sinha, N. 1968. "Chhos Srid Gnyis Ldan." *Bulletin of Tibetology* 5. 3, 13–27.

Social Forestry Division. 2010. *National Strategy for Community Forestry: The Way Ahead*. Thimphu, Bhutan: Royal Government of Bhutan.

Thinley, J.Y. 2007. "What Is Gross National Happiness?" In *Rethinking Development: The Proceedings of the Second International Conference on Gross National Happiness*, ed. Centre for Bhutan Studies, pp. 3–11. Thimphu, Bhutan: Centre for Bhutan Studies.

Uddin, S., R. Taplin and X. Yu. 2007. "Energy, Environment and Development in Bhutan." *Renewable and Sustainable Energy Reviews* 11. 9, 2083–2103.

Ura, K. 2008. "Gross National Happiness Web Site." Available at http://www.grossnationalhappiness.com/Default.aspx (accessed July 31, 2011).

——. 2009. *A Proposal for GNH Value Education in Schools*. Thimphu, Bhutan: Gross National Happiness Commission.

Wangchuk, S. 2001. *Indigenous Natural Resource Management Institutions of Bhutan*. Thimphu, Bhutan: DSB Publications.

Watanbe, T. and D. Rothacher. 1996. "The 1994 Lugge Tsho Glacial Lake Outburst Flood, Bhutan Himalaya." *Mountain Research and Development* 16. 1, 77–81.

Xu, J., R.E. Grumbine, A. Shrestha, M. Eriksson, X. Yang, Y. Wang and A. Wilkes. 2009. "The Melting Himalayas: Cascading Effects of Climate Change on Water, Biodiversity, and Livelihoods." *Conservation Biology* 23. 3, 520–530.

Zurcher, D. and K. Choden. 2004. *Bhutan: Land of Spirituality and Modernization—Role of Water in Daily Life*. Elgin, IL: New Dawn Press.

Zurick, D. 2006. "Gross National Happiness and Environmental Status in Bhutan." *Geographical Review* 96. 4, 657–681.

5 Pursuing diplomacy overseas, fostering adaptation at home

The Church of Bangladesh's proactive responses to climate change

James Pender

Introduction

While scientists have been increasingly able to measure the effects of climate change on the physical environment, less is known about what its effects will be on the social landscape. Bangladesh serves as a valuable case study of this important topic not only because climate change is expected to impact the country severely as a result of its particular latitude, geography and topography, but also because these changes have manifested themselves earlier here than in most countries. As a result, the process of climate change adaptation also began earlier here, aided by a strong civil society, numerous non-governmental organizations (NGOs) and a history of natural disasters. While secular efforts have been substantial, one important and relatively overlooked participant in the response has been the Church of Bangladesh, an independent Protestant church that has been active in the region since the mid-twentieth century. As a tiny minority community in a predominately Muslim nation, the Church of Bangladesh has punched above its weight in terms of advocacy, mitigation and adaptation in its service of the country's poor. Examining how culture, history and local theological perspectives have shaped the Church's work will highlight the important positive role religious institutions can play as part of a broader global response to climate change. Given that Bangladesh has contributed little to the problem, lessons can also be learned from how the Church of Bangladesh has responded to climate change as a justice issue.[1]

Methodology

The material in this chapter is primarily based on the author's own experience, supplemented by a review of project proposals, reports and other publications of the Church of Bangladesh. I have lived in Bangladesh for nine years, having been sent to work with the Church of Bangladesh as an ecumenical missionary supported by The Church Mission Society (U.K.), United Society, Council for World Mission, the Church of Scotland, and the Methodist Church (U.K.). I have been based in three locations: urban Dhaka, rural Meherpur and the small city of Rajshahi.

As a "development advisor" to the Church of Bangladesh Social Development Programme (CBSDP), my role is to support the social development work of the Church of Bangladesh, which ranges from grassroots work with village women to assisting with organizational development. Initial focuses were on anti-human trafficking and arsenic mitigation, but as concern over the issue of climate change grew among the leaders of the Church I offered to use my training as an environmentalist to help develop the Church of Bangladesh's response to this issue through resourcing with information, approaches and appropriate techniques. As I am actually part of the story I tell—not quite an insider but not really an outsider either—this undoubtedly leads to some biases but also yields insights which would otherwise not be possible. Very little of the work I describe here was work I did, yet I was involved in some way in most of it through supporting Bangladeshi colleagues and facilitating action—hopefully a good example of how a modern missionary can support local-led initiatives of a national church. In one sense this chapter could be seen as my reflections on how the work has developed, as my time of service of mission work draws to a close, and reflections on why the Church of Bangladesh engaged with social and environmental issues so readily while Christians in my home country, the U.K., often seem reluctant to do so.

Bangladesh's situation in regard to climate change

Few places in the world are projected to experience as wide and severe a range of effects from climate change as is Bangladesh. First, average temperatures are projected to rise. Research in Bangladesh between 1948 and 2004 has already found the minimum temperatures of months to be rising, particularly in the winter (Quadir 2009). In addition, the IPCC's Fourth Assessment Report estimates that warming in South Asia is likely to be above the global average (Christensen et al. 2007).

Second, the weather will increase in variability and disasters will become more frequent. A variety of different studies point to average rainfall increasing in Bangladesh during the summer monsoon by around 2 to 7 percent by the 2050s and predict slightly less rainfall in winter months by around 3 to 4 percent by the 2050s (Tanner et al. 2007). There may also be: more extreme hot and cold spells; less precipitation when it is most needed for agriculture, and more in the monsoon season when it already causes floods; melting glaciers in the source areas of Bangladesh's rivers, which will in turn alter the hydrological cycle; more powerful tornadoes and cyclones; and a sea level rise displacing some communities, turning freshwater saline and facilitating more powerful storm surges. Floods are expected to affect 18 percent of currently flooded land more severely and over 12 to 16 percent more of the country than at present (Ahmed 2006). Cyclones could have up to 20 percent faster wind speeds and 32 percent higher storm surges by the 2050s (Tanner et al. 2007). A further consequence of the increase in weather variability will be increased illness as a result of stagnant or flood water and a possible extension of the range of vector-borne diseases such as malaria (Walter and Simms n.d.). The changing weather patterns will also seriously impact agriculture, with a potential reduction of 17 to 28 percent for rice and 31 to 68 percent for wheat production in Bangladesh (Karim, Hussain and Ahmed 1999). The impact will be intensified by the fact that Bangladesh is both one of the most populated and one of the poorest nations on earth (Pender 2010).

Bangladesh has also been ranked third most vulnerable country in the world to sea level rise (McGranahan, Balk and Anderson 2006). Sea level rise and associated saltwater intrusion may contaminate fish ponds, cropland and drinking water, forcing around 43 million off their land by 2080 (Mohal and Hossain 2007).

It is in this challenging context that the Church of Bangladesh became closely involved with climate change advocacy, mitigation and adaptation efforts.

Bangladesh history and the Church of Bangladesh

History of the Church and the nation

The Church of Bangladesh was born on the Indian subcontinent through the work of various Protestant missions during the British colonial period. Although the East India Company initially vigorously opposed the arrival of missionaries lest they expose its abuses of power or hinder trade by upsetting Indian political and religious leaders (Mangalwadi 2001), in later times the missions were undoubtedly favored by the colonial authorities, evidence for which can be seen by the extensive compounds and churches of the Church of Bangladesh in the old centers of cities like Dhaka and Rajshahi near the courts and centers of administration.

Although the missions that brought Christianity may well have been Western, most of the conversions were not carried out by Westerners at all but by trained native people, as is being increasingly recognized by historians of missionary efforts around the world (e.g., Hanciles 2008). This pattern of native-led evangelization was also evident in the work of the Oxford Mission in Bengal (present-day Bangladesh and the state of West Bengal, India). There the combination of selfless commitment to service and high intellect modeled by its British Oxford University graduate missionaries attracted a number of high-caste Hindu Bengali converts who in fact became its most successful evangelists (Robinson 2002).

As a result of having well-educated, confident and cultured Bengali evangelists, the traditions and practices of the Christians continued to exhibit many attributes common to traditional Hindu and Muslim worship in Bengal. This includes the removal of shoes and sitting on the floor during worship, and the traditional folk style of singing. Expressing faith through acts of service or charity is also an important characteristic of Christianity in Bangladesh, deriving from the earlier established religions of Islam (where this principle is known as *zakāt*) and Hinduism (where it is considered a route to salvation, or *dharma*). (Although acts of service or charity are an important aspect of the Christian Gospel, they have not normally been seen as an "expression of worship" to the same degree by the Western Church in recent centuries.)

There is also a significant appreciation and reverence for nature in traditional society, which is likely to have influenced and encouraged environmental action by the Church of Bangladesh. This appreciation manifests itself most strongly in the poems and folk songs of the Bengali literary tradition. For example, the work of Rabindranath Tagore and the mystic Lalon Shah invariably talks about the natural world and how we can encounter our creator through appreciation of its beauty, its natural cycles, and daily toil within its confines. In "Stream of Life," for example, Tagore writes:

> The same stream of life that runs through my veins night and day
> runs through the world and dances in rhythmic measures.
> It is the same life that shoots in joy through the dust of the earth
> in numberless blades of grass
> and breaks into tumultuous waves of leaves and flowers.
> It is the same life that is rocked in the ocean-cradle of birth
> and of death, in ebb and in flow.
> I feel my limbs are made glorious by the touch of this world of life.
> And my pride is from the life-throb of ages dancing in my blood this moment.
>
> (1913: 86)

The nature-venerating poetry of Tagore and others is included in a section of the Church of Bangladesh hymn book.

The turmoil associated with the struggle for freedom from colonialism and the break-up of the British Indian Empire earlier in the century was also formative in the life of the nation and in the history of the churches in Bangladesh. During this period the situation for Christians changed dramatically. After 1947 when the Indian Empire was split into the independent nations of India and Pakistan, Christians suddenly became a small minority. This was particularly the case in East Pakistan (now Bangladesh), where today only around 0.3 percent of the population are Christians of any denomination (even now the Church of Bangladesh numbers only around 20,000). In response, the English Presbyterian Mission, the Oxford Mission and the Church Missionary Society decided to unite their small congregations to create the Church of Pakistan. Following a liberation war, the Church was renamed the Church of Bangladesh, and brought under indigenous leadership (Baroi 2007a).

Splitting up the British Indian Empire was a traumatic affair causing much inter-sectarian violence (Collins and Lapierre 1975). The leaders of the new nation of Pakistan soon became antagonistic towards their minorities, especially in East Pakistan, where the mostly-Hindu minority initially numbered around a third of the population. Discontentment with the perceived second-class treatment of the eastern part of Pakistan mounted, inflamed by the declaration that Urdu would be the state language there, even though the common language was Bengali and very few people could speak Urdu. This gave rise to the Bengali nationalist movement, which pressed various demands, and eventually even for independence. In the meantime, persecution of minorities became more intense. The Pakistani authorities in Islamabad blamed intellectuals and religious minorities for the protests and persecuted them vigorously (Muhith 1978; Plunkett et al. 2000). The defining moment for Bangladesh was the "Language Movement" and its culmination the "Liberation War" from Pakistan in 1971, when the proud Bengali people fought for their right to use their mother tongue Bangla, or Bengali. The bitter and bloody war of independence incidentally brought some inter-community unity as Hindus, Christians and Muslims fought shoulder to shoulder against the Pakistani government. After the genocide, atrocities, and death of 3 million Bengalis in nine months of fighting, a secular constitution was established. Under the new government, however, evangelism and proselytizing became difficult and at times dangerous, and were discouraged by state institutions. This led churches to reconsider how they could express their faith in the new context within which they found themselves.

Consequence of this history: the Church adopts "works of service" approach

With the end of colonialism also went the Church's privileged position. Now that evangelism could no longer be the prime expression of its witness, the Church had to redirect its focus. Having experienced persecution from the state of Pakistan, church leaders were initially cautious, yet eventually the immense needs not only of their own members but of their countrypeople at large pulled them outwards. In particular, the devastation caused by the war had caused famine, starvation and hardship for the people of the newly liberated Bangladesh, leading to a great need for relief and aid support (Muhith 1978; National Museum, Dhaka pers. obs. 2004). As organized institutions, churches (including the Church of Bangladesh) became important conduits for channeling external assistance, especially that from faith-based organizations abroad with which they already had international links.

The Bengali cultural practices of expression of faith through acts of compassion and a love of the natural world, as well as the new political realities of Bangladesh (particularly the discouragement of proselytism and the humanitarian crisis), led Church leaders to decide in 1984 that the Church of Bangladesh should seek to focus on sharing the Gospel through works of service (for all faith, caste and ethnic communities). This was accomplished through the establishment of the Church of Bangladesh Social Development Program (CBSDP) in 1984 (CBSDP 2004). Under the direction of its first indigenous national leader, Bishop Barnabas Mondol, the Church adopted John 10:10 for the vision statement of the CBSDP: "I have come that they may have life, and may have it in all its fullness," with the explanation: "In the light of Christ's teachings, CBSDP desires and seeks fulfilled communities, living in justice and peace with all of God's creation" (CBSDP 2011). This included the principle that "The environment will be safeguarded in order to sustain present and future generations' livelihoods and life requirements, such as food and clean air to breathe, as well as to be faithful stewards of all God's Creation" (CBSDP 2004). The inclusion of phrases like "safeguarding the environment" and "living at peace with creation" in its constitutions demonstrated the Church's conviction that development needed to be sustainable, perhaps growing out of a strong understanding of rural issues, as most of its Church members have livelihoods fully or partly sustained by agriculture.

Given that as recently as 2005 it was estimated that 40 percent of the population was living below the national poverty line, and 49.6 percent on less than $1.25 per day (ADB 2011), humanitarian work has continued to be necessary. Today CBSDP works in 16 districts through 14 area offices with 365 staff operating programs in: social empowerment; climate change, food security and disaster management; community health; education and training; and microcredit. Between 40,000 and 50,000 people are beneficiaries of these programs, including over 17,000 women's development group members. Thus the beneficiaries of the Church of Bangladesh's efforts to alleviate poverty and assist the most vulnerable actually outnumber the members of its congregations!

The Church's response to climate change

The Church's first environmental work

The Church has engaged in environmental work since at least 1982, when one component of CBSDP's interventions was "social forestry." In hindsight it can also be observed that many of CBSDP's activities since then have had the effect of mitigating climate change. This includes tree planting and helping communities adapt to increasingly severe weather.

The Church takes note of the climate issue

With the publication of the Stern Report in 2006 by an economist for the British government (Stern 2006), the world realized that climate change was not just a concern for environmentalists but would have an impact on all aspects of human society. This was also the year that the Church of Bangladesh entered the climate change arena with the visit of Bishop Michael S. Baroi, the Church's moderator, to Edinburgh. There he gave a speech on the impact of climate change in Bangladesh to the Church of Scotland General Assembly (Pender 2009). Since then two other Church of Bangladesh bishops, Paul Sarker and Samuel Mankhin, have also spoken frequently on the threat of climate change to Bangladesh as they have traveled abroad.

As a "united church" made up of components from various missions and church denominations, the Church of Bangladesh has played an active part in a number of international church groupings such as the Anglican Communion and Council for World Mission. It also has international partners ranging from Reformed churches, to Presbyterians, to Methodists, to Anglicans, to the United Church of Christ. Thus the Church of Bangladesh is well placed to play an important advocacy role for action on climate change within a good number of church organizations. As many Bangladeshis are living in poverty and Bangladesh's emissions of greenhouse gases are low, the leaders of the Church of Bangladesh recognize that the solution to climate change is largely out of their hands and lies with developed countries or rapidly industrializing ones, mostly in the West. While the Church of Bangladesh can therefore do little of international significance in terms of mitigation in its own country, it does have many friends and partners in the Christian as well as development communities in the West, particularly in Europe and North America, which are among the worst culprits in terms of greenhouse gas emissions.

What the Church has done

Externally

In 2006 Bishop Michael S. Baroi spoke powerfully of the need for Western countries to take action on climate change and its devastating threat to life in Bangladesh both at the Church of Scotland General Assembly and at a conference of over 100 bishops of the Episcopal Church of the U.S.A. In these speeches he urged them to influence their members, societies and governments. In a subsequent letter to the editor of *Green Christian* magazine published in Summer 2007 by the U.K.-based organization Christian Ecology Link, he wrote:

> I really want to be sure that my grandchildren, and their children after them, will be able to enjoy the beauty of my country that I have enjoyed, and be able to have enough land to live, and enough land for food. May I humbly plead to you all to continue the fight for a greener environment, continue the fight against climate change for the sake of my country, my people and for the world at large.
>
> (Baroi 2007b)

These efforts to highlight the threats of climate change to Bangladesh have continued with subsequent bishops and lay leaders of the Church of Bangladesh at the many meetings that they attend around the world.

To date the Church of Bangladesh has had perhaps the greatest impact in Scotland. The General Assembly of the Church of Scotland was so moved by Bishop Baroi's speech that they created a new project called "Responding to Climate Change." The project reflected growing concerns in the Church of Scotland and its congregations about the impact of climate change not only in Scotland but around the world. It encourages people in Scotland to take action to reduce their emissions by providing practical resources for greening churches, including an "eco-congregation" scheme which accredits churches for their achievements in reducing their impacts on the environment. It also advocates for national and international action through engagement with regional and national government policy makers.

The Church of Bangladesh Social Development Programme has also enabled poor villagers such as members of CBSDP-facilitated village organizations (committees formed from poor women beneficiaries of the CBSDP in a particular location) to get involved in advocacy when the opportunity presents itself. In one case when a British Methodist minister studying climate change for his sabbatical went to Govipur village organization, the women told him in no uncertain terms to take a message back to the U.K. that "people there should stop pollution, look after the environment and stop emitting greenhouse gases, for Western countries are increasing greenhouse gases but it is we in Bangladesh that are suffering!" The minister shared this message at the UK Methodist Church Conference in 2007 and was later quoted on BCC TV News for Reporting Scotland in the lead-up to the 2010 United Nations Framework Convention on Climate Change summit in Cancun. A visit by the moderator of the Church of Scotland to Bangladesh during 2010 also caused him to share the climate-related concerns raised by people he had met on the same TV program. Perhaps even more significantly, the moderator raised the issue a few weeks later when he visited the UK prime minister David Cameron. During the meeting he gave Cameron a summary of the report *Climate Change, Its Impacts and Possible Community-Based Responses in Bangladesh* (Pender 2010), which had been produced by the Church of Bangladesh. The full report has been widely circulated, having been published on the websites of Norwegian Church Aid, the Church of Scotland and the University of Sussex's Institute of Development Studies' ELDIS. It was also used by a joint Bangladesh–British parliamentary sub-committee on climate change and Bangladesh, the Ministry of the Environment in Norway, NORAD and the director of the Bangladesh Government Climate Change Cell. An additional 500 copies have been translated into Bengali and distributed to organizations within Bangladesh, which is significant, as it constitutes the first comprehensive report on climate change published in the Bengali language.

The Church of Bangladesh has additionally taken leadership on the issue within its regional ecumenical relationships. In September 2008 it hosted a consultation of the Council for World Mission South Asia Region (CWM-SARC) titled "Climate Change and Water: Challenges and Responses of South Asian Churches," which brought together over 40 people, including delegates from churches in North and South India, Sri Lanka and Nepal, as well as Bangladesh. A focal point of the discussion was how to be effectively involved in advocacy and to carry out climate change adaptation. It also stressed that delegates must take action personally, in their localities, and examine climate change from a theological perspective. At the end of the consultation a jointly agreed commitment was made to implement practical measures to address climate change within individual churches, corporately and within wider society (Pender 2009).

Internally

Beyond the international arena, CBSDP has also had some success in the local arena, where since 2007 it has been engaging in climate change mitigation activities. Although the country's low emissions mean the onus is not on it to mitigate, it was soon discovered that, once people heard about climate change, for various personal and utilitarian motives they were very keen to take action themselves. Church leaders decided that to deny them that possibility was to disempower them, rendering them essentially helpless to engage with the crisis. Fortunately there is a wide range of measures that would not only reduce emissions but also confer a variety of community benefits. This includes encouraging the use of organic fertilizers, planting fruit, timber and medicinal trees, and installing smokeless fuel-efficient stoves.

CBSDP has had a particularly significant impact in regard to tree plantation in the hot, moderately drought-prone Meherpur district. In 2008 alone 40,500 trees were directly distributed to development group members and other beneficiaries of the CBSDP, while technical support was given to 40 village nursery holders, who themselves supplied 300,000 trees to the community during the year. These trees were mainly fruit-producing species, though timber and medicinal trees were also popular. In addition to sequestering carbon, they can provide shade, food, medicine, timber or income. Beyond homestead plantation, in Meherpur district CBSDP has planted over 67.5 kilometers of roadsides with trees that both sequester carbon and benefit agriculture, as the trees prevent winds that erode soil and increase evapo-transpiration rates (Pender 2009).

Another carbon-reducing project CBSDP promotes is the installation of smokeless and fuel-efficient stoves. It installed 150 in homes in Rajshahi division between 2009 and 2011. Such stoves are beneficial to the environment, because they reduce the denudation of local vegetation by burning fuel more efficiently. They are even more important in mitigation, because in addition to reducing greenhouse gas emissions they also reduce "black carbon" emissions. Black carbon is a major cause of retreating glaciers, snow and ice melt in the Himalayas, because when it resettles on ice or snow it changes the snow's albedo, thereby increasing melting (*Developments* 2009; Moral 2002). Smokeless stoves are also beneficial, because they reduce family respiratory illness from smoke inhalation, reduce the work of (mostly) women in collecting fuel, and benefit family budgets when fuel needs to be purchased.

While mitigation may be an "optional" focus in Bangladesh, adaptation measures are urgently needed and essential if communities are to continue to survive in the face of climate change. Among the villages where CBSDP operates, those that are most vulnerable to climate change are in the drought-prone Barind tract (with its poor clay soils in the northwest of Bangladesh in Rajshahi division) and those in the floodplain of the Brahmaputra (or Jamuna) River (in Barisal and Gopalganj districts). Therefore this is where CBSDP has so far focused most of its actions regarding climate change adaptation and disaster risk reduction.

To assess what most needed to be done, during 2010 CBSDP conducted "environmental assessments" on its project activities around the country, which gathered data on the impact of climate change, environmental degradation and natural disasters via secondary sources and village-based focus group discussion.

The assessment for the districts of Natore and Rajshahi districts of Rajshahi division showed that the groundwater table was dropping fast, that droughts were increasing and that natural resources were declining. As the people with whom the CBSDP works are the poorest of the poor, with little land, the adaptation measures it has elected have been small in scale, such as measures that people can undertake around their homes, particularly in their small vegetable gardens. These have included various ways of producing compost that improve soil texture and moisture retention, the use of old plastic bottles for drip irrigation in order to provide a constant supply of water to plants throughout the day, and the use of old sacks filled with compost for growing vegetables on infertile or compacted land around homes. With support from the University of Rajshahi CBSDP has also been collecting, conserving, distributing and promoting over a dozen species of naturally drought-resilient wild plants that were formerly collected from uncultivated land and used as vegetables by local communities but are now scarce.

In the floodplains of Gopalganj and Barisal districts where severe flooding regularly occurs, climate change adaptation is closely tied to disaster risk reduction (which is itself sometimes considered part of climate change adaptation). Much of the CBSDP's work there has involved raising awareness and promoting safe practice in the event of the floods and other disasters that are predicted to accompany climate change. This has included: building 17 flood or cyclone shelters; setting up village disaster management committees in 50 villages; land raising in key places such as roads or around flood shelters, churches or temples; raising tube-wells above the level of floods so they are not submerged; and clearing a length of canal so that flood water may drain away from the land unimpeded.

Other climate change adaptations include the promotion of *baira*, an ancient form of hydroponics. Also known as "floating gardens," *baira* are among the best-known agricultural adaptation techniques in Bangladesh. This technique developed centuries ago in the seasonally flooded areas in the south of Bangladesh, though more recently researchers have modified the design to obtain optimum yields. Until 40 years ago rice straw was the main material used to make *baira*, but since then water hyacinth plant material has taken its place. To make a *baira*, a bamboo pole the length of the required floating garden is first placed over a floating bed of water hyacinth; then more water hyacinth and other rotting aquatic plants are pulled together. After a week or so more decomposable plants (such as water hyacinth) from the water bodies are piled on top. Using bamboo poles to keep the *baira* in place, either a ball or a cushion made from aquatic plants or coconut husks is then assembled to make a bed ready for seed planting. Both summer and winter vegetable seedlings are cultivated on *baira*, which simply rise and sink along with flood waters. When finished the material is used as compost and the produce sold (IUCN 2005). Because *baira* provide a means of producing food even during periods of excessive precipitation, they are an important means of adapting to climate change.

Since 2010 the Barisal–Gopalganj office of the CBSDP has also been implementing a new approach to climate change adaptation in Bangladesh through its Tearfund-supported project "Mobilizing the Church in responding to the threats posed by climate change." As in many churches around the world, the Church of Bangladesh congregations in Barisal Deanery have become rather self-centered, focusing on internal politics, meetings and religious practices revolving around their church buildings, despite the outward-looking approach of the Church of Bangladesh as a whole. To counter this inward focus, this project has trained "community animators" from 25 congregations, including church committee members, clergy and church workers. These animators are tasked with raising awareness, starting village disaster management committees and planning community tasks like raising the level of village connecting roads and growing and distributing tree seedlings or vegetables grown on church land. In addition to working in their own villages, each parish also chooses a neighboring non-Christian village to serve with the same initiatives. The project provides training and support so that climate change or disaster risk reduction messages and training can be integrated into the activities of the Sunday schools, women's guilds, golden (old) age groups, youth groups, and health and education programs of the Church of Bangladesh. Workshops for clergy and church leaders on the theology of environmental care, climate change action and social work have meanwhile sought to provide a base for deepening the spirituality of the Church of Bangladesh. Overall it is hoped that this project will result in a much more outward-looking Church, not just at the institutional or leadership level but all the way down to individual Christians in remote villages.

Discussion

The Church of Bangladesh has played a significant role in developing responses to climate change, because it sees the Church as a vehicle to serve the poor and needy in society, and because it is convinced that the Church should be at the heart of efforts to bring justice to the downtrodden in both humanity and nature. Although CBSDP's efforts in climate change adaptation and advocacy have been relatively small compared to those by large secular non-governmental organizations in Bangladesh, its work is widely respected within community-based adaptation to climate change networks. Furthermore while other faith-based development organizations or religious bodies may have engaged with either advocacy or adaptation in Bangladesh, they have not matched the scope of those of the Church of Bangladesh, which spans theology, church congregations and multi-faith beneficiary communities as well as practical action in climate change advocacy, mitigation and adaptation. It is expected that the Church will continue to walk in this tradition. In the words of Bishop Paul S. Sarker, moderator of the Church of Bangladesh:

> Consumerist human society has been destroying creation for some time in many ways and ignoring the consequences of it. One of the reasons may be that today we have less faith in God, and that we depend on our own power. Faith in God helps us to understand our responsibility to take care of the creation and the people of this wonderful earth as stewards. We are living in a planet that can't be compartmentalized. We are one human family living together in God's world, for we not only depend on each other but also learn from each other. Thus to protect this planet from destruction we need to take necessary and bold action together as one international family without delay!
>
> (cited in Pender 2010: 2)

Whether Christian, Muslim, Hindu, Buddhist or animist, many in Bangladesh and around the world have gotten involved in action on climate change and the environment as a result of faith convictions. As former UK prime minister Tony Blair has said, "faith matters—whether you are religious or not. It matters because it inspires people to act . . . yes some of the worst actions of recent times have been committed by people of faith. But also some of the best" (Blair 2010). As this case study has demonstrated, the Church of Bangladesh has been able to channel theological and disaster relief-related concerns toward significant action on the front of climate change. While it is only a small part of what must in the end be a global effort, it amply demonstrates the impact that a faith-based organization can have when the right mixture of leadership, concern and resources is assembled. The story of how the Church of Bangladesh began working to counter the threat of climate change is therefore important in highlighting the work of faith-based organizations and non-Western Christian perspectives.

Note

1 The country emits only 0.2 tons per capita of greenhouse gases (Zia-Begum 2002), and just 0.24 percent of the global emissions of carbon dioxide (the main greenhouse gas) (Quadir 2009).

References

ADB (Asian Development Bank). 2011. *Asian Development Bank and Bangladesh: Fact Sheet*. Dhaka, Bangladesh: Asian Development Bank.

Ahmed, A.U. 2006. *Bangladesh Climate Change Impacts and Vulnerability: A Synthesis*. Dhaka, Bangladesh: Climate Change Cell, Bangladesh Department of Environment.

Baroi, M.S., ed. 2007a. *The Church of Bangladesh Profile: Serving Humanity Is Proclaiming the Good News*. Dhaka, Bangladesh: Church of Bangladesh.

——. 2007b. "Letter to the Editor." *Green Christian* 63, 19.

Blair, T. 2010. "Religion and Reducing Poverty: A Question of Faith." *Developments* 48, 37–38.

CBSDP (Church of Bangladesh Social Development Programme). 2004. *Constitution: Church of Bangladesh Social Development Programme*. Dhaka, Bangladesh: Church of Bangladesh.

——. 2011. "Revised Constitution: Church of Bangladesh Social Development Programme." Draft. Church of Bangladesh, Dhaka, Bangladesh.

Christensen, J.H., B. Hewitson, A. Busuioc, A. Chen, X. Gao, I. Held, R. Jones, R.K. Kolli, W.T. Kwon, R. Laprise, V. Magaña Rueda, L. Mearns, C.G. Menéndez, J. Räisänen, A. Rinke, A. Sarr and P. Whetton. 2007. "Regional Climate Projections." In *Climate Change 2007: The Physical Science Basis. Contribution of Working Group I to the Fourth Assessment Report of the Intergovernmental Panel on Climate Change*, ed. S. Solomon, D. Qin, M. Manning, Z. Chen, M. Marquis, K.B. Averyt, M. Tignor and H.L. Miller, pp. 881–882. Cambridge, U.K.: Cambridge University Press.

Collins, L. and D. Lapierre. 1975. *Freedom at Midnight*. New Delhi, India: Vikas.

Developments. 2009. "Clearing the Air: How to Give Your Lungs and the Planet a Break at the Same Time." *Developments* 46, 14.

Hanciles, J.J. 2008. *In the Shadow of the Elephant: Bishop Crowther and the African Missionary Movement*. Oxford, U.K.: Church Mission Society.

IUCN Bangladesh. 2005. *Baira: The Floating Gardens for Sustainable Livelihood*. Dhaka, Bangladesh: IUCN Bangladesh Country Office.

Karim, Z., S.K.G. Hussain and A.U. Ahmed. 1999. "Climate Change Vulnerability of Crop Agriculture." In *Vulnerability and Adaptation to Climate Change for Bangladesh*, ed. S. Huq, Z. Karim, M. Asaduzzaman and F. Mahtab, pp. 39–54. Dordrecht, Netherlands: Kluwer Academic.

Mangalwadi, V. 2001. *The Quest for Freedom and Dignity: Caste, Conversion and Cultural Revolution*. Mumbai, India: GLS.

McGranahan, G., D. Balk and B. Anderson. 2006. "Low Coastal Zone Settlements." *Tiempo* 59, 23–26.

Mohal, N. and M.M.A. Hossain. 2007. "Investigating the Impact of Relative Sea Level Rise on Coastal Communities and Their Livelihoods in Bangladesh." Draft final report. Institute of Water Modeling (IWM) and Center for Environmental and Geographic Information Services (CEGIS), Dhaka, Bangladesh.

Moral, Shishir. 2002. "Energy Resources." In *Bangladesh Environment: Facing the 21st Century*, 2nd edn., ed. P. Gain, pp. 162–180. Dhaka, Bangladesh: Society for Environment and Human Development (SEHD).

Muhith, A.M.A. 1978. *Bangladesh: Emergence of a Nation*. Dhaka, Bangladesh: Bangladesh Books International.

Pender, J.S. 2009. *The Church of Bangladesh and Climate Change*. Dhaka, Bangladesh: Church of Bangladesh.

——. 2010. *Climate Change, Its Impacts and Possible Community-Based Responses in Bangladesh*, 2nd edn. Oslo, Edinburgh and Dhaka, Bangladesh: Norwegian Church Aid, Church of Scotland and Church of Bangladesh. Available at http://www.cbsdp.org (CoS and Bengali language versions) or http://bit.ly/James_Pender (NCA version).

Plunkett, R., A. Newton, B. Wagenhauser and J. Murray. 2000. *Bangladesh*, 4th edn. Victoria, Australia: Lonely Planet.

Quadir, D.A. 2009. "Observed Climatic Variations." *Daily Star*, 18th Anniversary Supplement, February 25, 60–62.

Robinson, A. 2002. *The History of St. Andrew's Mission, Haluaghat*. Dhaka, Bangladesh: Church of Bangladesh.

Stern, N. 2006. *Report of the Stern Review: The Economics of Climate Change*. London: HM Treasury.

Tagore, R. 1913. *Gitanjali: Song Offerings*. London: Macmillan; Goodreads.com e-book.

Tanner, T.M., A. Hassan, K.M.N. Islam, D. Conway, R. Mechler, A.U. Ahmed and M. Alam. 2007. *ORCHID: Piloting Climate Risk Screening in DFID Bangladesh*. Brighton, U.K.: Institute of Development Studies, University of Sussex.

Walter, J. and A. Simms. n.d. *The End of Development? Global Warming, Disasters and the Great Reversal of Human Progress*. London and Dhaka, Bangladesh: New Economics Foundation and Bangladesh Centre for Advanced Studies.

Zia-Begum, K. 2002. "Threat to Lives of Many and Lifestyles of Some." *Clime Asia*, COP8 Special Issue, 4–5. Available at http://www.cansouthasia.net/?Itemid=157.

6 From theology to a praxis of "eco-jihad"

The role of religious civil society organizations in combating climate change in Indonesia

Ulil Amri

Introduction

Climate change will greatly affect many aspects of life across social, political, economic, and cultural domains. Climate change is likely to increase the frequency of disasters such as floods, droughts, and sea level rise. In turn, such conditions produce significant economic loss, ignite conflicts, and alter the demographic composition of the population in affected regions. As it is a global issue, many countries have begun to collaborate in order to design strategies and form institutions to tackle the problems which may result from climate change. Indonesia is among them. The country has established Dewan Nasional Perubahan Iklim (DNPI—National Climate Change Council) in 2008 to reduce the effects of climate change in the country (DNPI 2012).

There are five main tasks of this organization, including: formulating national policy, strategy, and action to combat climate change; coordinating strategy implementation in combating climate change through adaptation, mitigation, technology, and finance strategies; setting carbon trade mechanisms; conducting monitoring and evaluation of the climate change policies; and strengthening Indonesia's bargaining position with industrial countries in order to make them more responsible in combating climate change effects (DNPI 2012). Besides that, the country also committed to reducing its carbon emission by 26 percent by 2020. In addition to national and international institutions devoted to addressing climate change, the issue has gained formal institutional support from civil society.

In Indonesia, two of the oldest and largest religious civil society organizations, Muhammadiyah[1] and Nahdlatul Ulama,[2] have joined efforts to promote climate change awareness and activism. This chapter will elaborate the role of both organizations in combating climate change in Indonesia. It argues that both organizations have played a pivotal role in the effort to combat climate change and that such a role is opening a new field of praxis called "eco-jihad." The term *jihad* comes from the Arabic language and means "striving" or "struggling." However, the term has a complex set of meanings. Generally, there are two types of interpretation of *jihad*: classical and contextual. The first one directly refers to the Qur'an and the Hadiths, which mention *jihad* as "warfare with spiritual significance": "Fight them until there is no [more] *fitnah* [tumult] and [until] worship is [acknowledged to be] for Allah. But if they cease, then there is to be no aggression except against the oppressors" (Qur'an 2:193). Therefore, this interpretation is often associated with a politically offensive action such as war (Moniruzzaman 2008) because, in the Qur'an and Hadiths, Muslims are commanded to fight until there is no more persecution and get involved in *jihad* until the Day of Judgment (2008: 3). In contrast, the second interpretation defines *jihad* in a contextual way as a commitment of Muslims to struggle for Islamic universal principles such as peace, justice, and human welfare (Abu Sulayman 1987), and now against environmental degradation (Moniruzzaman 2008).

The chapter starts by presenting a brief history of climate change and the engagement of civil society organizations with it. It then describes the histories of Muhammadiyah and Nahdlatul Ulama and discusses their theologies, which draw mainly from the Qur'an and Hadiths, the two most important sources of Islamic jurisprudence.[3] It places these theological frames in relation to organizational praxis,[4] which includes theory and theology-informed practice about both adaptation and mitigation steps. Specific to the praxis side, schools, universities, *pesantrens* (Islamic boarding schools), mosques, and *pengajians* (Islamic teaching forums) related to the two organizations are important in combating environmental degradation in Indonesian society. By investigating the work of religious civil society organizations in such efforts, we can see the contemporary intertwinement between religious praxis and ecology.

This chapter is based on a year of research conducted in 2010 in three provinces of Indonesia. The three provinces are Jakarta, West Java, and Yogyakarta. Jakarta was chosen because it is the capital city of Indonesia and where Muhammadiyah and Nahdlatul Ulama's headquarters are placed. Jakarta was also chosen as a representative of urban areas. The second province was West Java, including Bogor and Garut regencies. They are chosen as representatives of rural areas. Bogor has been especially selected as a green city by the Muslim Association for Climate Change Action. Meanwhile Garut is one of the most rapidly deforested areas in Indonesia. The third area was Yogyakarta province (Yogyakarta municipality and Bantul region). It was chosen because, in this region, Muhammadiyah and Nahdlatul Ulama are actively involved in combating climate change through planting trees and creating bioenergy projects.

As qualitative research which focused on people's thought and action in combating climate change in Indonesia, the research questions are also about how Islamic beliefs play a role in that change. During semi-structured interviews, I asked general questions like "What do you know about climate change?" "What do your beliefs [Islam] say about environmental degradation?" and "What do you do to combat these environmental changes?" Their responses were followed by new, exploratory questions based on their answers to each question.

Thirty informants were interviewed. They were selected from the three regions and consisted of three groups: organizations' officials; *pesantrens* leaders (Nahdlatul Ulama) and school principals (Muhammadiyah); and adherents including *santri* (Nahdlatul Ulama) and students (Muhammadiyah). In Jakarta, I interviewed Nahdlatul Ulama and Muhammadiyah officials and *pesantren* leaders who have knowledge about environmental issues and a commitment to combat climate change. In Bogor, Garut, Yogyakarta, and Bantul I focused on interviewing *pesantrens* leaders and their adherents. Analysis consisted of a coding technique that facilitated interpretation of informants' responses.

Besides interviews, I also analyzed primary documents which contain both organizations' involvements in environmental and climate change issues. Such documents included books, reports, statements, congress decisions, and relevant internet sources in order to enrich the interview data. Information from outside the three provinces chosen is mainly from the documents and internet sources. Since Muhammadiyah and Nahdlatul Ulama make a point to document their activities, these documents were valuable in getting the big picture of religious civil society organizations' efforts regarding combating climate change.

Climate change and the role of religious civil society

Thirty-seven countries were signatories to the Kyoto Protocol, agreeing to reduce their greenhouse gas contributions (United Nations Framework Convention on Climate Change 2010). Since that time, the countries have conducted follow-up meetings almost every year to discuss practical and technical schemes to implement the agreement. Non-government organizations (NGOs, also known as civil society organizations or CSOs) have become involved as well. According to Hoffman (2008), the engagement of these organizations began in 1992 during the United Nations Conference on the Environment and Development.

The involvement of CSOs can be seen from two sides: policymaking and action (implementation). On the policy side, research by Betsill and Correll (2001) showed three arenas where CSOs played a role: activity, access, and resources. First, CSOs have become active participants in designing and advocating for policy on several international environment forums. For example, during United Nation Convention to Combat Desertification and Kyoto Negotiation meetings in 1994 and 1997 respectively, CSOs formed side events to raise important issues to be brought into the final day's meeting. They have also tried to persuade country leaders to adopt their policy proposals. Secondly, CSOs have substantial access to the international environment forums, such as getting involved in the negotiation process either directly or indirectly or using informal relationships with country delegates. Thirdly, CSOs have valuable technical skills to enable work with a wide public.

On the action side, CSOs have worked on adaptation strategies in order to minimize the impact of climate change (Blanco 2006). For example, they have designed projects which are aimed at improving local community livelihoods with respect to sustainable environmental resource practices in the face of changing climates. Specific techniques generated by local communities in dealing with climate change are discussed by Wisner (2010), who argued that people use their own knowledge combined with external ideas and tools in order to face the changing conditions. If each community has a different way of thinking about and dealing with climate change-related issues, its religious beliefs and practices will in many cases play an important role.

Religious CSOs have been particularly active during these international forums (Posas 2007; see also Chapter 18 by Kerber, this volume) and on-the-ground adaptation action (Wisner 2010). Besides getting involved in international forums, they have articulated eco-theological frameworks and arranged actions or movements in response to climate change and their beliefs. While there are a number of transnational and national religiously based organizations engaged on climate change, the question remains: How effective have they been? To pursue this question, we will now examine the role of two religious CSOs in Indonesia.

Indonesia's religious civil society organizations

With approximately 202 million Muslims (or 86 percent of the country's population), Indonesia contains the world's largest Muslim population. Sunni Muslims are the majority, while Shi'ites are only about 1 million (Buehler 2009: 51). Three-quarters of the Muslims are affiliated with the two CSOs, Muhammadiyah and Nahdlatul Ulama (Mujani and Liddle 2004: 122).

Muhammadiyah is a modern-reformist Islamic CSO (Nakamura 1980; Noer 1973; Saleh 2001) which was established by Ahmad Dahlan in 1912 in Yogyakarta. In its early period of development, Muhammadiyah focused primarily on purifying Islamic teachings in the region from local superstitions and traditions. To do so, Muhammadiyah built modern Islamic schools which combined Islamic reformist values with a Western educational system (Fuad 2002). Later, Muhammadiyah widened its mission to provide economic empowerment and health services, and more recently to protect the environment by carrying out environmental education and activism. Today, Muhammadiyah has at least 25 million members, with 6,118 mosques, 5,519 schools, 172 universities, and 457 hospitals spread all over the country (Muhammadiyah 2012).

Muhammadiyah manages its schools, universities, and hospitals professionally. The revenue of these institutions goes to finance Muhammadiyah's existing programs and/or new programs. In some cases, Muhammadiyah encourages its members to self-fund their programs. Nevertheless, Muhammadiyah often receives funds from other sources, including government and private institutions from both domestic and foreign countries. In the case of climate change, Muhammadiyah is largely supported by the Indonesian government and international donors like the Australian Agency for International Development.

The second religious CSO, Nahdlatul Ulama, is known as a traditional-pluralist Islamic CSO (Barton and Fealy 1996; Bruinessen 1994; Bush 2009). It was established in 1926 by Hasyim Asy'ari and is committed to guarding Islamic plural and traditional ways of life. Nahdlatul Ulama's main base was in the village, where many traditional Islamic boarding schools (*pesantren*) existed. About 17,000 *pesantrens* are associated with Nahdlatul Ulama. Each *pesantren* has hundreds of students (*santri*) and is headed by a cleric (*kyai* or *ulama*). The word *kyai* is derived from Javanese and means a respected person; *ulama* is Arabic and means an Islamic scholar. In practice, both words are used interchangeably. Loyalty to the *kyai* is a main characteristic of *pesantrens*. In Nahdlatul Ulama, a *pesantren* is owned by the *kyai*, not the organization. Nevertheless, most of the *kyais* affiliate with Nahdlatul Ulama.

Nahdlatul Ulama's income mainly comes from its members or adherents, about 40 million people, most of whom donate some amount of money to the organization. The vast majority of these members are economically categorized as Indonesia's lower middle class. Nahdlatul Ulama also receives donations from government and international donors like the United States Agency for International Development and the World Wide Fund for Nature.

Muhammadiyah and Nahdlatul Ulama's eco-theology

Islamic eco-theology is mainly based on the canonical sources of Islam, especially the Qur'an and Hadiths. Respectively, these are the first and the second sources of Islamic jurisprudence. Muslims believe that the Qur'an contains God's words as direction for humanity, while the Hadiths contains tools for understanding the Qur'an. Ozdemir argued that eco-theology is "the logical outcome of a Qur'anic understanding of nature and humankind" (2003: 28). The Prophet Muhammad (peace be upon him, or PBUH) was the person who first exemplified the Qur'anic worldview; his sayings and actions are regarded as the Hadiths. His attitude of treating animals and nature informed Muslims throughout the ages. Today, according to Ozdemir, both the Qur'an and the Hadiths "are awaiting rediscovery and analysis" by Muslims in order to protect the earth from further crisis (2003: 28).

Nevertheless, since Muslims have been living in many corners of the globe, Muslim environmentalists gain their inspiration or reference not only from the Qur'an and the Hadiths, but also from many other sources (Foltz 2003: 252). The culture of the place where Muslims live has an important effect on Muslim environmentalism (eco-theology and praxis). In Indonesia, organizations like Muhammadiyah and Nahdlatul Ulama have developed an Islamic eco-theology based on the combination of canonical sources and the particularly Indonesian Muslim culture. According to Fealy and White, Islam in Indonesia "bore none of the severity and rigidity attributed to Middle Eastern forms, earning it praise for its moderation and tolerance" (2008: 1).

Muhammadiyah and Nahdlatul Ulama, as the oldest and biggest Islamic organizations in the country, are believed to have contributed significantly to this cultural characteristic of Indonesian Islam. Both organizations teach orthodox Islamic doctrines but practice them moderately (Hefner 2002). Moreover, when setting forth their views of eco-theology and action to combat environmental degradation, both organizations utilize such characteristics as a reference, besides the Qur'an and the Hadiths.

Muhammadiyah's eco-theology refers mainly to the Qur'an, and is centered on the relationship among God, humankind, and the environment. God is put in the top position, whereas humankind and the environment both have an equal position under Him. In explaining this, Muhammadiyah refers to Al-An'am 165, which says that a human being is not only God's creature, but also His representative (*khalifah*) in the world. As creatures, human beings are obliged to worship their creator, obey His commandments, and avoid His prohibitions. Further, as representatives, they are responsible for managing, protecting, and always keeping the world in balance. According to the Islamic viewpoint as understood by Muhammadiyah, these two roles are obligatory. If an individual fails, he or she will be judged by God in the hereafter (Lembaga Lingkungan Hidup Muhammadiyah 2007).

Regarding the relationship between humans and the environment, Muhammadiyah views both as God's creations and interprets this to mean that humanity has to take care of the environment properly: "Human cannot be separated from environment. As part of environment, human and environment relation is a mutual relationship and . . . complementary . . . in which each has different role" (Lembaga Lingkungan Hidup Muhammadiyah 2007: 9).

Unfortunately, such a mutual relationship has been destroyed by humanity's arbitrary and greedy exploitation of the environment. Besides, humans have also arbitrarily used their authority as God's representatives. Instead of acting as protectors, they have become traitors. Consequently, according to Muhammadiyah, the environment reacts by causing catastrophes such as floods, droughts, and earthquakes. Therefore, Muhammadiyah's eco-theology calls for a reinterpretation of the human–environment relationship according to what has been mandated by God.

Environmental issues have already become a daily discussion topic for members of Nahdlatul Ulama, since the majority of them live in remote areas and most of them utilize the surrounding environment as a direct source of livelihood. Therefore, eco-theology has been implicitly embedded with Nahdlatul Ulama since its birth. Hasyim Asy'ari, the founder of Nahdlatul Ulama, for example, often delivered environmental issues in his teaching. According to Informant A in 2010, "*Kyai* Hasyim always encouraged people in his teaching to take care of the environment by planting more trees, using natural resources efficiently, and not throwing waste anywhere."

Nahdlatul Ulama uses both the Qur'an and the Hadiths as sources to construct the theology. There are dozens of verses of the Qur'an and Hadiths which contain environmental issues. Among the verses most cited by the organization are Al-Baqarah 11–12 about the prohibition against damaging the environment, verse 27 about the disadvantages inflicted on those who commit such an offence, verse 60 about how the very rich earth contains a lot of natural resources, and verses 220 and 250 about the role of God in protecting the earth from damage. Besides Al Baqarah, there is Al-Maidah 64, which explains God's wrath on people who cause destruction on earth while, as Al-A'raf 56 mentions, God is actively taking care of it. Furthermore, the most cited Hadiths are from Imam Bukhari and Ahmad. Bukhari mentioned that once upon a time the Prophet Muhammad (PBUH) said that, if someone has land, he or she should plant trees or food crops on it. Meanwhile Ahmad told that the Prophet also urged someone to plant a tree if he or she had seed in hand.

Besides that, Alie Yafie, a respected *ulama* from Nahdlatul Ulama, has written a book titled *Merintis Fiqh Lingkungan Hidup* (Initiating environmental *fiqh*[5]) (2006). The book aims to show the way of Islam in protecting the environment from further destruction, including climate change. According to his analysis, the problem is mainly caused by global economic development: "The dominant development model today is based on economic globalization . . . the result is that our world has experienced a big transformation. . . . Some of the characteristics are the unlimited economic and technological growth which in turn bring impact to our life . . . [and] water, soil, and air pollution" (2006: 49–56).

Nahdlatul Ulama's eco-theology represents an effort from within the organization to recontextualize the message of God to the current situation so that humans protect and conserve the environment. What is unique about this theology is that it combines the Qur'an, which contains general principles of living on earth, with the Hadiths, which provide many examples and practical guidelines of how humans should treat the environment. Thus, the Hadiths here are positioned as the Qur'an's supplementary information in order to make God's message clearer. Similarly to Muhammadiyah, Nahdlatul Ulama also attempts to reinterpret God's message to remedy the current situation where environmental degradation is becoming very serious.

In Muhammadiyah the experts who are competent to talk about climate change work in a body named Lembaga Lingkungan Hidup (Institute for the Environment—LLH), now changed to Majelis Lingkungan Hidup (Environment Council—MLH), whereas in Nahdlatul Ulama they work in Gerakan Nasional Kehutanan dan Lingkungan Hidup (National Movement for Forest and Environment—GNKL) and Lembaga Penanggulangan Bencana dan Perubahan Iklim (Institute for Disaster and Climate Change Mitigation—LPBI).

In relation to climate change, Muhammadiyah made a historical decision during its annual congress (*muktamar*) in 2010 by putting climate change and environmental degradation as a major agenda for its future. Muhammadiyah's understanding is that climate change and environmental degradation will affect human life, including poor health and a weakening economy. Unfortunately, public awareness of such a condition is still poor. Therefore, Muhammadiyah urged all Indonesian people to communicate the condition by using all possible channels, such as educational institutions, mass media, and social organizations. Furthermore, the organization's *muktamar* decision called for: "[The] national government [together with Muhammadiyah] to obey international conventions on carbon emission reduction, carbon trading, and energy consumption reduction according to the Kyoto Protocol, Bali Declaration, Copenhagen [Climate Change Summit], and others." Muhammadiyah also urged all faith communities to conduct interfaith action in combating climate change through community-based activities (Muhammadiyah 2010: 146–147).

Meanwhile, Nahdlatul Ulama had released national advice (*tausiyah*) about protecting the environment from the threat of climate change. The organization worried that current ecological crises in the form of air and water pollution, deforestation, and lost biodiversity could become worse. Therefore, through its *tausiyah*, Nahdlatul Ulama urged:

> First, the Indonesian government MUST do real action against environmental degradation . . . second, Indonesian government MUST uphold to the law, regarding deforestation and environmental degradation as criminal actions . . . third, Nahdlatul Ulama adherents as well as all citizens must reject and oppose forest and environmental destruction . . . fourth, Nahdlatul Ulama adherents as well all citizens MUST conduct eco-jihad through reforestation programs, protection of trees, forest guardianship, and environmental conservation.
>
> (Nahdlatul Ulama 2007: n.p.)

The praxis of combating climate change

As mentioned above, the climate change issue stimulates not only the government but also civil society organizations to take action. In the Indonesian context, both Muhammadiyah and Nahdlatul Ulama showed that they have become convinced of the need to combat a changing climate. They both take action by mobilizing resources to inform adherents about the impact of environmental degradation, combating deforestation and conducting reforestation, creating alternative livelihood and energy options, and managing natural resources effectively.

On the one hand, as socio-religious organizations with a range of focuses, Muhammadiyah and Nahdlatul Ulama are not as environmentally focused as the single-focus environmental CSOs. Nevertheless, their actions to protect the earth from further degradation (which will be elaborated on shortly) signal that their role is becoming more important and active now. It is a role that may become even more impactful in the years ahead, because these two organizations have greater resources and more support from the grassroots than do environmental CSOs. Muhammadiyah and Nahdlatul Ulama's formal action on the environment dates back to 2000 (after the establishment of LLH) and 1994 (after Nahdlatul Ulama conducted *muktamar*) respectively, but some of their adherents had started to combat environmental degradation and climate change long before that date.

Sounding the alarm on environmental degradation

The most prevalent action conducted by Muhammadiyah and Nahdlatul Ulama is raising awareness of the impact to the whole country of a degraded environment. Few people in Indonesia are aware of climate change, what it is, why and how it happens, and the current and likely impacts on their lives. That Indonesia is the largest archipelagic country in the world, with approximately 17,500 islands, makes it difficult to reach all regions with awareness programs. Therefore, both Muhammadiyah and Nahdlatul Ulama have mobilized resources through mosques and Islamic teaching forums (*pengajian*).

Both organizations believe that the mosque has been a very powerful channel for socio-cultural communication among Muslims, as it is often used to spread and to transform any idea into action. Muhammadiyah began utilizing the mosque as a venue for conveying environmental issues in 2006 after it launched the *mubaligh lingkungan* (environment specialist preachers) program. Nahdlatul Ulama began in 1994 after its 29th *muktamar*, which decided that environmental pollution should be categorized as a forbidden action (Musa 2012). For even longer, said Informant A, many *kyais* had already inserted into their teaching in the mosque consideration of environmental issues such as keeping the environment clean, restraining from cutting trees, and not throwing waste anywhere.

For a long time, the mosque has been the primary venue for propagating the Islamic way of life (Munson 2001). In this case, the *kyais* or *ulamas* usually begin by mentioning relevant Qur'anic verses, followed by some explanation and practical examples to connect to the adherents' (*jamaah*) daily lives. The teachers close with conclusions or recommendations about what actions should be taken to protect the environment. Although this classical form of teaching does not provide the adherents time to ask questions, the mosque has become a venue for transmitting basic information about environmental degradation or climate change and potential or ongoing impacts. This activity takes place in many mosques which are located near Muhammadiyah and Nahdlatul Ulama's schools, universities, *pesantrens*, and *kyais*' or *ulamas*' homes. Those adherents who wish to find more information and discussions about the topic may join another teaching forum named a *pengajian*, which can be held either in or out of the mosque.

In a *pengajian*, people have an opportunity to put questions about the topic in detail to the *ulama*. However, the *ulama* may not be able to answer technical questions related to environmental degradation or climate change, so only theological aspects of it are usually addressed unless the *pengajian* is attended by other experts. The *pengajian* is attended by men and women of all ages. A potential further effect of this forum is that people who have joined a *pengajian* may transfer their knowledge to other people at home and their neighborhood so it can be publicly and widely known later on. Another informant explained his experience:

> I usually give *pengajian* in mosque about the importance of protecting environment to people in my village. Most of the audiences are housewives. They are a special audience . . . because of their ability to spread the *pengajian* message to wider audiences. . . . When they go home, they spread message to husbands and children. Often, they also spread it to neighbors near their home.
>
> (Informant B, 2011)

Besides communicating about the issue, the mosques and *pengajians* also recommend action, such as changing consumption behavior and reducing greenhouse gases through economizing on electric use at home. Nevertheless, some barriers stand in the way of better communication. An important barrier is the limitation of clerics' basic knowledge about climate change. Another barrier is the unavailability of experts in both organizations in rural areas. Therefore, it seems that the action is most effective in the urban areas, since more knowledgeable clerics are available in these areas. One informant even said that the action of combating climate change is still elitist or only discussed at the top level of organizations and still lacking adequate dissemination of information to the larger community. Meanwhile, many other people who are not well informed about the action, or even the issue of climate change itself, are left behind.

Besides mosques and *pengajians*, both organizations mobilize schools and *pesantrens*. Both Muhammadiyah and Nahdlatul Ulama realize that educational institutions have a pivotal role in creating a better future, as they educate and build the character of students. The two organizations also believe that religious ethics can mitigate the problems. Based on the assumption that the contemporary Indonesian education system is strongly associated with modern secular values and to a certain degree ignores the moral and spiritual aspects of life, the curriculum of both Muhammadiyah schools and Nahdlatul Ulama *pesantrens* incorporates Islamic values and ethics into its education system. Following the adoption of environmental and climate change issues in Muhammadiyah and Nahdlatul Ulama's eco-theology lately, schools and *pesantrens* are slowly also adopting such issues.

A common expectation is that emphasizing Islamic ethics on the environment in schools and *pesantrens* can result in the quick spread of the message on environmental degradation throughout the country so that people are more aware of its impact in the future. Clearly more research to test this assumption is warranted. Unlike the case for mosques and *pengajians*, raising awareness on environmental degradation and its climate change impact through schools and *pesantrens* seems more structured in terms of who gives information and to whom it is given, what kind of information is given, and other details of the educational program.

Combating deforestation and conducting reforestation

Indonesia is one of the world's countries most vulnerable to global environmental change. Since 1950, this country has lost approximately 40 percent of its existing forests. According to Global Forest Watch (2011), Indonesia's forest cover fell from 162 million to 98 million hectares over this time. Since the 1990s, Indonesia's rate of deforestation has been 2 million hectares per year. The country's deforestation is believed to accelerate the greenhouse gas concentrations in the atmosphere, which in turn changes the earth's climate. In response, both Muhammadiyah and Nahdlatul Ulama have acted to combat deforestation and to conduct reforestation and afforestation. These actions involve GNKL, LPBI, MLH, schools, universities, and *pesantrens* as well.

One of the main concerns is illegal logging. Although this activity may bring benefit to its adherents in the short term in the form of cash, Nahdlatul Ulama believes that such activity will cause more damage than benefits in the long run. To protect the forest from exploitation, GNKL, LPBI, and *pesantrens* urge their *jamaah* or *santri* not to cut trees and to keep an eye on those who intend to do illegal logging.

> Since 1999, we have collaborated with many *pesantrens* under the banner of Asosiasi Pesantren Nahdlatul Ulama [Pesantren Association of Nahdlatul

> Ulama] in conserving the environment. Our main concern was to stop illegal logging, deforestation, because we believed that the biggest ecological crisis in our country comes from this sector. As we know, Indonesia's forest is the second biggest after Brazil. Then, we put *pesantrens* at the front line, because their position is very close to the forest. Besides, theologically, the Qur'an underlined the importance of forest conservation, as it mentioned the word "*Sajaroh*" [tree] many times. If the Qur'an mentions something many times, this shows its great importance. So, we must therefore take it into account.
>
> (Informant A, 2010)

In order to make the action of combating deforestation effective, Nahdlatul Ulama appoints *pesantrens* as the main base of action. *Pesantrens* then collaborate with villagers to guard and protect forests. *Pesantrens* also urge the villagers to think about and to compare whether illegal logging brings benefit to them and the generation after them or not. This kind of collaboration seems not difficult to establish, since people in forest areas have reciprocal relationships with the *pesantrens*. The *pesantrens*, on the one hand, need assistance from the villagers to combat illegal logging, while on the other hand the villagers need spiritual enlightenment from *pesantrens*.

Nahdlatul Ulama also actively combats deforestation caused by mining activities. This organization argues that any mining activity will damage the environment, not only land, forest, water, and air, but also human life. For example, through GNKL, Nahdlatul Ulama challenged the local government to abrogate a mining permit which had been given to companies and individuals in Silo, Jember district of East Java province (Nahdlatul Ulama Online 2009). This kind of action might be unexpected, because in many places people tend to support the opening of mining sites in order to get benefit from the site. However, for Nahdlatul Ulama, despite the mine's short-run profit, the long-run mining activity was believed to devastate the environment. Unfortunately, even though they were supported by local leaders and hundreds of villagers, the local government with support from local scientists and legislators continued to give permits to explore the site.

Besides combating deforestation, Muhammadiyah and Nahdlatul Ulama conduct reforestation. Muhammadiyah, for instance, mobilizes its schools and universities. The school and university students are invited to plant trees in both their schools and the surrounding areas. The aim of this program is to build student awareness of protecting the environment. Students from Sekolah Menengah Pertama 1 Muhammadiyah (Junior High School/SMP 1 Muhammadiyah) in Yogyakarta, for example, planted trees on the foothills of Mount Merapi after the eruption at the end of 2010. In the same year, Muhammadiyah University of Purworejo, Central Java, planted 2,000 *jati* (teak hardwood) trees in Sokoagung village.

Muhammadiyah's reforestation focus is in urban areas, since the majority of its adherents live in these areas, but it does not discourage its followers if they want to participate in the rural areas as well. According to a Muhammadiyah leader, many urban areas in Indonesia are suffering from a lack of trees. Economic growth and rapid industrialization are admitted to be two of the causal factors. Therefore, Muhammadiyah collaborates with the central, provincial, and regional governments to provide tree seeds for people so that they can plant the seeds near their homes. By doing so, Muhammadiyah shows that its organization invites all people not only to be aware of current environmental problems caused by climate change and their impact through *pengajians* but also to participate directly by planting trees.

While Muhammadiyah conducts reforestation mainly in urban areas, Nahdlatul Ulama does similar action in the rural areas. This organization holds that combating climate change must not end up in mosques and *pengajians*, but should be translated into actions. With thousands of *pesantrens* spread in many rural areas, Nahdlatul Ulama has planted millions of trees in those areas since 2006 (Tempointeraktif 2007). It has committed to supporting the government in planting 1 billion trees each year (Agroindonesia 2011). Nahdlatul Ulama focuses its afforestation projects in barren lands, planting specific species of trees like *acacia auriculifarmis* and *tectana grandis* effective at absorbing carbon dioxide in the atmosphere.

Creating alternative livelihoods and energy, and managing resources

Illegal logging, which mainly takes place in rural areas, is one of the main causes of deforestation in Indonesia. Illegal logging has historically been an important source of livelihood. Therefore, in order to stop this illegal logging, an alternative livelihood must be provided. Through the *pesantrens* Nahdlatul Ulama has tried to create such alternatives by planting empty, non-forested areas with new crops of fruits and vegetables or using the lands to build stock farming.

By planting fruits and vegetables and building stock farming, *pesantrens* aim to empower local people not to depend on forest resources alone. These people are then equipped with technical and practical tools to work on such fields successfully. For the people, these fields look very promising, as fruits, vegetables, and stock farming have had considerably high economic value. On the one hand, these alternative livelihoods help people to earn money safely, and on the other hand the people back the *pesantrens*' efforts to protect the environment. Those *pesantrens* which have already had plantation and stock farming programs continue to widely extend their territory of plantation or stock farming. It is not difficult for them to do this, as each *pesantren* has had quite wide open land and extensive landholdings.

Pesantren involvement in boosting the rural economy started a long time ago. Known to be financially independent institutions in providing education, *pesantrens* got involved in various locally scaled businesses to finance their activities and provide jobs for people surrounding the *pesantrens*. It is through the *kyai*'s direction that many sources of livelihood such as agribusiness, agriculture, and fisheries have been created. Many people believe that the *kyai*'s leadership brings not only physical welfare, such as sufficient income for many people in the rural areas, but also spiritual welfare, such as God's blessing.[6]

Meanwhile, Muhammadiyah has focused on alternative energy projects and managing resources efficiently. Since 2011, Muhammadiyah offices, schools and universities have been mandated by the Central Board of Muhammadiyah through MLH to use energy efficiently in their daily activities. They are also mandated to reduce the use of paper, water, and electricity. This program is called Gerakan Aksi Hijau di Kantor (Green Action in the Office). In implementing the program, Muhammadiyah asks its adherents to print on both sides of the paper, use recycled paper, and reduce air-conditioner use. At the same time Muhammadiyah invites wider public participation in carrying out these actions.

Another program is called Shodaqoh Sampah (Charity from Waste). Muhammadiyah followers collect waste from their households. They separate organic and non-organic waste, and leave the organic by assuming that it will decompose naturally in a communal waste space. A small proportion of the non-organic waste is recycled by the branch office of Muhammadiyah to become new products, such as handicrafts, which are sold to the public. Any revenue from this activity goes to the poor.

It would not seem difficult for Muhammadiyah to do this, since it is a centralized organization in which hundreds of local branches are under its control so that any instruction that comes from the Central Board must be obeyed. However, it seems that the instructions do not work well in practice, because not all Muhammadiyah schools and universities directly implement such mandates. One reason is that Muhammadiyah's concern with environmental issues just started in 2000. Such a recent starts means that its program has not reached the lowest grassroots level yet. A second reason is that education and economic empowerment are still the dominant programs in Muhammadiyah, and environmental programs are seen as of lower importance. In response to this state of affairs, Muhammadiyah is prioritizing environmental issues alongside education and socio-economic empowerment within the five years 2011–15. It intends to establish more infrastructure at the central, provincial, district, sub-district, and village level, and to increase the number of *mubaligh lingkungan* (interview with Informant C, 2011).

So far, Muhammadiyah has made good progress. Its achievement in the waste management project is particularly exceptional. All of its schools and universities are mandated to collaborate with local people to change garbage into useful materials. They started this project in 2008 and focused on urban areas, which are full of waste. Muhammadiyah operates the only vocational school which educates students to be waste management professionals. Muhammadiyah has committed itself to continue developing more vocational programs to train environmental professionals. An informant explained:

> This is the first school in Muhammadiyah and maybe in the country which is concerned with alternative energy. This school represents Muhammadiyah's effort in mitigating climate change. We develop a specific field concentration named alternative energy which aims to create energy from household waste collected from surrounding areas. All supporting technologies are constructed from simple and cheap materials gained from local Bantul. These

> technologies are developed by collaboration between teachers and students. The final products are many, including bioethanol and charcoal. We hope these products can be produced massively so that the public can utilize them.
>
> (Informant D, 2011)

As one of the biggest civil society organizations, with hundreds of universities spread all over the country, Muhammadiyah is challenged to make a significant contribution toward providing solutions for replacing fossil energy. So far, although Indonesia has abundant alternative energy sources, it has not pursued this option. Therefore, Muhammadiyah has taken this opportunity seriously by mobilizing its universities to create such energy by learning from and collaborating with other institutions which already have supporting technologies. At present, Muhammadiyah has mobilized two universities, the Muhammadiyah universities of Malang and Yogyakarta, in cooperation with Dutch institutions to produce alternative bioenergy, which focuses on developing biofuel, biomass, and biogas. In addition, one of these universities is now planning to open a new department of environmental theology (Jurusan Teologi Lingkungan). This department will be responsible for teaching about managing the environment based on Islamic religious ethics. According to Informant D, it is expected that these two universities will stimulate Muhammadiyah's universities in other regions to do the same.

Combating environmental degradation: a new *jihad*

The Central Board of Muhammadiyah does not mention explicitly that its role in combating environmental degradation is a form of *jihad*. However, its proponents did use this term to describe what they do in protecting the environment to combat environmental degradation:

> To me, protecting the environment is a kind of *jihad*. *Jihad* means a serious endeavor. Without seriousness, we cannot protect the environment from destruction. For example, we would like to create a clean environment in a classroom. We cannot do it without being serious about it. Therefore, before class teaching begins, I always check each classroom seriously to make sure that all classes have been cleaned.
>
> (Informant B, 2011)

On the other hand, in Nahdlatul Ulama the term "eco-jihad" seems more common. It is used frequently by both the top level of the organization and its followers. Nahdlatul Ulama's GNKL *tausiyah* in 2007, which called for national attention to the current ecological crisis, used the term. The organization mentioned that, because of this crisis, it is urgent to conduct collective *jihad bi'iyah* (eco-jihad). According to GNKL, this is compulsory not only for Nahdlatul Ulama followers but for all Indonesian people, because environmental degradation is not related to a person, a single community, or a single religion, but all people (Nahdlatul Ulama 2007).

Besides GNKL, the Nahdlatul Ulama branch of Central Java called for a similar *jihad.* During an event called Ceramah Lingkungan (Environmental Islamic Teaching) a *kyai* invited his followers to get involved in eco-jihad in a forest area of Muria, Pati district (Gerakan Pemuda Ansor 2009.) As is widely known in the country, the Muria forest has been dramatically degraded as a result of illegal logging for many years. The *kyai* mentioned that, without forests, humans will suffer a significant loss, because forests underpin our life on earth. Further, he stated that a forest is also God's creature, like humans, so its life is also important. This teaching forum was attended not only by Nahdlatul Ulama followers but also by local government officials.

Pesantrens also use the term *jihad.* For example, one informant from a *pesantren* called upon *jihad* to protect the environment by saying:

> We must worry about the current situation where environmental degradation has become a real threat. Therefore, we need eco-jihad and eco-mujahid [eco-jihadists]. . . . In this pesantren, we already have eco-mujahid consisting of 100 male students and 100 female students who are united in the department of environment of this *pesantren*. Their main task is to campaign as well as to protect the *pesantren*'s environment. The *pesantren* also assigns them to do social work outside such as planting trees and campaigning on the importance of protecting the environment.
>
> (Informant E, 2011)

Many *santris* have embodied *jihad* in their action in implementing an eco-friendly lifestyle in the *pesantren*. They have campaigned for an eco-lifestyle to the villagers who live near the *pesantren*, such as how to recycle *air wudhu* (water used for ritual ablution before praying) and planting trees. These are some of the regular programs set by the *pesantren* for its *santris*. So far, many villagers have demonstrated their interest by joining in planting trees and participating in protecting the environment in their place (Green Radio 2011).

Conclusion

As a global issue, climate change has attracted worldwide attention. Governments and civil society organizations, including religious CSOs, have pivotal roles in tackling the issue. As the Indonesian case shows, some religious CSOs have started to engage with the issue of climate change. Muhammadiyah and Nahdlatul Ulama, the two oldest and largest religious CSOs in Indonesia, have introduced an eco-theology followed by praxis to tackle climate change. To do so, they have begun to mobilize their institutional resources such as schools, universities, mosques, *pesantrens*, and *pengajians*, and socio-cultural resources such as ties between adherents and a Muslim environmental praxis. Strong support from the grassroots is the main strength of these organizations, in comparison with CSOs in the country that are solely focused on the environment.

The eco-theology of both organizations is mainly based on canonical sources including the Qur'an and the Hadiths. With the presence of this theology, there are common characteristics of Muhammadiyah and Nahdlatul Ulama's eco-theology, which is to reinterpret the Islamic message of God on the human–nature relationship. Both organizations have expressed the command of God to always maintain and protect the environment. These organizations believe that the current environmental problems are rooted in human behavior and actions on earth which fail to actualize Islamic teachings.

Furthermore, Muhammadiyah and Nahdlatul Ulama are trying to put their developing eco-theology into praxis. Both organizations focus on bringing attention to climate change and its impact on the wider public, combating deforestation and conducting reforestation, creating alternative livelihood and energy options, and managing resources efficiently. Both organizations collaborate with local people, for example, in planting fruits and vegetables, building stock farming, and recycling waste. This reciprocal relationship helps not only both organizations to combat climate change, but also the local people to find alternative and more environmentally sustainable livelihoods. This effort to combat environmental degradation and its impact opens a new field called "eco-jihad." This kind of *jihad* is based on the combination of literal and contextual interpretations of *jihad*. While the Central Board of Muhammadiyah does not specifically consider its effort to be *jihad*, its followers claim that it is. Meanwhile, both leaders and followers of Nahdlatul Ulama acknowledge their action in protecting the environment as *jihad bi'iyah* (eco-jihad), because environmental degradation has become a serious threat to human life on earth.

However, there are some constraints that hamper the praxis. For example, when Muhammadiyah and Nahdlatul Ulama attempted to sound the alarm of environmental degradation throughout the country, they found the limitations of the *ulamas*' basic knowledge on the issue of climate and environmental change. Coupled with the unavailability of experts of both organizations in rural areas, this has slowed the campaign's progress in such areas.

Another constraint is that it is perceived that there is an element of elitism in the campaign of climate change, as it is restricted to the concern of a few organization leaders and has not really reached the broad masses of the adherents. Many *kyais* and *jamaah* in rural areas do not know the dynamics of the campaign. The outer islands of Indonesia, which would be severely impacted by climate change, are particularly untouched by the campaign. Besides, for Muhammadiyah in particular, although it is a centralized organization, it seems that the instructions from the Central Board about combating environmental degradation through changing school and university practices are not implemented well at the lower levels.

Indonesian CSOs' effort in combating climate change is therefore still far from ideal. There is much work to do, such as enhancing basic knowledge of the CSOs' officials on environmental issues and raising awareness of the climate change threat to the whole country. Although Muhammadiyah has introduced the *mubaligh lingkungan* program, it is not enough. Furthermore, it seems that the issue of climate change is only familiar in certain areas of Indonesia (especially Java) but not in the outer regions. In fact, the other islands are also severely impacted by the effect of climate change. Therefore, both organizations need to engage broad masses from the other islands in order to prove that combating climate change is not solely an elite concern but also affects all members of the organizations. Finally, there should be more efforts from Muhammadiyah and Nahdlatul Ulama to practice a praxis of eco-jihad in raising an awareness of climate change, combating deforestation and conducting reforestation, and creating alternative livelihood and energy sources in the country in the future.

Glossary of terms

Fiqh: Islamic jurisprudence
Fitnah: tumult
Jamaah: adherents
Jihad: warfare with spiritual significance
Jihad bi'iyah: eco-jihad
Khalifah: representative of God on earth
Kyai: Islamic scholar or cleric
Mubaligh lingkungan: environment specialist preacher
Muktamar: National congress
Pengajian: Islamic teaching forum
Pesantren: Islamic boarding school
Santri: student of *pesantren*
Tausiyah: advice
Ulama: Islamic scholar or cleric

Notes

1 Muhammadiyah literally means "follower of Muhammad."
2 Nahdlatul Ulama literally means "awakening of *ulama* [Islamic scholars or clerics]."
3 The Qur'an and the Hadiths are the first and the second source of Islamic jurisprudence respectively. Muslims believe that the Qur'an contains God's words as direction for humanity, while the Hadiths contain tools for understanding the Qur'an.
4 In his famous book *A Theology of Liberation*, Gutierrez (1974) defined theology as a critical reflection which requires a person to get involved in the socio-historical process of his or her social environment. In thus explaining praxis, Gutierrez did not simply define it as the application of theology, but as active socio-historical participation to build a new society which is free from any injustice. Similarly, Ogden (2005) called for liberation work through redemption and emancipation. In the context of faith-based organizations' roles in combating climate change, this chapter employs both scholars' notions of theology and praxis.
5 The term *fiqh* refers to Islamic jurisprudence.
6 Many people believe, especially in rural areas, that the *ulama* is very close to God as a result of his spiritual knowledge and that God blesses him very much. Therefore, following the *ulama* means getting close to the blessing of God.

References

Abu Sulayman, A.H. 1987. *The Islamic Theory of International Relations: New Directions for Islamic Methodology and Thought.* Herndon, VA: International Institute of Islamic Thought.

Agroindonesia. 2011. "Satu Miliar Pohon Jilid II di Bukit Merah Putih." Available at http://agroindonesia.co.id/2011/12/01/satu-miliar-pohon-jilid-ii-di-bukit-merah-putih (accessed September 12, 2012).

Barton, G. and G. Fealy. 1996. *Nahdlatul Ulama, Traditional Islam, and Modernity in Indonesia.* Melbourne: Monash Asia Institute.

Betsill, M. and E. Correll. 2001. "A Comparative Look at NGO Influence in International Environmental Negotiations: Desertification and Climate Change." *Global Environmental Politics* 1.4, 86–106.

Blanco, A.R. 2006. "Local Initiatives and Adaptation to Climate Change." *Disasters* 30.1, 140–147.

Bruinessen, M.V. 1994. *NU: Tradisi, Relasi-Relasi Kuasa, Pencarian Wacana Baru*, Yogyakarta: LkiS.

Buehler, M. 2009. "Islam and Democracy in Indonesia." *Insight Turkey* 11.4, 51–63.

Bush, R. 2009. *Nahdlatul Ulama and the Struggle for Power within Islam and Politics in Indonesia.* Singapore: Institute of Southeast Asian Studies.

DNPI (Dewan Nasional Perubahan Iklim). 2012. "About DNPI." Available at http://www.dnpi.go.id/index.php?option=com_content&view=article&id=33&Itemid=3 (accessed September 12, 2012).

Fealy, G. and S. White, eds. 2008. *Expressing Islam: Religious Life and Politics in Indonesia.* Singapore: Institute of Southeast Asian Studies.

Foltz, R. 2003. "Islamic Environmentalism: A Matter of Interpretation." In *Islam and Ecology: A Bestowed Trust*, ed. R. Foltz, F.M. Denny and B. Azizan, pp. 249–280. Cambridge, MA: Harvard University Press.

Fuad, M. 2002. "Civil Society in Indonesia: The Potential and Limits of Muhammadiyah." *SOJOURN: Journal of Social Issues in Southeast Asia* 17.2, 133–163.

Gerakan Pemuda Ansor. 2009. *Nahdlatul Ulama Serukan Jihad Melawan Perusak Hutan.* Jakarta. Available at http://gp-ansor.org/14117-12102009.html (accessed June 2, 2011).

Global Forest Watch. 2011. *Indonesia's Forest in Brief.* Jakarta. Available at http://www.globalforestwatch.org/english/indonesia/forests.htm (accessed March 6, 2011).

Green Radio. 2011. *Jihad Lingkungan dari Jabon Mekar*. Jakarta. Available at http://www.greenradio.fm/news/latest/4339-jihad-lingkungan-dari-jabon-mekar- (accessed June 20, 2011).

Gutierrez, G. 1974. *A Theology of Liberation: History, Politics, and Salvation.* London: SCM Press.

Hefner, R.W. 2002. "Globalization, Governance, and the Crisis of Indonesian Islam." Conference on Globalization, Self-Capacity and Muslim Self-Determination, University of California Santa Cruz.

Hoffman, M.J. 2008. "Where the States Are: Environmental NGOs and the UN Climate Change Negotiations." In *Transnational Activism in the UN and the EU*, ed. J. Joachim and B. Locher, pp. 29–42. London: Routledge.

Lembaga Lingkungan Hidup Muhammadiyah. 2007. *Teologi Lingkungan*. Jakarta: Kementerian Negara Lingkungan Hidup and Lembaga Lingkungan Hidup Muhammadiyah.

Moniruzzaman, M. 2008. "Jihad and Terrorism: An Alternative Explanation." *Journal of Religion and Society* 10, 1–13.

Muhammadiyah. 2010. "Tanfidz Keputusan Muktamar Satu Abad Muhammadiyah." Available at http://www.muhammadiyah.or.id/muhfile/download/Tanfidz%20Muhammadiyah/Tanfidz%20Muhammadiyah%20Sept%202010.PDF (accessed September 11, 2012).

——. 2012. "Database Persyarikatan." Available at http://www.muhammadiyah.or.id/id/content-8-det-amal-usaha.html (accessed September 16, 2012).

Mujani, S. and W.R. Liddle. 2004. "Politics, Islam, and Public Opinion." *Journal of Democracy* 15, 109–123.

Munson, Z. 2001. "Islamic Mobilization: Social Movement Theory and the Egyptian Muslim Brotherhood." *Sociological Quarterly* 42.4, 487–510.

Musa, A.M. 2012. "Teologi Lingkungan." *Seputar Indonesia Daily Newspaper*, August 23. Available at http://www.seputar-indonesia.com/edisicetak/content/view/500895/44/.

Nahdlatul Ulama. 2007. *Pengurus Besar Nahdlatul Ulama Keluarkan Tausiyah tentang Pelestarian Hutan dan Lingkungan Hidup*. Jakarta. Available at http://www.nu.or.id/a,public-m,dinamic-s,detail-ids,1-id,9538-lang,id-c,warta-t,PBNU+Keluarkan+Tausiyah+Tentang+Pelestarian+Hutan+dan+Lingkungan+Hidup-.phpx (accessed September 11, 2012).

Nahdlatul Ulama Online. 2009. *GNKL NU Minta Eksplorasi Mangan Dihentikan*. Jakarta. Available at http://www.nu.or.id/a,public-m,dinamic-s,detail-ids,1-id,17159-lang,id-c,warta-t,GNKL+NU+Jember+Minta+Eksplorasi+Tambang+Mangaan+Dihentikan-.phpx (accessed May 26, 2013).

Nakamura, M. 1980. "The Reformist Ideology of Muhammadiyah." In *Indonesia: The Making of a Culture*, ed. James J. Fox, pp. 273–286, Indonesia, Australian Perspectives, vol. 1. Canberra: Australian National University.

Noer, D. 1973. *The Modernist Muslim Movement in Indonesia, 1900–1942*. Singapore: Oxford University Press.

Ogden, S.M. 2005. *Faith and Freedom: Toward a Theology of Liberation*, rev. and enlarged edn. Eugene, OR: Wipf and Stock.

Ozdemir, I. 2003. "Toward an Understanding of Environmental Ethics from a Qur'anic Perspective." In *Islam and Ecology: A Bestowed Trust*, ed. R. Foltz, F.M. Denny and B. Azizan, pp. 3–38. Cambridge, MA: Harvard University Press.

Posas, P.J. 2007. "Roles of Religion and Ethics in Addressing Climate Change." *Ethics in Science and Environmental Politics* 2007, 31–49.

Saleh, F. 2001. *Modern Trends in Islamic Theological Discourses in Twentieth Century Indonesia: A Critical Survey*. Boston, MA: Brill.

Tempointeraktif. 2007. "NU Jember Tanam 1 Juta Pohon." Available at http://www.infoanda.com/linksfollow.php?lh=BwcGXFJaCwUA (accessed September 12, 2012).

United Nations Framework Convention on Climate Change. 2010. *Kyoto Protocol*. Jakarta. Available at http://unfccc.int/kyoto_protocol/items/2830.php (accessed March 14, 2011).

Wisner, Ben. 2010. "Climate Change and Cultural Diversity." *International Social Science Journal* 61.199, 131–140.

Yafie, A. 2006. *Merintis Fiqh Lingkungan Hidup*, Jakarta: Ufuk Press.

7 Churches building resiliency to climate change in Solomon Islands

Andreana Reale

Introduction

Solomon Islands, like other small island developing states in the Pacific region, is feeling the effects of climate change. Solomon Islands is a small Pacific island country that sits east of Papua New Guinea, consisting of nearly 1,000 islands, with geographies ranging from tiny remote atolls to large mountainous islands. The culture is predominately Melanesian, with a population of just over half a million. Its people, scattered across remote villages that are often separated by many days of travel, are familiar with the threat of climate- and weather-induced hazards such as cyclones, flooding and drought, having lived with and adapted to climate variability for hundreds of years. The emerging threat of climate change exacerbates these hazards and brings new ones, such as rising sea levels, more frequent and destructive king tides, erosion and consequent salt water incursion.

In a country whose government has limited relevance to most villagers, it is churches, as well as other informal power structures, that provide a level of governance within villages in the midst of many challenges, including climate change. Like other Pacific island countries, Solomon Islands is a highly Christianized country, with 92 percent of the population affiliated with a Christian church (Bureau of Democracy, Human Rights, and Labor 2007). The majority of these people attend church weekly. This chapter explores the way churches have filled a governance void, playing an important role in climate change resilience and adaptation.

The hazards brought about by climate change only become disasters when coupled with underlying vulnerabilities—an understanding integral to the disaster risk reduction framework that is used in this chapter (Wisner et al. 2004). Many people are forced to live or sustain livelihoods in areas highly exposed to the effects of climate change, and an insecure food supply and lack of other resources will cause a village to struggle to cope on its own. Many villages in Solomon Islands are extremely isolated, and the government department responsible for disaster relief, the National Disaster Management Office (NDMO), based in the capital Honiara, is often undermined by inadequate funding and resources (Yates and Anderson-Berry 2004).

Yet Solomon Islanders cannot be characterized as passive victims in the midst of climate change. People have been coping with extreme weather and climate variability for many generations, and they continue to do so. Solomon Islanders possess significant *resilience* —that is, qualities that enable people to cope with, adapt to and recover from hazards (Pelling and Uitto 2001). Solomon Islanders are also increasing their *adaptive capacity* —that is, the extent to which a system can modify its circumstances (i.e., adapt) to move to a less vulnerable condition (Luers et al. 2003). Building both resilience and adaptive capacity are projected as crucial features of response in Solomon Islands to effects of a changing climate.

One site of resilience and adaptive capacity is the many churches that operate in villages across Solomon Islands. Churches, in this context, are the institutionalized gatherings of Christians that form for the purposes of worship, prayer, mutual assistance, teaching and support. These churches are operational at a grassroots level, and are often part of a wider denomination that includes centralized leadership, decision making or coordination.

While the government in Solomon Islands is not generally active in the everyday lives of the majority of Solomon Islanders who live in remote villages, the various church denominations across the country form a steady structure. This structure attends to the spiritual, social and physical needs of its people—from its role in prayer, song, rituals and rites through to its role in conflict mediation and economic development. The Christian faith forms a significant part of people's identity. The part churches play in disaster risk reduction and climate change adaptation is of great importance, since they are involved across governance scales from the village church structure to the urban headquarters and ecumenical organizations to the linkages that various denominations create with overseas counterparts.

Many factors contribute to resilience and adaptive capacity in the face of climate change and variability. Community coherence, health and educational outcomes, level of economic development, availability of external relief and presence of disaster plans and preparedness are among the factors identified in other research (Cutter, Boruff and Shirley 2003; Wisner et al. 2004). Solomon Islands illustrates some of the ways in which religious bodies can be involved in all levels of climate change resilience and adaptive capacity, beyond the discrete areas of development (e.g., Belshaw, Sugden and Calderisi 2001; Dickson-Waiko 1999; Kamaara 2000), disaster relief or recovery (e.g., Falk 2010; Fountain, Kindon and Murray 2004; Gillard and Paton 1999). The effects of climate change should not be seen in terms related purely to the hazard, but in conjunction with underlying social, political, economic and environmental processes.

While Solomon Islands churches are extremely important to the social, political, economic and environmental fabric, I caution from the outset about the danger of over-burdening them. Religious bodies cannot and should not be used to "fix" all of Solomon Islands' problems. Instead, their work is part of a mix of efforts by a range of NGOs and governments in building the infrastructure, institutional capacity and social capital to respond to stressors created by climate change.

Methodology

This chapter brings together important secondary sources (largely peer-reviewed research and government and NGO reports) on the role of the church in disaster risk reduction and climate change adaptation in Solomon Islands, and is complemented by an ethnographic methodology that may be described as "rapid" in nature (Beebe 2001). At the time of writing this chapter, I was an Australian university-based researcher, and integrated volunteer work with participant observation and informal interviews. I spent a month in Solomon Islands in July and August of 2010, largely in Honiara (the capital), where I volunteered with a church-based organization providing schooling for at-risk children. I also spent time in a rural village.

The fact that I am a person of Christian faith enabled me to stay with the family of the husband-and-wife pastoral team running the church and associated ministry. I was thus well embedded in a church community. During my stay I engaged in conversations with key informants about the role of churches in climate hazards, including the director and other personnel of the NDMO, and office bearers and program managers from Solomon Islands Christian Association (SICA) and the Church of Melanesia. These conversations were made possible because of my place as a guest within the church community.

At the same time, there were inherent limitations in my "positionality" as a researcher. In particular, I acknowledge my own tendency, as a white person, to see "white" behavior and culture as the norm, and hence to view, at times, other cultures negatively (see for example Milner 2007; Tillman 2002). I am grateful to some of my Solomon Islands friends and informants for pointing out my bias. I also acknowledge that because I did not conduct comprehensive research within villages the observations in this chapter are preliminary in nature, and would benefit from further research.

The role of churches in Solomon Islands society

Western minds, formed in a context where the church is firmly separated from the state and religion related to the private realm, may assume that the key role of religion is the formation of personal belief or spirituality. Yet, in other times and in other places, religion is not so easily contained, operating, as Ninian Smart (1996) put it, in various "spheres" of life, including the political, the economic, the legal and the social. Christianity in Solomon Islands is far from a private affair, extending into all of these areas through networks of extremely active and thoroughly indigenized churches.

The state constitutes a weak presence in the lives of most Solomon Islanders (Hegarty et al. 2004). Like other oceanic nations, Solomon Islands is a colonial creation with no overarching indigenous polities (Douglas 2007). The modern system of government is a foreign invention that is not familiar to most of its "citizens" (Hegarty et al. 2004). The governments of Solomon Islands—both nationally and at a provincial level—are hampered by elected leaders who are beholden to the micro-interests of their constituencies (Hegarty et al. 2004). The government is impeded further by an exceedingly slow economy. The result is a state that is not active in the everyday lives of most of its "citizens," and thus has little relevance to local villages—except perhaps at election time—and many informal systems of governance fill the void. These alternative governance and authority systems include the "chiefly" system (traditional village-based governance), the *wantok* system (literally "one talk"—kinship relations between people of the same language group) and churches.

Churches represent a kind of authority within communities that is not shared by government. Unlike the far-removed government structure, there is at least one church in virtually every village. Church structures are intimately connected with the modern-day chiefly system, and churches have a significant role in the day-to-day running of villages. They are involved in managing village life and village-level decision making, and this, coupled with their involvement in the everyday lives of their members, forms a steady structure of leadership and village-based governance. This structure of governance is pertinent to the development process in many communities (Rhodes 2007).

Church leaders exercise considerable decision-making influence and authority at a village and sometimes provincial level (Cox and Morrison 2004). As the principal of a school in Ranongga island put it, the church pastors are her "secret weapon" in ensuring that her students are doing well, as they are recognized authorities in the village. While villagers may not follow the directions of the chairperson of the school committee, they will not ignore the pastor (McDougall 2008). Churches are deeply influential and deeply respected.

My research showed that churches are generally part of village "committees" which are involved in village-level decision making and the general running of the village. The pastor, minister or priest will usually play an important role in the committee, as do other church members and the village chief. Villages that are well organized have members from all village churches represented on the committee (although the presence of a number of churches in a village can add some significant challenges to village coordination—see the Ranonggan case study below). These committees are often utilized by agencies and religious bodies conducting development work, and, as we will later see, efforts to adapt to climate change and reduce disaster risk.

Underlying the success of the church is the role of women. Solomon Islander Alice Aruhe'eta Pollard (2000) has stated that in church activities women outnumber men and are so powerful and important that, without them, church-related activities in villages would die (2000: 44, 50). Church-affiliated women's groups take a prominent position in all levels of Solomon Islands society. They are expanding rapidly and are now well established in most villages (Pollard 2003). The groups are the legacy of early Protestant missionaries who organized women's gatherings for prayer, Bible literacy, domestic training and moral improvement, and have now become indigenized (Douglas 2003). They are of key significance to women and their communities, and often move beyond the domestic sphere by providing important educational services or tackling political and social issues such as logging and domestic violence (Scheyvens 2003).

It is pertinent also to note the integral role churches play in Solomon Islands civil society. The organizations that form civil society in Solomon Islands are primarily NGOs, community-based organizations, women's organizations, trade unions and churches (Douglas 2007; Prasad and Snell 2004). These institutions are being mobilized as alternative structures to fill the political void (Douglas 2003). Most of these organizations have deep connections with churches. Churches often run their own aid and development organizations, such as the Church of Melanesia's Inclusive Communities Program, and churches run a myriad of local activities that assist local villages and communities, such as the Hope Academy school for disadvantaged children in Honiara, where I volunteered, run by a local Pentecostal church. Churches are also significantly involved in services delivery, such as health and education, where they fill a void left by government (Bird 2007; McDougall 2008; Rhodes 2007).

Solomon Islands churches have a strong tradition in ecumenicalism—that is, churches from differing denominations working in collaboration. SICA is perhaps the most successful of these efforts, with a large headquarters in Honiara and membership consisting of six churches (mainline Protestant and Roman Catholic). Smaller Pentecostal churches have formed the Solomon Islands Full Gospel Association (SIFGA). It is worth noting that, in Solomon Islands, reference to "the church" usually implies SICA churches. When the government associates with "the church," it is generally SICA churches. Perhaps a flaw in an otherwise strong practice of ecumenicalism is the dominance of SICA, which, according to some of my informants, has often left smaller churches feeling left out and under-resourced.

Churches worked well together during the civil unrest, known locally as the "tensions," that occurred between 1998 and 2003. The tensions resulted in the displacement of 35,000 Solomon Islanders, the death of about 200, and an armed takeover of the Honiara government in June 2000 (Hameiri 2007).

The networking power of church-related women's groups was demonstrated in the weeks following the coup. The SICA Federation of Women called women together, conducted combined services and prayers, and collected offerings to help victims, as well as established programs to feed displaced families and provide continued humanitarian assistance (Pollard 2003). Women's groups made prayer visits to militants on opposite sides of the battlefield, encouraging them to lay down arms and forgive each other (McDougall 2008).

The crisis years catalyzed many courageous peacemaking efforts from Christian organizations. It has been said that every church in every level of engagement played some role in the reconciliation and peace process (Bird 2007; see also McDougall 2008). Churches were also involved in some of the formal political peace process, although in some cases were overlooked—highlighting the "informal" role to which they are often perceived to be confined (Douglas 2007).

Christian churches are the only modern bureaucratic organizations in Solomon Islands that are connected at a deep and broad level to rural areas—an advantage that the state has never achieved (McDougall 2008). "For all the complexity and conundrums," writes McDougall, "if any governance initiative is to move beyond

Honiara or provincial centres, it must in some way engage with churches" (2008: 16). Understanding this, faith-based groups such as World Vision and Scripture Union are associated with SICA to undertake development projects (Bird 2007). Similarly, aid arms of overseas churches link with counterpart churches in Solomon Islands for development purposes (Bird 2007), and transnational Christian agencies such as World Vision are also active in Solomon Islands. Below, I will discuss the collaborative work between AusAID, SICA and the NDMO.

The role that churches play in governance, women's empowerment, civil society, health, education, peacemaking, networking and development is crucial for the general wellbeing of Solomon Islands, but also contributes directly to the resilience and adaptive capacity of Solomon Islands to the effects of climate change. For effective development and adaptive capacity, the Solomon Islands government, NGOs and bilateral or multilateral development programs will need to further deepen their relationships with local churches and their denominational hierarchies or centers.

Climate change in Solomon Islands

Like other parts of the world, Solomon Islands is experiencing the effects of climate change. The region has seen annual mean surface temperatures increase between 0.5 °C and 0.8 °C per century from 1901 to 2005 (Trenberth et al. 2007). In addition, there has been a long-term decrease of 20 to 40 percent in annual mean precipitation from 1901 to 2005 (Trenberth et al. 2007). Interview data from three remote islands revealed irregularly occurring dry periods, periods with continuous and heavy rain, windy periods and changes in weather variability, which are likely the effects of climate change (Rasmussen et al. 2009).

Small islands have a special set of climate change-related challenges. Rising sea levels, induced by climate change and exacerbated by other environmental problems like mangrove erosion, bring "king tides" into low-lying islands (such as atolls) that flood homes, pollute fresh water sources and poison crops. Changes in rainfall can pose special problems for small islands with limited access to fresh water (Julca and Paddison 2010). Changes in strength and frequency of cyclones pose a significant threat to livelihoods and human security, and future changes in temperatures and acidity of coastal waters can threaten reefs and their biological richness, as well as an important protective function for vulnerable coasts (Rasmussen et al. 2009). In addition, Solomon Islands is exposed to high levels of climate variation, for example El Niño and La Niña, which cause droughts and floods (Rasmussen et al. 2009).

Vulnerability, resilience and adaptive capacity

In other words, Solomon Islands has a high level of exposure to the current and potential effects of climate change. Yet exposure to hazard only becomes disastrous in the face of underlying vulnerabilities, which undermine resilience and adaptive capacity. The disaster risk reduction framework (see for example Cutter,

Boruff and Shirley 2003; Handmer 2003; Pelling and Uitto 2001; Wisner et al. 2004) helps us to understand climate-induced disasters (either fast-onset such as cyclones or slow-onset such as water scarcity) not as simple natural events, but as the product of a range of social, economic, political and environmental factors. A disaster risk reduction approach aims to reduce vulnerability and increase resilience and adaptive capacity by addressing the underlying factors that cause disasters, rather than focusing solely on the hazard event and disaster response.

In Solomon Islands, many people are vulnerable to hazards (climate-induced and otherwise) in the following ways:

- The semi-subsistence lifestyle of the majority of Solomon Islanders means that the cash required for essential items such as fuel and kerosene is in short supply. For example, a typical monthly cash income for households on the Weather Coast of Guadalcanal is SBD$100, yet for some villages fuel for travel to the nearest town centre in Avuavu could cost SBD$1500 (Kastom Gaden Association 2005b). The lack of cash supply in the context of a cash economy can adversely impact on disaster recovery and adaptive capacity.
- Many parts of Solomon Islands suffer from chronic food insecurity. Semi-subsistence lifestyles are undermined in part by a new reliance on imported crops such as rice and a decrease in the biodiversity of food gardens (see for example Kastom Gaden Association 2005a, 2005b). Food insecurity can be exacerbated by climate change, which can affect agriculture through erosion, salt water incursion, cyclones and storm surges, as well as heat stress and drought, and impact on fisheries (Barnett 2011). Crops lacking in biological diversity can put communities at risk of food shortages, if those crops succumb to climate-related hazards.
- Social unrest, especially the "tensions" between 1998 and 2003, can lead to social dysfunction, which erodes a community's ability to cope with and adapt to hazards.
- The extreme isolation of many communities makes timely disaster relief very difficult. Around 85 percent of Solomon Islanders live not in provincial towns or capitals, but in rural villages around the vast scattering of islands that make up this country. It can take a week to reach some communities from the capital, Honiara. Ships travel to some remote villages as irregularly as twice or three times a year.
- Many communities lack basic infrastructure such as roads, airstrips, radios and health facilities, which increase their isolation and overall development.
- A lack of formal governance at a provincial or national level has led to under-development and reduced adaptive capacity.

Yet Solomon Islanders possess significant capacity to adapt to climate change and climate variability, and have done so for centuries. Communities that continue traditional disaster mitigation methods can be more resilient in the face of a hazard, for example planting fast-maturing, short-cycle crops like sweet potatoes immediately after a cyclone, or using traditional food preservation techniques, to ensure food security (Rasmussen et al. 2009). Similarly, traditional building techniques can reduce cyclone damage (Rasmussen et al. 2009). Remittances and assistance from relatives in urban areas can increase the ability of communities to "bounce back" from disasters and adapt to new hazards, and educational attainment can also assist in this (see the Ranonggan example below). Although formal governance is weak in Solomon Islands, there are a number of levels of strong, coherent informal governance, including the chiefly system and churches, as discussed. All of these factors contribute to resilient communities in the face of climate change shocks.

A disaster risk reduction framework captures a range of social, economic, political and environmental factors that underpin resiliency amidst hazard events. Climate change, colossal as it is, remains one factor amongst many (Mercer 2010). It will often exacerbate pre-existing environmental problems, such as the destruction of mangroves which protect coastal communities from storm surges, or the environmental damage caused by the mining and logging industries. Hence it is important that overall resiliency be strengthened. The disaster risk reduction approach also recognizes the ways in which communities have been adapting for centuries to climate trends and extremes (see for example Campbell 2006; Gaillard 2007). The emerging approach of climate change adaptation, which focuses on future uncertainty and new risks, has much merit, and is being increasingly incorporated into disaster risk reduction (Mercer 2010; Tearfund 2008).

Disaster relief

While disaster relief is not the focus of a disaster risk reduction approach, it is nonetheless a factor in overall community resilience. Extreme remoteness and chronic underdevelopment make disaster relief very difficult indeed. Key infrastructure, especially in rural areas, is in short supply—including roads, wharfs and telecommunication (Manilofia 2006).

Cyclone Zoë in December 2002 illustrated the difficulties faced by Solomon Islands disaster relief providers (Yates and Anderson-Berry 2004). The cyclone severely impacted the extremely remote islands of Tikopia and Anuta, which are 1,025 kilometers south-southeast and 1,110 kilometers southeast of Honiara, respectively. Zoë was a Category 5 cyclone that hammered the islands for three days—right in the midst of the social unrest of the tensions. It was nine days before either of the islands received any external assistance—the response delayed not only by the islands' isolation but also by international political protocols and failing national infrastructure (the social unrest meant that the NDMO office in Honiara was forced to function without electricity and telecommunications). While miraculously everybody survived, the damage to the islands was severe. Moreover, minimal infrastructure on the islands also made response difficult: there were no airstrips and no jetties, so ships had to offload via canoe or dinghy.

Disaster response in Solomon Islands would be challenging to even the most well-resourced government, and for the resources available the Solomon Islands NDMO performs well. However, as with disaster management strategies in other

Pacific island countries, Solomon Islands government disaster management requires significant capacity strengthening in terms of physical infrastructure, finances and human resources (McKenzie, Prasad and Kaloumaira 2005).

The culmination of these challenges has led the NDMO to focus less on disaster relief and more on increasing the capacity of communities to cope with disasters, climate-induced and otherwise, more or less on their own. I explore this important topic below.

Churches and disaster risk reduction

Church involvement in disaster relief

Religious groups are often involved directly in disaster relief. In Solomon Islands, churches often have assets that can be helpful to the government in times of disaster, such as ships or useful church personnel. They often have the strongest building in a village and thus are used for shelter in a cyclone. Local churches often have the capacity to organize the disaster response: after the 2007 western Solomon Islands earthquake-tsunami, for example, the United Church on the island of Ranongga organized villages on the less damaged eastern side to gather food for the devastated western coast (McDougall, Barry and Pio 2008).

Denominational headquarters, based in Honiara, sometimes coordinate their own disaster relief, particularly if an affected area is dominated by people of a particular denomination. The Church of Melanesia (Anglican), for example, provided relief aid in the wake of the 2007 western Solomon Islands earthquake-tsunami to Western Province, which has a large Anglican population. Its efforts were funded by local and overseas donations, and partnerships with other church-based NGOs. Relief aid was distributed through village-based disaster committees, of which churches were a part, regardless of denomination. For rehabilitation, on the other hand, the Church of Melanesia worked mainly with Anglicans, helping to reconstruct homes, schools and churches.

The church network can be a powerful force, but can sometimes be forgotten and left out of discussions. On the other hand, according to one informant, individual churches do not always communicate adequately with the NDMO, causing duplication of resources in some areas and subsequent gaps in other areas. When churches and governments work well together, the results can be compelling. For example, I was told that Caritas (the Catholic aid agency) worked effectively with the NDMO on a water sanitation project after the 2007 earthquake-tsunami in western Solomon Islands (Choiseul and Shortland Islands): the NDMO bought water tanks while Caritas installed them. SICA often sets up disaster accounts, and distributes through the NDMO or individual churches. Where there is successful partnership, churches and church organizations can be very effective in disaster relief.

At the time of research in 2010, a new version of the Solomon Islands government National Disaster Risk Management Plan was forthcoming, which would incorporate churches directly into its structure for disaster response and recovery.

As we will see below, the Solomon Islands government, through the NDMO, is working closely with churches for disaster mitigation as well.

Church involvement in community-based disaster risk reduction and adaptive capacity

Given the geographical challenges and chronic infrastructure problems of Solomon Islands, external relief on its own, as recognized by the NDMO, is not an effective way to reduce the impact of disasters in Solomon Islands. Instead, it is more effective to bolster resilience and adaptive capacity to climate hazards at a local level—a focus that has come to be known as "community-based disaster risk reduction." Faith communities provide key resources for this.

Community-based disaster risk reduction focuses on local strategies that are designed to reduce the impact of a hazard through preparedness and hazard mitigation at local levels (Australian Red Cross 2008). Community-based disaster risk reduction takes into account broad socio-economic conditions that affect resilience and adaptive capacity, such as health, education, institutionalized social capital, food security and traditional knowledge.

Community-based disaster risk reduction has found acceptance around the world. These self-reliant strategies are not new, but often form part of the fabric of the indigenous knowledge within communities. One such strategy, practiced in some Solomon Islands communities, involves traditional food preservation techniques such as the production of *nambo* (dried breadfruit) that can be used in the wake of a cyclone when other food sources are scarce. Another strategy is place-based methods of recognizing when a hazard is imminent, such as reading the movement of birds or insects. The latter knowledge base is essential for communities that do not have radio access.

The 2007 earthquake and subsequent landslides on the island of Ranongga provided an example of the role of churches in community-based resilience and adaptive capacity. Although an earthquake is not a climate event, it is relevant in terms of the effectiveness of pre-existing resilience in the aftermath of such an event. The earthquake lifted the island a meter above the high water mark, causing major damage to the shoreline, settlement and garden areas. Buildings that were not destroyed were tilted and structurally unsound; landslides and subsequent rains destroyed many bush gardens (McDougall, Barry and Pio 2008). Typical of disasters in Solomon Islands, external help was limited. Nevertheless, many villages coped well.

The research in the affected areas concluded that churches provided institutional resources and social capital which on the whole increased resiliency to the hazard. For example, soon after the earthquake, the United Church in Ranongga organized for villages on the less devastated eastern coast on the island to send locally grown food to villages on the damaged western coast. The United Church also distributed cash and rice to affected villages.

In addition, the research showed that the presence of churches in villages affected community organization. It is important to note that the presence of churches does *not* always result in good community organization. Rather, the *type* of church makeup in a village impacted heavily on how well the community organized itself in the wake of the earthquake. Communities with a single village church were better able to organize themselves and recovered more quickly, as compared with communities that contained a number of churches (McDougall, Barry and Pio 2008).

Since missionization, Solomon Islands villages have found their social, geographic and spiritual centre in a single church (McDougall 2008). In more recent years, Solomon Islands villagers have become more divided along denominational or even religious lines (McDougall 2008). Since communal village work that goes beyond the scope of extended families has generally been coordinated by the church, multi-denominational or multi-religious villages are often less coordinated than villages that are centered around just one church (McDougall 2008). This has been particularly true in the context of disaster recovery, when a coordinated approach is necessary for the rebuilding process.

This dynamic was clearly demonstrated in the wake of the Ranongga earthquake. Here, I will contrast the situation in the village of Modo with that of Sabala. The village of Modo had just one church: the Seventh Day Adventist Church. As such, its village committee was not divided along denominational lines, and the people of the village formed a single congregation. There were clear sites of power and lines of authority. In Modo, the community functioned collectively, and as a whole made the decision on the advice of government geologists to relocate to a safer area. They were assisted by the government and NGOs. The result was a thriving new settlement with a primary school, pre-school and kindergarten quickly established and running. Villagers described an "all for one" approach; families with more land gave to others with no land nearby, and men worked together to saw wood needed for homes and community buildings. Community funds were used to buy fuel for the chainsaw, raised through community markets selling fish and garden produce. A community disaster committee formed to provide leadership and coordination in the recovery effort.

The nearby village of Sabala was similarly affected by the earthquake. However, unlike Modo, it has three different churches. Despite other examples of ecumenicalism in Solomon Islands, these churches did not work well together and did not function collectively in the aftermath of the disaster. This hampered Sabala's recovery from the disaster. People moving away from the shoreline found it very difficult to clear land, rebuild and have enough time to produce food, and thus faced food shortages. Additionally, people from Sabala felt they had been overlooked by external disaster assessors and did not get their fair share of relief supplies.

Yet other factors that existed prior to the earthquake also impacted disaster recovery. Education is a key factor, which an overwhelming advantage for Modo. The worldwide Adventist church emphasizes education and runs some of the best schools in the country. Seventh Day Adventist villages such as Modo tend to have higher education levels than other villages in Solomon Islands. Higher levels of education proved advantageous for Modo's disaster recovery, as they enabled people to negotiate the bureaucracy involved in obtaining relief assistance.

Education can assist disaster recovery in less obvious ways. Seventh Day Adventist islanders, for example, often have relatives with skilled positions living in urban areas who can be called on for assistance in times of adversity (McDougall, Barry and Pio 2008). Many send regular remittances, which are used for essential items such as fuel and kerosene; perhaps this cash source was also responsible for the presence of an outboard motor in Modo. The outboard motor, in turn, enabled the Modo villagers to buy supplies and seek external assistance from the nearest urban centre. This kind of assistance is typical of the interdependency characteristic of the *wantok* system in Solomon Islands (Handmer and Choong 2009).

Yet, in Sabala, there was not even a primary school. Sabala villagers were less likely to have educated relatives in urban areas to send remittances. Unlike Modo, the Sabala village did not possess an outboard motor for transport and thus was not able to seek external assistance in the same way. The village was generally less developed than Modo.

At this point I will note that there is always a tension between the benefits that churches bring to community cohesiveness and their consumption of individual resources. The Ranongga earthquake research found that many villagers prioritized rebuilding the church over and above their own recovery, in the hopes of experiencing blessing. From an outsider's perspective, it is impossible to determine whether this was an appropriate tradeoff. To the people who make this decision, clearly the role of the church in community cohesion and provision of spiritual goods was worth the individual cost.

Specific disaster mitigation and preparedness are also important to a community's coping capacity in the face of climate change. Again, Solomon Islands churches have a large role to play here. The Pacific Community-Focused Integrated Disaster Risk Reduction (PCIDRR) program, which I was shown by personnel at the NDMO in Honiara, is an example of church–government partnership for community-based disaster risk reduction. The program is funded by AusAID and jointly administrated by SICA and the National Council of Churches in Australia, which has partnered with Solomon Islands for a long time. These program partners work closely with the NDMO, utilizing NDMO expertise. The goal is to equip communities to respond to disasters on their own, by engaging with the existing skills, knowledge and governance within a particular community.

Facilitators from SICA and NDMO assist each village to identify hazards and potential ways to reduce the risk of the hazard. From the general village committee a village disaster risk committee is formed. This group creates a village disaster plan, identifying the risks and helping to enact the solutions, including devising a community evacuation plan and running simulation exercises that link to a nationally devised warning system. The PCIDRR essentially bridges the gap between the work that the NDMO conducts at a provincial level and the village level.

Examples of community-based disaster risk reduction strategies identified during a PCIDRR process included the following:

- A dead tree was identified as hazardous in the event of a cyclone; the solution identified was to cut it down.
- A village identified that it would be difficult for children or elderly or disabled people to cross a certain stream that was blocking the path inland in the event of a cyclone or tsunami; the community identified the need for a bridge.
- A river that regularly flooded, destroying crops that villagers had planted in the fertile river soil, did so because it was blocked by fallen logs upstream, which then created a dam that burst in the rainy season. The community identified the need to remove the logs.

This latter task required only bush knives, axes and the labor of people—resources that each household possessed. Most of the solutions do not require a large input of external resources; they are not focused on building infrastructure but instead utilize the assets that already exist in the village. This "strength-based approach" appears to be finding success amongst NGOs elsewhere in Solomon Islands.

A report of the AusAID Pacific Enhanced Humanitarian Response Initiative (PEHRI, 2010) contended that the PCIDRR program was overly ambitious, with too many villages targeted and not enough resources to ensure effective disaster risk reduction strategies in each one. Nonetheless, the concept of engaging with churches to facilitate disaster risk reduction and increase adaptive capacity in the face of climate change is an important one, because of their role in the social fabric and governance of Solomon Islands.

Churches in Solomon Islands are also contending with slow-onset climate change-induced hazards, such as sea level rise and temperature change, and salt-water incursion. This climate change effect most heavily impacts the low-lying atoll islands. Ontong Java Atoll, or Lord Howe Islands, in the Malaita province is one such area. The people of Ontong Java rely on swamp taro, coconut and fish to survive. During a cyclone in 2008 the island was inundated with seawater. Afterwards, the swamp taro began to weaken and die. The effects were far reaching: in a society where food is integral to social connection, a lack of food not only causes hunger but negatively affects social cohesion. The people of Ontong Java required new, salt-resistant crops.

The Church of Melanesia responded with the Ontong Java Climate Change Program, described to me by the man who runs it. It involves the introduction of what he called "atoll permaculture" to the islands. The idea is to create a "forest of food"—introducing around ten species of salt-resistant crops. At the time of interview, some of these crops had already been planted, and it has been reported that they are doing well. Again, the program utilizes the village-based church structure to implement this project. The program partners with non-church organizations like the NDMO, Kastom Gadan and the Ministry of Agriculture, and is funded by an American Christian development organization.

The geography of Solomon Islands, combined with its minimal infrastructure and limited ability to deliver relief aid, ensures that community-based disaster risk reduction is the most promising way to address climate change hazards. Solomon

Islands churches have a key role to play in both of these areas, and are most effective when they partner with non-church parties such as government and NGOs.

Conclusions

The research herein aims to demonstrate the role of religious institutions in building resilience and adaptive capacity in communities and countries at risk from the effects of climate change by exploring the role of churches in Solomon Islands. Resiliency, as demonstrated, involves more than responding to discrete hazards, but is determined by pre-existing social, economic and political factors. Churches in Solomon Islands bolster resilience and adaptive capacity through their provision of social capital, institutional resources and governance. This was demonstrated in the example of the Seventh Day Adventist Church in the Modo village of Ranongga. On the other hand, a counter-example is Sabala, where the churches may have had an impact on social capital and governance that reduced disaster recovery. This suggests that it is not the presence of churches per se that impacts positively on disaster resilience and adaptive capacity, but rather the *type* of institution. Thus it would be unwise to rely uncritically on churches for bolstering resilience and adaptive capacity. A better recommendation is to see them as important actors to consider and to resource.

Churches are entrenched deeply within the social fabric of Solomon Islands. They provide multi-level governance and leadership, as well as conducting important health, education, development and disaster risk reduction work. For these reasons, it is very important that churches be supported and kept within the conversation, and equally important that churches partner effectively with governments and NGOs. The alternative would be the deliberate and costly replacement of an existing civil societal infrastructure with another structure.

There is a need for research that further explores the role of faith groups in building resiliency to climate change and other hazards (Lunn 2009). The case of Solomon Islands shows that it would be productive for research to attend to the way that religion is responding—and can potentially respond—to the challenges of climate change in building overall resiliency and adaptive capacity.

Acknowledgments

There were many people who made this research possible, including: Ruth and Hennessey Maetala and their family in Solomon Islands, who embraced me and introduced me to many others, as well as taught me many things and commented on drafts of this chapter; Rebecca Monson, who introduced me to Ruth and thus started me on this journey; Regina Gatu, who taught me Pijin at her kitchen table in Melbourne; Loti Yates and other staff at the NDMO in Honiara, who explained many things and provided helpful feedback on drafts; John Handmer and Blythe McLennan at the Centre for Risk and Community Safety at RMIT University, who have supported me in this research and provided helpful feedback; and the many warm, welcoming people of Solomon Islands, especially the New Hope United

Pentecostal Church in Kombivatu, Honiara, who supported me in my stay and my research.

References

Australian Red Cross. 2008. "Traditional Knowledge and Red Cross Disaster Preparedness in the Pacific." Available at http://216.14.195.135/media/Traditional_knowledge_and_Red_Cross_disaster_preparedness_in_the_Pacific_(3).pdf (accessed January 17, 2011).

Barnett, J. 2011. "Dangerous Climate Change in the Pacific Islands: Food Production and Food Security." *Regional Environmental Change* 11 Suppl. 1, 229–237.

Beebe, James. 2001. *Rapid Assessment Process: An Introduction*. Walnut Creek, CA: AltaMira.

Belshaw, D.G.R., C. Sugden and R. Calderisi. 2001. *Faith in Development: Partnership between the World Bank and the Churches of Africa*. Washington, DC and Oxford, U.K.: World Bank and Regnum Books.

Bird, C. 2007. "Blowing the Conch Shell: A Baseline Survey of Churches Engagement in Service Provision and Governance in the Solomon Islands." AusAID. Available at http://www.ausaid.gov.au/publications/pdf/solomons_church_survey.pdf (accessed February 19, 2010).

Bureau of Democracy, Human Rights, and Labor. 2007. "International Religious Freedom Report 2007." U.S. Department of State. Available at http://www.state.gov/j/drl/rls/irf/2007/90154.htm (accessed October 18, 2012).

Campbell, J.R. 2006. *Traditional Disaster Reduction in Pacific Island Communities*. Lower Hutt, New Zealand: Institute of Geological and Nuclear Sciences.

Cox, J. and J. Morrison. 2004. "Solomon Islands Provincial Governance Information Paper." AusAID. Available at http://www.ausaid.gov.au/publications/pdf/sols_provincial_gov.pdf (accessed March 3, 2010).

Cutter, S.L., B.J. Boruff and W.L. Shirley. 2003. "Social Vulnerability to Environmental Hazards." *Social Science Quarterly* 84.2, 242–261.

Dickson-Waiko, A. 1999. "Civil Society and Development, Non Government Organizations and Churches." *Development Bulletin* 50, October, 44–46.

Douglas, B. 2003. "Christianity, Tradition, and Everyday Modernity: Towards an Anatomy of Women's Groupings in Melanesia." *Oceania* 74, 6–23.

——. 2007. "Christian Custom and the Church as Structure in 'Weak States' in Melanesia." In *Civil Society, Religion and Global Governance: Paradigms of Power and Persuasion*, ed. H. James, pp. 158–174. London: Routledge.

Falk, M.L. 2010. "Recovery and Buddhist Practices in the Aftermath of the Tsunami in Southern Thailand." *Religion* 40, 96–103.

Fountain, P., S. Kindon and W. Murray. 2004. "Christianity, Calamity, and Culture: The Involvement of Christian Churches in the 1998 Aitape Tsunami Disaster Relief." *Contemporary Pacific* 16.2, 321–355.

Gaillard, J.C. 2007. "Resilience of Traditional Societies in Facing Natural Hazards." *Disaster Prevention and Management* 16.4, 522–544.

Gillard, M. and D. Paton. 1999. "Disaster Stress Following a Hurricane: The Role of Religious Differences in the Fijian Islands." *Australasian Journal of Disaster and Trauma Studies*, 1999.2.

Hameiri, S. 2007. "The Trouble with Ramsi: Reexamining the Roots of Conflict in Solomon Islands." *Contemporary Pacific* 19.2, 409–441.

Handmer, J. 2003. "We Are All Vulnerable." *Australian Journal of Emergency Management* 18.3, 55–60.

Handmer, J. and W. Choong. 2009. "Resilience: Wantoks, Transnational Traders and Global Politics." In *Rethinking Insecurity, War and Violence: Beyond Savage Globalization?*, ed. D. Grenfell and P.W. James, pp. 208–220. New York: Routledge.

Hegarty, D., R. May, A. Regan, S. Dinnen, H. Nelson and R. Duncan. 2004. *Rebuilding State and Nation in Solomon Islands: Some Options for the Regional Assistance Mission*. State, Society and Governance in Melanesia Discussion Paper 2004/2. Canberra: Research School of Pacific and Asian Studies, Australian National University.

Julca, A. and O. Paddison. 2010. "Vulnerabilities and Migration in Small Island Developing States in the Context of Climate Change." *Natural Hazards* 55.3, 717–728.

Kamaara, Eunice. 2000. "The Role of the Christian Church in Socio-Economic and Political Development in Kenya." *Journal of Third World Studies* 17.1, 165–176.

Kastom Gaden Association. 2005a. "Extreme Living, Extreme Need." Available at http://pacific-edge.info/wp content/uploads/2009/07/wc_makira_06.pdf (accessed February 24, 2010).

——. 2005b. "People on the Edge." Available at http://kastomgaden.org/wp-content/uploads/2010/03/gualalcanal_nov05 (accessed January 17, 2011).

Luers, A.L., D.B. Lobell, L.S. Sklar, C.L. Addams and P.A. Matson. 2003. "A Method for Quantifying Vulnerability, Applied to the Agricultural System of the Yaqui Valley, Mexico." *Global Environmental Change* 13, 255–267.

Lunn, J. 2009. "The Role of Religion, Spirituality and Faith in Development: A Critical Theory Approach." *Third World Quarterly* 30.5, 937–951.

Manilofia, R. 2006. "Disaster Communications in Solomon Islands." ITU/ESCAP Disaster Communications Workshop, December 12–15, Bangkok, Thailand. Available at http://www.itu.int/ITU-D/emergencytelecoms/events/ThailandWorkshop/final1/Session%203/SESSION%203%20%5BSolomon%20Islands%5D%20Country%20Paper.pdf (accessed February 5, 2010).

McDougall, D. 2008. "Religious Institutions as Alternative Structures in Post-Conflict Solomon Islands: Cases from Western Province." State, Society and Governance in Melanesia Discussion Paper 2008/5. Canberra: Research School of Pacific and Asian Studies, Australian National University.

McDougall, D., I. Barry and S. Pio. 2008. "Disaster and Recovery on Ranongga: Six Months after the Earthquake in the Western Solomons." Independent report. Available at http://westernsolomons.uib.no/docs/McDougall,%20Debra/McDougall%20(2007)%20Ranongga%20Disaster%20Recovery%20(report).pdf (accessed February 24, 2010).

McKenzie, E., B. Prasad and A. Kaloumaira. 2005. "Economic Impact of Natural Disasters on Development in the Pacific," vol. 1. AusAID. Available at http://www.ausaid.gov.au/publications/pdf/impact_pacific_report.pdf (accessed August 24, 2010).

Mercer, J.2010. "Disaster Risk Reduction or Climate Change Adaptation: Are We Reinventing the Wheel?" *Journal of International Development* 22.2, 247–264.

Milner, H.R.2007. "Race, Culture, and Researcher Positionality: Working through Dangers Seen, Unseen, and Unforeseen." *Educational Researcher* 36.7, 388–400.

PEHRI (Pacific Enhanced Humanitarian Response Initiative). 2010. "Independent Progress Report: Mid Term Evaluation of the AusAID Pacific Enhanced Humanitarian Response Initiative."

Pelling, M. and J.I. Uitto. 2001. "Small Island Developing States: Natural Disaster Vulnerability and Global Change." *Global Environmental Change Part B: Environmental Hazards* 3.2, 49–62.

Pollard, A.A. 2000. *Givers of Wisdom, Labourers without Gain: Essays on Women in Solomon Islands*. Suva, Fiji: Institute of Pacific Studies, University of South Pacific.

——. 2003. "Women's Organizations, Voluntarism, and Self-Financing in Solomon Islands: A Participant Perspective." *Oceania* 74, 44–60.

Prasad, S. and D. Snell. 2004. "The Sword of Justice: South Pacific Trade Unions and NGOs during a Decade of Lost Development." *Development in Practice* 14.1, 267–279.

Rasmussen, K., W. May, T. Birk, M. Mataki, O. Mertz and D. Yee. 2009. "Climate Change on Three Polynesian Outliers in the Solomon Islands: Impacts, Vulnerability and Adaptation." *Geografisk Tidsskrift: Danish Journal of Geography* 109.1, 1–13.

Rhodes, D. 2007. "Analysis of the 'Community Sector' in Solomon Islands." AusAID. Available at www.ausaid.gov.au/publications/pdf/Community_Sector_in_SI_May_08.pdf (accessed February 23, 2010).

Scheyvens, R. 2003. "Church Women's Groups and the Empowerment of Women in Solomon Islands." *Oceania* 74, 24–43.

Smart, N. 1996. *Dimensions of the Sacred: An Anatomy of the World's Beliefs*. London: HarperCollins.

Tearfund. 2008. "Linking Climate Change Adaptation and Disaster Risk Reduction." Available at http://www.tearfund.org/webdocs/Website/Campaigning/CCA_and_DRR_web.pdf (accessed May 3, 2011).

Tillman, L. 2002. "Culturally Sensitive Research Approaches: An African-American Perspective." *Educational Researcher* 31.9, 3–12.

Trenberth, K.E., P.D. Jones, P. Ambenje, R. Bojariu, D. Easterling, A. Klein Tank, D. Parker, F. Rahimzadeh, J.A. Renwick, M. Rusticucci, B. Soden and P. Zhai. 2007. "Observations: Surface and Atmospheric Climate Change." In *Climate Change 2007: The Physical Science Basis*, ed. S. Solomon, D. Qin, M. Manning, Z. Chen, M. Marquis, K.B. Averyt, M. Tignor and H.L. Miller, pp. 236–336. Contribution of Working Group I to the Fourth Assessment Report of the Intergovernmental Panel on Climate Change. Cambridge, U.K.: Cambridge University Press.

Wisner, B., P. Blaikie, T. Cannon and I. Davis. 2004. *At Risk: Natural Hazards, People's Vulnerability, and Disasters*. London: Routledge.

Yates, L. and L. Anderson-Berry. 2004. "The Societal and Environmental Impacts of Cyclone Zoë and the Effectiveness of the Tropical Cyclone Warning Systems in Tikopia and Anuta, Solomon Islands, December 26–29, 2003." *Australian Journal of Emergency Management* 19.1, 16–20.

8 Prophecies and climate change in the Mam Altiplano of Guatemala

Julie Hermesse

Introduction

In 2005, tropical storm Stan made landfall on the shores of southwestern Guatemala.[1] Located in western Guatemala, the town of San Martín Sacatepéquez (SMS), the object of this study, was devastated by tropical storm Stan's passage, primarily on the material and agricultural level.[2] Mud flows, caused by landslides from the mountains above SMS, buried the town from October 3 to 6, 2005. Most of San Martín's inhabitants, also called *tinecos* (an abbreviation of *San Martinecos*) diagnosed climate change as being the cause of meteorological events and climatic disturbances.

The 2005 cyclone season's historic power (in addition to Stan, Katrina, Rita and Wilma passed through that year) also raised questions within the scientific community regarding whether the occurrence and violence of these tropical cyclones were related to climate change. Climatologists almost unanimously reject the assumption of an increase in the number of hurricanes due to climate change. Instead, they concentrate their questions on the link between the rise in temperatures in the past 30 years and the increase in the hurricanes' power and destructive capacity (André, Royer and Chauvin 2008; Webster et al. 2005).

Whereas scientists develop theories regarding extreme climatic phenomena, the populations who are the victims have their own interpretations of the events, based on their cultural systems. Anthropological interest in climate has developed greatly over the preceding decade. As Steve Rayner underlines, this new interest surely has something to do with the increasing worry about climatic changes and variability (Rayner 2003: 277).

Based on long-term ethnographical research in Guatemala's Altiplano region, this chapter delves into local sources of knowledge about climate change. My primary finding is that the *tinecos* analyze Stan in the light of prophecies. The prophetic representations, based as much on the Christian religion as on Mayan spirituality, differ from one another both in what they suggest is the origin of the prophesied events and in the results they predict. I argue that in the context studied there exist two ideal-typical interpretations of the etiology of natural phenomena like Stan and, more broadly, climate changes: one regards these phenomena as being of divine origin and the other points the finger at anthropogenic influences.

Research site and methodology

Located 24 kilometers from the regional city of Quetzaltenango, San Martín Sacatepéquez's municipal territory extends from the mountainous chain of the Sierra Madre to its Piedmont. San Martín has a surface area of 100 square kilometers, and is composed of "cold lands" of the Altiplano and "hot lands" of the ecological plateau called the *boca costa.* The Altiplano communities differ from the Piedmont's in many ways. In order to preserve the research project's coherence, I focused my attention exclusively on the populations residing in the Altiplano. In this rural and mountainous area of southwestern Guatemala, I concentrated my research on the urbanized center of San Martín, where 5,078 of the municipality's 29,116 inhabitants reside.[3]

The identity referents characterizing the area are the following: a rural, peasant minifundist economy, a population of Mayan origin, high demographic density and elevated poverty indices.[4] In San Martín, 85 percent of the town's inhabitants are of Mayan Mam ethnic origin (INE 2003). The remaining percentage corresponds to the mixed-race fringe of the population, commonly called *ladinos* in Guatemala. There are 23 national ethnic groups that cohabit the territory, 21 of which are Mayan, each of them possessing their own language and dialectal variations. The Mam ethnic and linguistic area covers the regions of Quetzaltenango, Huehuetenango and San Marcos.

Whereas the *ladino* population, with some rare exceptions, is monolingual, Mams are generally bilingual, speaking both Mam and Castilian (Spanish). Mastery of Castilian however varies according to generations, levels of schooling, and distance from the urban center. Aware that learning a language also means discovering a symbolic universe, at the beginning of my research I concentrated my efforts on learning Mam. However, my knowledge of Mam remains rudimentary. The young people found speaking their dialect with me ridiculous. Generally, they would initially try to approach me in English. The older *tinecos* tested my knowledge of their language in exchanging greetings. After the usual questions, they quickly switched to Castilian. When meeting with monolingual Mam elders, I generally arranged to be introduced by a third person who was familiar to them and fluent in both languages. This individual helped me gain my informants' confidence while also helping ensure that I was understanding their remarks. The language generally used in my daily discussions with the *tinecos* was thus Castilian.

As with most Latin-American religious configurations, three religious poles coexist in Guatemala: Mayan spirituality, the Catholic religion since the sixteenth-century Spanish conquest and, recently, Protestantisms. Anchored in a thousand-year-old cosmogony, Mayan spirituality is a means of entering into a relation with "*lo sagrado*," the sacred, through rituals and daily practices. Reference to the Mayan ideological system, commonly called the Mayan cosmovision, is made to explain social and community relations as well as to interpret the world and its events. Mayan cosmovision justifies and explains relationships with the natural environment, and animates daily activities and behaviors like agricultural work, as well as allowing communication with what I call "figures of the invisible."[5] The shamans are specialists in ritual involving both life cycles and agricultural cycles.[6] The shamans of San Martín rarely assume their role publicly or socially. Their current marginalization confines them to a role that is relatively cut off from the local social and cultural scene of San Martín. Despite a drastic reduction in the number of shamans[7] according to most of the *tinecos*, I listed a total of over 50 shamans just in the hamlets of San Martín's cold lands.

Over the last 30 years, Protestantism, especially in its evangelical, Pentecostal form, has established itself in San Martín. Its growth is such that today approximately 50 percent of the inhabitants of the municipal center have converted to one of the Protestant currents.[8] In March 2009, I counted a Presbyterian church and 13 Pentecostal evangelical churches in San Martín's urban center. These rural and popular Protestantisms have developed in relative autonomy, although they are often supported by churches headquartered in Quetzaltenango or the capital. Natives of San Martín, or even Quetzaltenango, the current pastors have lent an autochthonous face to a religion with foreign origins.

Facing the proliferation of Protestant churches, the unique and massive Catholic church dominates the town's central square. Yet the architectural dominance of Catholic buildings is just a facade hiding internal tensions specific to the Guatemalan Catholic universe. Through a long syncretic process, the historical fusion between Catholicism and Mayan spirituality finds expression in daily life, despite the Catholic Church's opposition. For example, some religious brotherhoods are supported by actors identified as traditionalists (*tradicionalistas*) who defend local, syncretic customs (*costumbres*) that are based in Mayan spirituality.

From a methodological point of view, my research is based on a long-term ethnographical commitment, a technique which is necessary when analyzing the complex factors surrounding the catastrophes that climatic disasters provoke. The current chapter draws on 15 months of field work, undertaken between March 2006 and August 2010. In addition to conducting participant observation and informal discussions on a daily basis, I interviewed San Martín's inhabitants. The reflections developed here are the result of analyses of these encounters, which probed *tinecos*' understanding of what Stan and other climatic phenomena represented in their town. In order to remain as close as possible to the *tinecos*' remarks, in what follows I will use as many extracts from my interviews and discussions as possible.

Observation of climatic change by the Mams

According to recent studies, Guatemala figures among the countries most affected by climate changes. The *Global Climate Risk Index 2012*, published by the NGO Germanwatch, cites Guatemala as being the second most affected country (after Pakistan) when it comes to the climatic events of 2010, particularly hurricanes (Harmeling 2011). Future climatic scenarios for Guatemala, correlated with the country's vulnerability indices, predict a probable increase in floods and droughts (Castellano and Guerra 2009).

For populations whose principal source of income depends on their agricultural activities, direct climate observation is crucially important. As the anthropologists Strauss and Orlove note, these folk understandings of climatic changes are embedded within collective notions of time, as is demonstrated by the tendency for people to describe these changes by "not[ing] that the weather was different in the time of their parents, grandparents, and earlier generations" (2003: 9). Indeed, as lifetime observers of weather phenomena, the elders I encountered compared the current climatic situation with what they knew in their childhood and with what their grandfathers told them.[9]

In San Martín, the inhabitants recount the passage of other storms previous to Stan. Back then, they hasten to point out, there was no custom of naming "hurricanes." They say that hurricanes occurred during a period circumscribed between July and October—a period corresponding to the hurricane season defined by scientists. Appearing in the month of October 2005, Stan would not have appeared exceptional in the hurricane calendar. As such, most residents I encountered interpreted Stan as a natural phenomenon. However, Stan's extraordinary character and the context of many climatic events have also engendered local discourses which describe these climatic events in terms of climate change. According to a local named Álvaro:

> These phenomena are more powerful today. Before, there were storms, but it lasted a day, not more; there was not that much wind, the streams did not swell that much and the water did not leave the mountains. . . . Nobody was worried. Only last year, with Stan, we started to get worried.
>
> (Interview, April 21, 2006)

The elders tell us that Stan stood out because of its intensity and exceptional character. Nobody in living memory could remember a like event.

Besides Stan, other phenomena are also observed as signs of climate change. First of all, the alternation of the dry and rainy seasons, heretofore relatively punctual in the Central American climatic calendar, seems to be upset. As Carmen explained, "It has become impossible to say when the rainy season will start and will come to an end. Before, the rains came and came to an end on a given date." The unfolding of the climatic calendar shows signs of being disturbed. For, she continued:

> The climate is changing now. This year [2010], it rained too much. There were already two hurricanes during May and June: Alex and Agatha. Whereas last year, rather than being in winter [i.e., the rainy season], the months of July and August, well, it was like mid-summer. And, as in summer, the weather was nice during the day and very cold nights, and in fact there was a frost in July!

The climatic seasons are not what they used to be: it rains when it is not supposed to and does not rain when rain is expected. Whereas the rainy season of 2009 was characterized by a drought, the 2010 rainy season set a rainfall record with the occurrence of tropical storms Agatha, Alex and Frank. In August 2010, a farmer, Pedro, complained about climate change: "Now it rains, but it rains too much! My vegetables are still in the earth; the onions and potatoes are rotting. The year before, we had already lost our harvests because of drought." Efraín added: "We sure feel climate change here. Now we've lost almost all our potatoes because of rain. It rains too much, and that's not normal."

The media certainly play a major role in spreading information on the subject. Some of the *tinecos*, often professionals who have benefited from education outside the community, have looked into the reasons for climate change and are familiar with the specialized literature. Yet Pablo deplored his town's lack of knowledge on climate change. In his view, "People here notice that there are variations in the climate, but they do not know why." According to him, most of the inhabitants of San Martín refer to God in talking about climate change:

> In San Martín, the inhabitants placed the blame on God. They said, "If God wills it, then that's the way it has to be." They are not really worried about the situation. And that's the case for both the Catholics and the evangelicals. Because, according to them, we cannot go against God's decision. For example, little information has been disseminated on the effect of greenhouse gases.
>
> (Interview, July 23, 2010)

If most of the inhabitants of San Martín diagnose climate change as being the cause of weather events and climatic disturbances, it is interesting to consider the sources of these discourses as well as the explanations the *tinecos* give as to the origin of climate change. I found that the causes at the basis of climate change were a frequent topic of discussion within the community.

Christian perspectives on climate change: Apocalypse now?

Many residents of San Martín told me that the Supreme Being, the Transcendent Creator, had given Stan its breath. The *tinecos* who were members of evangelical denominations explained Stan's divine origin in terms of particularly entrenched millenarian and apocalyptic ideas: Christ will return to earth and will destroy the terrestrial world to establish a new world free from all sin. According to Marco, the Monte Sinaï Church's evangelical pastor, for example:

> The end of the world is written in the word of God. God says that the sky and the earth will disappear but only his words will remain. When the disciples asked Jesus Christ what would be the signs of his return, God announced that there would be earthquakes, wars, famine, plagues, high winds, hurricanes. . . . We are Christians and we trust in God's word.
>
> (Interview, April 30, 2006)

For these evangelical converts, Stan is thus interpreted as a sign of Christ's coming, prefiguring the end of the world. One *tineco*, Cesar, told us that during hurricane Stan his neighbors who fled their homes to seek refuge shouted and cried in the quagmire of water and earth: "We're going to die today; the world will come to an end because God has decided to finish us off."

According to many evangelicals, being converted to Christ provides an assurance of having one's place amongst the "saved." Confident in that divine protection, Florinda explained why her husband, Adolfo, did not want to join the *tinecos* who evacuated San Martín during Stan's passage in the following manner: "People from the town of Concepción came to get us in a four-wheel-drive vehicle but my husband answered: 'God is going to liberate us and defend us here. If you want to leave, go ahead, but I'm staying put here because Christ has promised to help me.'" No believer who has the faith should fear or flee events like Stan, Adolfo said subsequently.

For the *tinecos*, humans should ask themselves what their sin is, behind the divine sending of Stan. Cesar, a Bethania Church convert, considered that "God sends hurricanes so that we will repent, because too many people live in disorder." Florinda explained that disorder was due to a failure to worship God: "We do not respect God anymore. We no longer believe in God. But, for me, He's the judge." Many *tinecos* with whom I spoke said that God had manifested himself through a hurricane to remind people of him, as well as to install a reign of order and respect. Adolfo, an evangelical believer, described God as "a father who sees to it that his children are obedient and not ill bred. That's why God punishes us. We're punished so that we'll see God's power."

For the Catholics, God is certainly not completely foreign to the tropical storm. Yet most of the Catholics I interviewed did not adopt such a literal interpretation of the Bible. "They [Catholics] place less accent on the end of the world," explained the shaman Marcelo. For their part, the evangelicals justify the lack of attention the Catholics pay to signs of the Apocalypse by their ignorance of the Bible.

Non-evangelicals consider the argument of divine protection by conversion as part of the evangelical churches' proselytism strategy. From that perspective, the former priest in San Martín described the evangelical God as a retributive God who punishes people because they are not sufficiently converted to Protestantism and do not accept Christ. Consequently, for the shaman Marcelo, the evangelicals are only worried about the need to save themselves spiritually. In his view, the pastors manipulate their believers using sermons propagating the fear of not being saved:

> The pastors scare the population when there's a situation like the one caused by Stan, saying, "Come with us for the end of the world is at hand! Accept Jesus Christ! Repent! Save your soul! Stop thinking about material things, turn yourself towards God alone! For we can see signs of Christ's coming in the world's decomposition. The hour has come for the Son of God to come to earth a second time." So people are afraid and forget about their futures.
>
> (Interview, April 28, 2006)

Carla too thought that the evangelicals in San Martín did not worry enough about material things: "What happens here and now matters little to them." Although Carla was converted to an evangelical church in North America, she did not join a church on her return to Guatemala because, she said, "I did not share their way of thinking." She vigorously rejected the evangelicals' preaching the end of the world. For her, those words only sowed fear and confusion among the converts:

> At the time of Stan's passage, the evangelicals said "Oh my Lord, God's coming! It's the end of the world and this is our punishment." They scared people, who then felt confused, because a lot of people do not understand things and live in ignorance. For example, they would say: "Everything that's taking place, if there is an earthquake or a hurricane, it means that God is coming, that it's the end of the world." So people are afraid and some even want to kill themselves because they do not want to live and suffer through what's going to happen.
>
> (Interview July 17, 2007)

The recent passage of tropical storm Stan through Guatemala shows how influential an apocalyptic millenarian interpretation of events has been. The prevalence of such interpretations should not be surprising in light of scholarship connecting extreme natural phenomena with the rise of millenarian religious denominations (Barkun 1974).[10] Hence, for the particular case of Guatemala, it is interesting to look into the emergence of this apocalyptic thought of Protestant origin. In fact, we can trace the passion for this kind of preaching roughly to the time of the 1976 earthquake and the evangelical Protestant effervescence it caused (Hermesse 2010).

At dawn on February 4, 1976, a seismic rupture located on the Motagua fault, which crosses 80 percent of Guatemalan territory, devastated every one of the country's counties. The death toll was high: 23,000 Guatemalans lost their lives, 76,000 people were injured, and over 1 million lost their homes. On the whole, nearly 5 million Guatemalans (out of a population which, at the time, did not equal twice that) were affected by the violent tremor. For the anthropologist Sylvie Pédron-Colombani, "the misery was an 'excellent' springboard for establishing those new Churches" (1998: 9). Having come to aid the affected populations, evangelical missionaries from North America brilliantly made up for the deficit in official emergency aid, without restricting their care and services to converts alone. David Stoll presents the case of the 1976 Guatemalan earthquake as a perfect illustration of the modus operandi, which he calls "*el evangelismo del desastre*," or the evangelicalism of disaster (Stoll 1990: 25). North American Protestant organizations saw in the tragedy "the opportunity to help Guatemala physically and improve it spiritually" (Garrard-Burnett 1998: 120).

Statistics testify to the success of the evangelical missions' proselytism: in the few months following the earthquake, membership in the evangelical churches rose from 8 percent to 14 percent (Garrard-Burnett 1998: 121; Stoll 1990: 26). Moreover, even after the emergency was over and the aid services closed down, the

Protestant churches' growth continued unabated. In 1982, the annual growth rate of Protestant conversions in Guatemala was 23.6 percent.[11] The factors explaining the Protestant progression in Guatemala between the late 1970s and the early 1980s cannot be limited to the concrete answers Protestant congregations offered in meeting material needs or to a strategy of American infiltration. A key endogenous factor must be added to this discussion: the seduction exerted by the millenarian discourses the majority of the Pentecostal evangelical Protestant churches preached (Bastian 1990). In an increasingly impoverished country that is disaster stricken on the human, material and agricultural levels, intense questioning goes on about the meaning of life as well as relations between human beings and divine forces. Such questioning can easily lead to apocalyptic ideas, which (among other things) help explain current suffering, while also promising its imminent end.

My research into the aftermath of Stan in San Martín indicates that nowadays it would be unwarranted to say that the whole evangelical community is caught up in an eschatological fervor. Yet, if evangelicals are less likely to consider human acts to be the source of recent disasters, this does not mean that they are always dismissive of such explanations. José, an evangelical, does not accept that the will of God alone is the sole guide to events on earth. In his view, if the earth cannot be saved, it is because people have lacked the will to save it. Human will amplifies God's will: "The will of Christians counts for 90 percent and God only contributes 10 percent." Similarly, Pastor Martín commented that "We must begin to act and prepare ourselves." He continued along the same lines: "God blesses our lives, but we have to make an effort."

Such nuances in evangelical interpretations of human versus divine will are largely unknown or ignored by Catholics. For example, the former parish priest told me that, "For the evangelicals, the phenomena of nature emanate from God's power. Hence they say that there's no human responsibility, that we cannot do a thing." However, if some evangelicals mention the sins of humanity as being at the origin of Stan, there are also many who mention Stan as a divine message sent to wake us up to the destructive attitudes of people towards their natural environment. The two types of discourse are thus not systematically opposed, as former Pastor Rafael from the Bethania Church and the Dios de América Church artfully explained: "We evangelicals are aware of our impact on the whole system of nature, but we are also especially conscious of God's will. So we have to trust him and ask him to help us. The central message of the Gospel is Salvation."

Mayan prophecies on climate change

Aside from the Bible, regularly described by the evangelical *tinecos* as heralding future cataclysms to accompany the second coming of Christ, those inhabitants of San Martín who follow traditional Mayan spirituality also refer to Mayan prophecies as sources of knowledge on climate change.

Few researchers have the Mayan prophecies at the heart of their academic concerns. With the approach of the end of several cycles converging in 2012, this vacuum has given free rein to works, web pages, films and documentaries offering mystical and alarming messages that are disconnected from their original context within Mayan culture and religion. My purpose here in no way involves trying to convince anyone of the veracity of these prophecies but rather to consider these Mayan prophecies, still orally transmitted today in the Mam population studied, as revealing present-day worries. Indeed, it seems possible to me to analyze these Mayan prophecies without succumbing to the mechanical skepticism of our rationalist tradition or to the frenzy of fanatical esotericism. My point is rather to examine a prophetic tradition which has greatly influenced the local culture, no doubt in large part because of the Mayan texts' astonishing topicality.

For the *tinecos*, worry about a climatically disordered state is not new: it had already been announced orally by their ancestors. For the young traditionalist *tineco* Pascual, if global warming is a central topic at the heart of scientists' discussions today, his ancestors had already diagnosed it: "We did not need scientific research to report it. Our ancestors had already interpreted it. They warned us not to touch the mountains or the water, because they are sacred." Francisco agreed, saying that:

> The Mayans already knew that we would experience a climate change. As for the droughts, they knew that too. There have been many hurricanes and, well, they knew that too. And how did they know all that? I do not know. But I see that in the east of the country there's famine and there's been a hurricane. They had predicted it.

The old traditionalist Miguel also agreed that the hard times the *tinecos* face today had already been announced by the elders. When his elders met, they would say to Miguel: "There are hard times coming after our disappearance. Our children and our grandchildren will experience that. They will see it."

Besides the oral transmission of predictions, there are also many ancient Mayan texts where messages in the form of prophecies are recorded. We discover prophetic messages in the glyphs engraved on steles and on Mayan temples, as well as in the books of *Chilam Balam*, the *Popol Vuh*, the *Annals of Cackchiquels*, the play *Rabinal Achí* and the various codices which survived book burnings. The prophecies closest to the climatic realities, as well as chronicles of last events and mythological accounts, were written by Yucatan Mayans in the books of *Chilam Balam*. The *Chilam Balam*, written in the two centuries following the Spanish conquest, form an important part of Mayan literature. However, the various interpretations of ancient prophecies are prone to controversy. Various New Age currents re-appropriate the prophecies for themselves in offering interpretations of their meaning and those of crucial calendar dates.

For New Agers, the 2012 prophecies are amply described in the books of *Chilam Balam*. New Age authors like Barrera Vásquez and Rendón (2005), Yeitecpatl (2006) and Bermúdez (2007) estimated that the winter solstice of the year 2012 would mark the end of an era. Indeed the end of various short and long cycles in the Mayan calendar would coincide during the year 2012. In the northern hemisphere, the winter solstice of the year 2012 would mark the culmination of the long cycle of 5,126 years, begun in 3114 BCE. That date would coincide with the last days of the thirteenth *b'ak'tun*, a cycle of 144,000 days or approximately 394 years, as well as with those of the last *k'atun*, a short cycle of approximately 20 years. Though there is little doubt that these three cycles ended in 2012, determining the significance of these converging end points is much less straightforward. Many New Age authors interpret one of the prophecies of the books of *Chilam Balam* as announcing an era of darkness and terrestrial convulsions said to have begun seven years after the beginning of the last *k'atun*, or in 1999. Earthquakes, hurricanes and volcanic eruptions are predicted to increase appreciably thenceforth, and that would take place before the announced date of December 21, 2012.[12] According to them, these books forecast a heat wave which would raise the planet's temperature and produce unprecedented climatic, geological and social change. In other words, they believe climate change to have been anticipated by ancient Mayan prophets.

Certain cultural activists and Mayan shamans deplore the excesses of the New Age leaders' mystical interpretations, which they say constitute prophesying without really knowing Mayan cultural heritage.[13] Today these activists are reclaiming the 2012 prophecies as their own, and developing interpretations based on their own local ideological reflections. However, it is hard to determine whether or not these Mayan reclaimers have themselves been influenced by New Age currents for, as Sitler underlines (2006), "cross-pollination" between the contemporary Mayan and New Age participants in the 2012 movement is inevitable.

In the course of daily activities I participated in at San Martín, I regularly heard people refer to the "Mayan prophecies" about 2012. Yet the texts of *Chilam Balam* seem to be unknown to almost all the *tinecos* with whom I spoke. Rather than ancient literature—or contemporary for that matter—one of the important sources for disseminating these prophecies is television. For example, many *tinecos* describe their fascination with reports broadcast on the History Channel about Mayan prophecies. Those viewers' remarks converged: According to them, the Mayan prophecies are true and in the process of being accomplished. After having seen a program on Mayan prophecies, Yecenia commented the next day: "They explained why it's true, that something will happen in 2012." Similarly, Wilson explained how watching a program that talked about the Apocalypse in 2012 introduced him to the idea that "many religions regard that date as marking the end of the world." Referring to an apparent correlation between the destruction of the World Trade Center towers in 2001 and a Mayan prophecy involving birds, he added that "Mayan prophecies have all been fulfilled: the twin towers are dead; there are planes like large flying birds."

When speaking about Mayan prophecies, the *tinecos* defending Mayan spirituality and shamans nonetheless said that they did not believe in an end of the world in the year 2012. They resisted the common tendency to blend biblical notions of the Apocalypse with the end of cycles in the Mayan calendar. Victoriano, for example, was opposed to remarks heralding an end to the world in 2012. In his view, "It is false to believe that the world will end by divine mandate. . . . There have always been events like eruptions, et cetera. But the world will not end. The earth will not disappear." On the other hand, it is quite probable, he stated, that we may experience the end of humanity unless

> man analyzes his current lifestyle and reflects on the wrong he's doing and on the reasons why the earth is abnormally warm. . . . If man does not change his lifestyle, he will destroy his habitat and humanity will self-destruct. The earth will continue to live and a new living being, human or other, will appear.

Similarly to Victoriano, the shaman Álvaro sought other reasons for climate change than divine punishment for sin. According to Álvaro, the earth started to heat up in reaction to the misdeeds of humanity. Hence, the Creator has decided to challenge people because of the acts of destruction they have inflicted on the earth:

> In my opinion . . . Stan must be interpreted as a signal from our Creator that we are using up our natural resources. As is said in Mayan spirituality, I think that we men are enormously lacking in respect for Mother Nature in cutting down forests, and polluting the air and waters. That has brought on global warming, the earth's warming, which is the cause of the problem of Stan. All these things are manifestations of Mother Nature. They let us know that we'd better take care of it. I think it's not a question of a call from God because of our sins, but of what we've done with God's work, with Mother Nature.

In Álvaro's interpretation of climate change, God plays a role which cannot be eliminated. But, in unison with his fellow shamans, he rebels against the idea of an end of the world caused by God. If humanity comes to an end, the cause should be sought in the destructive acts of people, and not in some divine will.

According to inhabitants devoted to Mayan spirituality, Mayan prophecies thus warn about the dangers of anthropogenic damage to nature. But, in contrast to the fatalism sometimes associated with evangelical interpretations of prophecy, they do so in order to encourage people to change their behavior. Hence the prophecies should facilitate a powerful awakening, inciting us to change the course of events inscribed in the history of times. For the shamans I encountered, Mayan prophecies sought to attune human consciences to a more harmonious relationship with nature.

Conclusion: prophecies on climatic phenomena and their relationship to time

This chapter has observed that prophetic representations form part of both Mayan spirituality and evangelicalism. As our analysis has shown, we find two ideal-typical interpretations of catastrophic natural phenomena. The evangelicals mention the omnipotence of God at the origin of all natural events. For them, humanity lives in sin and forsakes God, who then makes his dissatisfaction felt, for example via meteorological phenomena. As for the shamans and people placing Mayan spirituality at the heart of their cultural heritage, they accuse humankind and its destructive acts of not taking care of "nature." Nature then makes humans understand their error (sometimes via God as intermediary) through hurricanes and, more broadly, by climate change.

Apocalyptic millenarian prophecies and Mayan prophecies offer their adherents distinct hopes: Mayan prophecies appeal to a terrestrial lived experience in which harmony between humans and the earth prevails, whereas the prophecies the evangelicals circulate project an idyllic future with the Lord in the hereafter. Thus, evangelical believers generally understand the disasters presaging Christ's return positively. Pédron-Colombani (2001) affirmed that the converts' messianic situation in waiting conditions their relationship to time. For the premillenarian evangelicals, the future is the most important time; it is one of transformation, personal improvement and the advent of God. In eschatological waiting, the premillenarian evangelicals refrain from personal involvement in history, instead encouraging "individual withdrawal, concentrating on one's personal engagement with Christ and individual sanctification" (García-Ruiz 1988: 7). The indigenous evangelicals are given the mission, not of saving the earth, but indeed of saving a maximum of human beings through conversion, in making them God's elected before the imminent end of the world. Converting and "accepting the Gospel" mean being personally placed under the protection of divine power. Unlike those who adhere to the evangelical belief system, wherein everything has a divine origin, the *tinecos* who express an attachment to Mayan spirituality strongly emphasize the anthropogenic role in climatic phenomena. The Mayan prophetic discourses manifest the urgency of a return to respect for nature.

But the distinctions drawn between divine and anthropogenic interpretations of climate change and between the groups circulating them must be nuanced. Non-evangelicals sometimes mention Stan's divine origin, just as certain evangelicals also acknowledge the impact of human beings on the development of extreme natural phenomena. *In fine*, the speakers' remarks agree on one point: extreme natural phenomena are a reaction to moral changes. They are payback either for humanity's sins or for neglect of the natural environment.

What fundamentally seems to distinguish the discourses of the *tinecos* as a function of their religious membership is the *outcome* of these extreme phenomena that they predict. In the Mayan worldview, destructions and creations alternate, with the image of the genesis of the "corn man" in the creation myth of *Popol Vuh*.[14] If 2012 coincided with the end of a major cycle, new cycles would begin. Thus Mayan cyclical time embodies a certain historical continuum (Boccara 1990). As for the evangelical conception of time, it is above all linear, following the example of biblical time, which considers that humanity has experienced only one creation and will experience its end when God destroys the terrestrial world. Hence an opposition exists between the Mayan cyclical and eternal conception of the world and the evangelicals' apocalyptical conception of it.

Do these distinct conceptions of time induce different attitudes toward climatic threats and other natural phenomena? Some *tinecos* I met extolled an ecological awakening and the urgency of saving the earth, whereas others encouraged withdrawal here on earth and, especially, the need to be saved by Christ. On the one hand, we might conclude from this that apocalyptic evangelical beliefs are a barrier to adapting actions to meeting climate changes. On the other hand, people associated with Mayan spirituality might be defined as actors raising local citizen awareness of the impacts of anthropogenic factors at work in climate change. Yet we need to supplement this analysis by observing the everyday practices the *tinecos* implement in managing their natural environment. In the sphere of religious practices, the shamans and their faithful regularly organize traditional ceremonies intended to protect what they call their Mother Nature.[15] But, in the field of natural environment management, my research in this agricultural town shows that the *tinecos*' practices regarding the earth are not distinguishable according to religious membership. In this context, where poverty indices are high, the first concern is survival on a daily basis rather than struggling against the causes and effects of climate change. Thus intensive and extensive agricultural practices tend to take precedence over an ecological safeguarding of the ecosystem.

Notes

1 Stan, reaching Mexico as a category 1 hurricane on the Saffir–Simpson scale, was outstanding for its slow movement and the vast quantity of rain produced in a short time (Pash and Roberts 2006).

2 The municipal authorities estimate that 21,500 of the 29,000 inhabitants of the town were directly or indirectly affected by Stan. According to the Centre for Research on the Epidemiology of Disasters (CRED) based at the Université catholique de Louvain, 455,314 Guatemalans were affected by the tropical depression (http://www.em-dat.net/disasters/Visualisation/profiles/countryprofile.php).

3 Data obtained from the population register of San Martín Sacatepéquez in July 2010.

4 In 2006, SEGEPLAN estimated that, in San Martín, 82.89 percent of inhabitants lived under the poverty threshold and that 28.09 percent lived under the threshold of extreme poverty. The minifundist peasant economy, in which farmers own only small parcels of land for cultivation, is representative of that rural poverty. Statistics gathered by the INE (2003) tell us that in San Martín the area of 87 percent of properties registered does not exceed 0.7 hectares.

5 I use this term to describe all the non-visible entities which compose the Mam universe, such as the vital principle of a mountain, deities, nahuals, the Heart of the sky, or an ancestor's spirit.

6 The shaman is commonly called *ajq' ij* or *ajkab'* in Mam. He assumes a role of collective interest because he has the gift of divination and a faculty for placing himself in direct contact with the divinities, spiritual figures and ancestors. He officiates at ceremonies in their honor as a sign of thanks or to ask for blessings. The shaman is still regularly referred to in Spanish as "*sacerdote Maya*" (Mayan priest) or in Mam as *chman*.

7 Catholic Action, a lay-run group that sought to increase Catholicism's influence on society, was responsible for marginalizing shamanism beginning in the 1940s. The expansion of evangelicalism since the late 1970s has further contributed to this trend.

8 This percentage represents the mean of figures provided by the various religious actors present in San Martín. It corroborates the data gathered by Carmen Martínez Bolaños in the *Diagnóstico Integral del Municipio de San Martín Sacatepéquez* in 2006. If there exists little in the way of quantitatively reliable data on the number of converts to Protestantism, it seems that Guatemala has the highest percentage of converts in all of Latin

America. The average figure generally advanced is between 25 and 30 percent of the total population (Freston 2008; Pédron-Colombani 2001).

9 Ethno-climatology is a mode of observation probably as old and as widespread as humanity itself, for every human being experiences variations in atmospheric conditions and meteorological phenomena (Katz and Lammel 2008).

10 More recently and in a distinct geographical place, one month after the January 2010 earthquake in Haiti, a priest from Port-au-Prince explained to a journalist the extent to which "the sects torture these unfortunates in telling them that God is punishing them because they've sinned. I constantly tell them that they're being manipulated. And it's only beginning" (quotation from "Dieu attendra, j'ai du boulot sur terre" [God will wait, I've got work to do on earth], *La Libre Belgique*, February 18, 2010: 18–19).

11 According to Garrard-Burnett, the annual growth rate is four times higher than during the 1970s (1998: 122). The evangelical proportion of the Guatemalan population has thus increased nearly sevenfold between 1960 and 1985 (Stoll 1990: 22).

12 I will not go into the ongoing debate among Mayanists on the subject of the end of the thirteenth *b'ak'tun* and its correspondence with the exact date of our Gregorian calendar.

13 On this subject, one might cite personalities such as: Carlos Barrios, a Mam shaman who has published various popularizations on shamanism and the Mayan calendar; Jose Murcía Batz, a defender of the idea of genetic transmission of ancestral knowledge; or the novelist Gaspar Pedro González, who predicts a "new era of light" after 2012. These authors share their vision, their dreams and their calculations to announce the end of numerous cycles and the beginning of a new era.

14 Book written in k'iche' dialect in the sixteenth century.

15 It is difficult to evaluate whether the term "Mother Nature" is indigenous or the fruit of foreign cultural inspiration. However, in the Mam language, we often find that elements referring to nature are composed of terms like "mother," "father," "grandfather" or "grandmother." For example, the Mam word "*Q'chu Tx'otx'*" is translated as "earth," but literally signifies "Mother Earth" or "Grandmother Earth." Similarly, "*Qman jba'l*" is translated from Mam as rain but literally signifies "Grandmother Rain." According to the *tinecos*, these lexical formulae are used to manifest a form of respect toward the elements composing nature.

References

André, J.-C., J.-F. Royer and F. Chauvin. 2008. "Les cyclones tropicaux et le changement climatique." *Comptes Rendus Geoscience* 340, 575–583.

Barkun, M. 1974. *Disaster and the Millennium*. New Haven, CT : Yale University Press.

Barrera Vásquez, A. and S. Rendón. 2005. *El libro de los libros de Chilam Balam*. México: Fondo de cultura económica.

Bastian, J.-P. 1990. *Historia del Protestantismo en America Latina*. México: Cupsa.

Bermúdez, D. 2007. *Profecías Mayas: Increíbles Revelaciones para Nuestra Época*. Buenos Aires: Editorial Kier.

Boccara, M. 1990. *La religion populaire des mayas: Entre métamorphose et sacrifice.* Paris : L'Harmattan.

Castellano, E. and A.Guerra. 2009. *El cambio climático y sus efectos sobre el desarrollo humano en Guatemala*. Guatemala : Programa de las Naciones Unidas para el Desarrollo.

Freston, P. 2008. *Evangelical Christianity and Democracy in Latin America.* Oxford, U.K.: Oxford University Press.

García-Ruiz, J. 1988. "L'Etat, le religieux et le contrôle de la population indigène au Guatemala." *Revue française de Sciences Politiques* 38.5, 758–769.

Garrard-Burnett, V. 1998. *Protestantism in Guatemala: Living in the New Jerusalem*. Austin: University of Texas Press.

Harmeling, S. 2011. *Global Climate Risk Index 2012: Who Suffers Most from Extreme Weather Events? Weather-Related Loss Events in 2010 and 1991 to 2010*. Bonn: Germanwatch.

Hermesse, J. 2010. "Diffusion et polarisation du protestantisme au Guatemala au lendemain du tremblement de terre de 1976." *Histoire et Missions Chrétiennes* 14, 35–58.

INE (Instituto Nacional de Estadística). 2003. *Censos Nacionales XI de Población y VI de Habitación 2002*. Guatemala: INE.

Katz, E. and A. Lammel. 2008. "Introducción: Elementos para una antropología del clima." In *Antropología del clima en México: Aires y lluvias*, ed. A. Lammel, M. Goloubinoff and E. Katz, pp. 27–52. Mexico: CEMCA/CIESAS/IRD.

Pash, R.J. and D.P. Roberts. 2006. *Tropical Cyclone Report: Hurricane Stan, 1–5 October 2005*. Available at http://www.nhc.noaa.gov/pdf/TCR-AL202005_Stan.pdf (accessed October 16, 2009).

Pédron-Colombani, S. 1998. *Le pentecôtisme au Guatemala: Conversion et identité*. Paris: Éditions du CNRS.

——. 2001. "Pentecôtisme et changement religieux au Guatemala." In *La modernité religieuse en perspective comparée: Europe latine–Amérique latine*, ed. J.-P. Bastian, pp. 199–211. Paris: Karthala.

Rayner, S. 2003. "Domesticating Nature: Commentary on the Anthropological Study of Weather and Climate Discourse." In *Weather, Climate, Culture*, ed. S. Strauss and B.S. Orlove, pp. 277–290. Oxford, U.K.: Berg.

Sitler, R.K. 2006. "The 2012 Phenomenon: New Age Appropriation of an Ancient Mayan Calendar." *Nova Religio: The Journal of Alternative and Emergent Religions* 9.3, 24–38.

Stoll, D. 1990. *Is Latin America Turning Protestant? The Politics of Evangelical Growth*. Berkeley: University of California Press.

Strauss, S. and B.S. Orlove. 2003. "The Anthropology of Weather and Climate." In *Weather, Climate, Culture*, ed. S. Strauss and B.S. Orlove, pp. 3–14. Oxford, U.K.: Berg.

Webster, P.J., G.J. Holland, J.A. Curry and H.-R. Chang. 2005. "Changes in Tropical Cyclone Number, Duration, and Intensity in a Warming Environment." *Science* 309, 1844–1846.

Yeitecpatl. 2006. *Las 13 profecías Mayas*. México: Editores Mexicanos Unidos.

9 Religious perspectives on climate change in the West Ivoirian mountainous region

Sadia Chérif and Joy H. Greenberg

Introduction

Climate change is one of the greatest environmental, social, and economic threats facing the planet today. It is already negatively impacting the livelihoods and living conditions of people in developing countries. The farming regions of sub-Saharan Africa, where agriculture provides the main economy on which rural populations depend, are especially at risk, as climate change causes unprecedented variability in rainfall and increases in temperatures (Christoplos et al. 2009; Handmer and Dovers 1999; Kurukulasuriya and Mendelsohn 2008). Agricultural yields are likely to decrease by as much as 50 percent by 2020 in some West African countries that are already afflicted with pervasive poverty (Hoffmann 2011). If climate change is not addressed more effectively, its impacts will impede progress toward viable development in sub-Saharan Africa and will make it impossible to attain the United Nations' Millennium Development Goals of ensuring environmental sustainability and ending poverty and hunger (United Nations 2010). In sum, climate change poses a grave risk to survival, development, and international security (Nzuma et al. 2010).

Farmers depend on sufficient rain for their crops. They depend on predictable weather patterns in order to plan when to plant and when to harvest. Climate change has, however, created unprecedented variability in the weather. Irregular rainfall and rising temperatures have made farming ever more challenging. Ultimately, it is expected that these changes will create extreme humanitarian crises (FAO 2009). Although organizations such as the Intergovernmental Panel on Climate Change (IPCC 2007) have proposed strategies for addressing climate change, there are serious issues with these approaches. IPCC strategies have proven ineffective so far and for numerous reasons, including, most obviously, that they are expensive and therefore unlikely to be implemented by farmers who are quite poor.

In this chapter we explore another reason for the lack of success to date: IPCC strategies do not take into account how local farmers understand what is happening to them. If they are to succeed, climate change adaptation strategies must take into account and must start from an understanding of how local farmers experience and think about the visible manifestations of climate change.

After offering some background and contextual information, we describe traditional religion at our study site, paying special attention to the environmental ethic embedded in it. Next, we describe some of the grave demographic, social, and cultural changes that had been increasingly stressing and disrupting traditional society even before the arrival of climate change. Population was exploding, putting new pressures on the land. Old ways were being abandoned, including the sustainable environmental behaviors the locals had long practiced. New faiths, whose teachings do not have an environmental ethic (or are hostile to having one), had gained a foothold and a following. Given the already dire situation, climate change could only make things worse.

Following a brief discussion of the impacts of climate change, we present evidence that shows both traditionalists and Christian Pentecostals misinterpreting the manifestations of climate change, with traditionalists blaming those who no longer adhere to old ways and Christians seeing signs of the end of times. Our goal, though, is not just to describe a hopeless dead end. We conclude with thoughts about possible ways out of the current situation.

Methods

This study used qualitative phenomenological research to analyze the local environmental narratives of farmers in mountainous western Côte d'Ivoire. This research design emphasizes individuals and their subjective experiences (Anadón 2006; Denzin and Lincoln 2000; Schütz [1932] 2008). The purpose is to arrive at an understanding of human social realities as they pertain to the environment and how these environmental perspectives are perceived by the social actors themselves (Poisson 1991). Social scientist Maribel Del Rio-Roberts (2009) suggested that focus groups are helpful when the topic of research concerns perceptions, opinions, and beliefs. For that reason, we used focus groups as our primary tool. We supplemented focus group data with individual interviews using a semi-directive interview guide and informal discussions with the local population. We asked interviewees to compare the annual rainfall of 2010 with that of the two previous decades. We asked them about changes in temperatures and about how changes in the weather were affecting agriculture.

The four main villages that compose the newly created municipality of Zagoué—Zagoué, Déoulé, Singouin, and Gouêtimba—were chosen for focus group sessions, one at each site. All four sessions and ten individual interviews were conducted in the local Goh language. Somewhat closely related to Toura or Wen, Goh is a variant of the Yacouba or Dan language, which is mainly spoken in the Department of Man in western Côte d'Ivoire, where Zagoué is located (see Figure 9.1). It is a little-known language, having, for the most part, escaped the notice of historians of the region (e.g., Loucou 1984). The choice to collect data in the Goh language turned out to be an especially useful way to study "local knowledge," which is constructed through communal events, symbols, and especially *shared language* (Geertz 1983).

To complete this study, residence was taken up in Déoulé for five weeks. It is from this "headquarters" that the three other villages of Zagoué were visited to conduct focus groups and informal interviews. The usual strategy for this consisted of, first, an unofficial visit with the leader of the village, around whom were often gathered

Figure 9.1 Map of Côte d'Ivoire showing 18 Mountains Region and Zagoué

Source: By Greg Greenberg, used with permission.

certain notables wanting to join the focus group. After an explanation of the object of the visit, a formal appointment generally was fixed for one of the rest days (Friday or Sunday) in the agricultural calendar so as to assure better participation.

The study site: geographic and economic context

Zagoué is located on a plateau in the 18 Mountains (18 Montagnes) Region of Côte d'Ivoire, where mostly rocky mountains reaching a height of more than 100 meters encircle the municipality. Zagoué has a tropical climate, with a long rainy season from March to October and a short dry season from November to February. Situated 9 kilometers from the town of Man, Zagoué municipality has a seasonal economy and residents who depend mainly on activities connected to agriculture. Its population is composed primarily of rural farmers. The town has some modern infrastructure, such as an elementary school, a hospital, and electricity.

Rice is the primary crop in this area and represents an essential source of income for the farmers (coffee and cocoa are also grown). Rice has been grown in the municipality since 1965, when the Ivoirian government launched the "crusade for rice" as a way of promoting food sustainability in the region (Chauveau and Dozon 1981: 651). Rice growing on mountain slopes begins with clearing the land. Trees are removed from the area to be cultivated as soon as possible after the harvest period, which runs from November to December. The removal of trees is accomplished by burning, which is done once or twice, depending upon local factors such as amount of wood, advancing rainy season, and abundance of the first rains, three to five of which are needed to ensure the extinction of ashes. Rice cultivation has thus contributed to deforestation and, hence, climate change (Ehui and Hertel 1989).

Traditional religion and its environmental ethic

Traditions such as fetishes, sacred groves, and rituals have been retained in their purest forms. For the Dan people of Zagoué, religious and ethnic identity relies on the ritual use of music and masks that reflect their animist tradition, known as *ge*. *Ge* is a religion that allots a soul to animals, plants, phenomena, and natural objects and for which all nature is animated by spirits called *genu* (the plural of *ge*). *Genu* are part of a pantheon of spirit intermediaries between God (*Zlan*) and humans. The spirits who manifest as *genu* originate in the wilderness—in certain mountains, trees, or streams (Reed 2001).

Zagoué animists believe that these protective spirits dwell in sacred forests and rivers as well as other revered places. Not only *places* are considered sacred; certain *species* are as well. There are prohibitions against harming certain native species related to these sacred sites. *Ge* compares both etymologically and theoretically with ancient Greek paganism, with *genu* resembling the *genius loci*, or local spirits, who were thought to inhabit nature.

The people of Zagoué also believe that they are dependent on nature and, hence, that it is essential for their continued wellbeing to treat nature well. Indeed, beliefs in the sanctity of nature and in the understanding that environmental health is the basis of human wellbeing are inseparable facets of a single worldview. Belief in protective spirits, in Goh named *lahi* ("the water which allows life"), expresses the feeling of human dependence on the rain that facilitates agriculture, ensuring food security. It is thought that propitiating protective spirits is necessary to counter problems like unproductive harvests, endemic diseases, and epidemics afflicting the local villages. By protecting sacred places, animist peasants contribute to environmental conservation despite pressures from agriculture and population growth that would otherwise encroach upon or outright destroy them (Cormier-Salem and

Bassett 2007; Nyamweru and Sheridan 2008). In this way, an environmental ethic is embedded in and implemented through traditional religious beliefs and practices.

Societal changes unrelated to climate change that erode environmental ethics

Climate change has arrived at a time in which other historical developments have already been at work undermining Ivoirian ecological sustainability.

Population pressures

A burgeoning population is causing considerable environmental stress. For the last 30 years, the province of Zagoué and other areas in western Côte d'Ivoire have been subjected to strong migrations from the north and south of the continent, where less accommodating climates make farming more difficult. Many have left their places of origin in hopes of improving their livelihoods, a migratory movement that seems to be a direct consequence of West Africa's reputation, unlike much of Africa, for having a forest zone favorable to agriculture.

People need to eat. More people require more land to farm. Agricultural economists Simeon K. Ehui and Thomas W. Hertel claim that the main impetus behind the clearing not only of the traditional sacred groves but all West African tropical forests is pressure from agriculture, which is "fueled by rapid population growth" (1989: 703). As a result, many nations are challenged by "how to feed an increasing population without irreparably damaging the natural resource base on which agricultural production depends" (1989: 703). As of 1989, Côte d'Ivoire, at 5 percent per year, had attained the fastest agricultural growth rate in sub-Saharan Africa, and its original 16 million hectares of tropical rainforest had been reduced by 3.4 million hectares (1989: 703). Although the rate of deforestation has slowed, thanks to conservation efforts by the government, remaining rainforest is now estimated to be at only 10 million hectares (Mongabay 2006). Besides the irreparable loss of trees, deforestation also deleteriously affects crop yields by making the land nonarable (Ehui and Hertel 1989: 703).

As environmental pressures increase, property ownership is becoming more contentious, particularly regarding sacred sites. Anthropologists Celia Nyamweru and Michael Sheridan cite case studies indicating that, today, "sacred ecological features are highly dynamic, deeply contested, and in rapid flux in African societies" as a result of competing interests (2008: 285).

The abandonment of tradition among youth

Even where sacred groves, taboos, and totems seem to be working to preserve habitats and species, as in Ghana, young people are often disinterested in following the old traditions. Such "ecological, social and symbolic dynamism has important implications for biodiversity conservation," Nyamweru and Sheridan maintain (2008: 285), for cultural traditions depend upon passage through generations to remain vital and relevant. Once a group loses its cultural traditions, it loses its "sacred ecology" (Berkes 1999).

The growing influence of other, "outside" religions

In addition to lack of interest by young people, the *ge* tradition has faced increasing challenges from outside religious groups. Islam came to Côte d'Ivoire following the conquest of the nation's western area by Samory Touré during the pre-colonial period. As a result of this conquest, foreign immigration doubled, especially from the heavily Islamic countries of Burkina Faso, Mali, Guinea and Niger. This incursion of Muslims resulted in conversions to the extent that Islam is now the most dominant religion in Côte d'Ivoire (38.6 percent of the population), followed by Christianity (32.8 percent), indigenous (11.9 percent), and none (16.7 percent) (CIA 2011).

Traditional spirituality has been tested even more since the appearance of Christianity in rural villages during the 1990s, when evangelists opened Pentecostal churches throughout Africa. In Côte d'Ivoire, the process of evangelization and conversion actually began earlier, during the years 1980–90. This is when demand for spirituality increased as a response to the stress caused by governmental programs that intended economic reform through the privatization of business but, in fact, further impoverished the people (Kouadio Ahou 2008). Because religions function in part to provide solace in the face of unendurable hardship and suffering, churches proliferated.

There now exist three religious belief systems in Zagoué: traditional animism, Islam, and Christianity. If attitudes toward nature are considered, however, there are only *two* religions, given that the animist religious precepts are respected by the Muslim community, with whom the animists are often aligned in opposition to Christianity. Villagers attribute this relative cohesion between practicing animists and Muslims to the successful integration of the latter. Unlike the Christians, Muslims generally do not resist or denigrate animist religious beliefs, but rather allow some degree of syncretism. In addition to their tolerance of animism, Muslims tend to be more accepted throughout West Africa because of Islam's simplicity and its coincidence with certain African sociocultural features, including group solidarity, polygamy, and circumcision (Charnay 1980: 140). Additionally appealing is its ingrained respect for elders, hierarchies, morals, and rules. For these reasons, Islam is regarded now as an "African" religion. This strong alliance between animism and Islam has contributed to the relatively poor acceptance of Christianity in the region.

The denigration of traditional beliefs by Christians

Traditionalist-Christian conflict is on the rise in West Africa (Cormier-Salem and Bassett 2007; Juhé-Beaulaton 2008). When Christians disrespect traditional animistic beliefs, the environmentally benign practices embedded in those traditional beliefs are also at risk. According to one Zagoué villager, "Since the arrival of the church in the village, women and children are no longer afraid of the masks. The rules and totems of the village are violated, and even the sacred forest says nothing to them." This villager refers to the Christian denigration of animist belief in the sanctity of certain natural places and of customs like the mask ceremonies. Once intended to scare followers into submission to the spirits, the masks have lost their power after being mocked repeatedly by Christians.[1]

As with the above-referenced mask ceremonies, the ritual of requesting rain, practiced mainly by women when the dry season is prolonged, has been challenged by Christians. This ritual consists of a festive procession around the village by women holding kitchen utensils that they beat together while chanting: "You, who caught the rain, if you eat what comes from the kitchen utensils, die." This chant is directed to the "raincatchers"—individuals living among them who are believed to be trapping the rain and keeping it from the farmers and their crops—with the intention of frightening them into allowing the rain to fall.

Recently, however, the women who devote themselves to the above ancestral practice feel mocked by Christians, who retort that only the Single God is able to bring rain to the village. According to one:

> Christianity recognized that it is God who makes fall the rain. It is God who allows the rain to come down from the sky towards the ground. The name *lahi* given to the rain corroborates my words, since this concept means the "water which saves humanity and all that exists on the ground." In other words, the creator of the rain is God only; rain does not depend on sacred rivers or forests, but is quite simply from God. Indeed, at the time when God created Adam and Eve at the beginning of humanity, there did not exist at that time any sacred places. Consequently, the drought is related to the fact that men no longer obey the rules of God or the Bible.

In addition, many Christians, by their own account, deny the mystic power of the rainmaker of Ganlié, a hamlet near Zagoué recognized by villagers to contain a "rainmaker" and "sunmaker" family. The rainmaker is solicited by animist peasants to make the rain return, but local Christians delight in pointing out that the traditional sacrifices and propitiations "demonstrate that they are not at all masters of the rain."

One local Christian convert told how he came to reject the traditional animist beliefs by first occupying and then building a plantation in a sacred forest:

> The place where I made my farm was regarded as a sacred forest. Indeed, it was known in the village that the spirits attacked any person who ventured there to clear this piece made of "bad" grass known under the name of Dépendant. Indeed, in this zone existed insects that attacked all men who entered. Of course, these insects attacked men everywhere, even in the non-sacred places such as plantations and farms. But I faced these insects and installed my cocoa plantation there. My success shows that sacred places and spirits do not exist.

The above comments describe the rationale used to disprove traditional beliefs: if the insects bite outside as well as inside sacred forests, then such places cannot be sacred. Other informants reported, and it was confirmed by local Christians, that some of them from the village caught and ate fish from the sacred rivers, which is traditionally taboo.

These examples illustrate that the practice of questioning longstanding sacred traditions and transgressing them has become so common in Zagoué that certain rituals are being progressively abandoned by the Goh community. Sacred places are profaned. The protection that the environment once enjoyed as a consequence or corollary of traditional beliefs and practices is being lost.

Climate change

The combination of population increase and the growing strength of religious belief systems that lack an ecological ethic has led to increased exploitation of natural resources, including what were once sacred forests, rivers, and species. Within this context, the local manifestations of climate change constitute an added stressor to an agricultural system and a traditional relationship to the land that have already been subjected to disruptive forces.

Climate change is creating conditions substantially different from those under which, in times past, agricultural activity succeeded. Relying on traditional knowledge for climate change adaptation is problematic, because changes in temperatures and rainfall fall outside of previous experience. The narratives and rituals that once served farmers well are no longer practical and may, in fact, be contributing to agricultural failure.

Successfully coping with the challenge of climate change would seem to require, first, understanding what is happening and, second, finding new ways to adapt. In fact, as we are about to describe, our research shows that the people of Zagoué certainly *see* the effects of climate change, but they *believe* that those effects are being caused by the religious conflicts we discussed above.

Competing explanations of climate change

When native farmers in the western mountainous region of Côte d'Ivoire explain the climatic changes they are experiencing, they point accusing fingers at the Christian community for not respecting the prohibitions and ancestral values honored by their tradition. It is believed that Christians profane the sacred forests by clearcutting them to install their plantations. One villager insisted the drought was due to the demolition of trees by Christians. He did not think that this was the only cause, however:

> We had sacred places here at the village, and when we made kola nut offerings it rained the same day. But today, certain villagers [*main nou a main nou*[2]] seem to think that these sacred places are not important because, according to them, nothing sacred lives there. So they cut down trees there and build

> their farms. At the time when I speak, these sacred places are exposed to the sun. Angry, the spirits [*Yi nan*] which lived in these sacred places have left them [*An toho am'ma lo ho*]. Also invisible, like God and therefore gods as well, these spirits are angry with us, and consequently the rain does not fall anymore.

The above comments demonstrate how traditionalists find it easier to blame Christian desecration of their sacred places for the drought than to take responsibility for finding an effective response to it.

Following this line of thought, another villager also holds Christians liable for the decline in rainfall at Zagoué:

> For me, the lack of rain is related to disrespect for the taboos of the village. For example, in our family, it is forbidden to eat certain foods, or totems [*Gba pon hon*], but modernized Christians consume them. Ultimately, the lack of rain is related to the disrespect by Christians for the village totems and taboos.

The groves are no longer protected by the *genu*, the rainmaker no longer creates rain, and the mask processions no longer enforce traditional beliefs.

The traditionalists blame the Pentecostals, who do not share their beliefs, for climate change. By the same token, the Pentecostals fault nonbelievers for crop failures that, to them, signify the coming Apocalypse. The reduction in rainfall, the drop in the agricultural outputs, the weak cloud cover even during the rainy season, the lack of water for crops at the end of the cycle, the destruction of harvests, and the displacement of the sowing periods are, *inter alia*, the principal signs which announce the Apocalypse—the end of the world. The scarcity of rain is viewed by the Pentecostals as the realization of God's prophecy, as revealed in the biblical Book of Revelation (Rev. 15–16). A Pentecostal believer told participants of his focus group that the only condition to ensure high precipitation would be conversion of all villagers to Christianity: "It is necessary that all the assembly joined together here accept this God as Creator of humanity. It is only under this condition that the rain will start falling abundantly again as it used to do several decades ago."

The above discussion indicates that both Pentecostal and traditional religions have their own explanations for the ways they experience climate change. From a Western scientific point of view, however, both misperceive the cause of the changes they see as the "wrong" beliefs and the blasphemous behavior of the other.

Is there a way forward?

In most West African countries like Côte d'Ivoire, agriculture remains the key economic and social factor and is of pivotal importance for assuring food security. However, climate change and overpopulation are irreversibly damaging the natural resource base on which agriculture depends.

Finding a way to curb the population growth that is applying extreme pressure to the environment is, obviously, a key piece of the answer. Of course, curbing population growth is easier said than done, for to speak of fertility is necessarily to address the pervasive poverty in Africa. This is because incomes rising above the subsistence level "bring about the social and economic changes that are necessary for fertility decline through changes in life styles and consumption patterns" (Kouamé, Schellekens and Rabenoro 2002: 274). With higher incomes, parents are able to consider opportunities for their children that increase the costs of raising them, which in turn is motivation for fewer children.

But, even if the interconnected problems of poverty and population can be solved, other difficulties stand in the way, as our research has shown. It seems clear from our narrative that a revisioning of the current, dysfunctional discourse into communally accepted, ecoethical narratives is crucial if West Africans are to address their looming ecological and humanitarian catastrophe successfully. A communal, ecological, mythic-religious narrative that embeds a reproduction ethic therefore seems imperative. Instead, however, local peasants of the municipality of Zagoué relate competing religious narratives on the drought, which many interpret as resulting from disrespectful religious practices. While Christians allot the lack of rain to the anger of a vengeful God, local animists believe that it represents retaliation by protective spirits for the desanctification of their sacred species, forests, and rivers by Christians.

In fact, none of the religious responses to climate change in Zagoué has been ultimately effective: the drought continues, endangering food supply and the future of its people. Rather than people bickering and pointing fingers at each other, it would seem more productive for religious and secular groups to become better educated regarding the causes of climate change and how best to respond to it.

Unfortunately, fundamentalist Christian sects like the Pentecostals are resolutely attached to literalist, anti-environmental interpretations of the Bible. They believe that the global ecocrisis is pre-ordained by God as punishment against nonbelievers, which supports their unwillingness to address the realities of climate change. Pentecostals are among the most anti-environmental of all religious groups, according to several studies. Social scientists Nalini Tarakeshwar et al. (2001) found that Protestant fundamentalists were among the least ecoethical, backing up similar results from Douglas Lee Eckberg and T. Jean Blocker (1996) and Michelle Wolkomir et al. (1997). According to Wolkomir et al. (1997: 343) and Jeanne Kay (1989: 228), conservative Protestant Christians—who are identified as Mormons, Southern Baptists, the Church of Nazarene and Pentecostal Holiness congregants, Seventh Day Adventists, and Jehovah's Witnesses—display the *most* literalist but *least* environmentally concerned views. These findings support the results of Eckberg and Blocker, who determined that "high sectarianism," or belief in a literal, dominionist interpretation of the Bible in which nature is desanctified, corresponded with the lowest "greenness" (1996: 348).

It is difficult to imagine Pentecostals ever being willing to find a common ground with the animist Goh or *any* community in the effort to create a mutually agreeable means of addressing their environmental problems. On the other hand, we can point to instances when African Christian churches have, in fact, redefined themselves and embraced ecoethical perspectives. For example, a coalition of African Independent Churches (AIC) and traditional African religions has been working to repair the drought-stricken landscape of southern Zimbabwe (Daneel 2005). AIC "are all-African churches founded *by* Africans *for* Africans" that have emerged in sub-Saharan countries and represent a majority of African Christians in Zimbabwe and South Africa (Daneel 2006: 537, emphasis in the original). By galvanizing the local peasants to defend their ancestral lands, the coalition has acted to reverse ecological failure. As a response to the despoliation of their sacred places, they have planted over 8 million trees, raised ecological awareness, and applied religious sanctions against harmful wildlife and water practices (Daneel 2005).

The AIC demonstrate that, under the right circumstances and given sufficient cause to do so, religions can change and rearticulate their teachings in ways that emphasize new messages. When there is a compelling reason, religions that otherwise espouse quite different views may be able to join in a common cause (Gottlieb 2006). Environmental crisis, when it rises to the level of an existential threat, can be such a cause. Such a metamorphosis is demonstrated by the traditional Zimbabweans above, whose practices, like those of the Ivoirian Goh, worked well for them until the pressures of "population, profit-oriented deforestation, overuse of water for commercial farms, soil erosion, and a decade-long drought created something profoundly new" (2006: 119). In other words, traditional belief conserved sacred forests only until the above stressors became so intense that the peasants felt forced to act or lose everything. Out of the tension arose a transformation of Zimbabwe's "age-old religio-ecological values into a modern programme of environmental reform" (Daneel 2001: 104). Mwari—once a rain god—became a "god of ecology"; and, although formerly appeased by the "observance of taboos against taking certain game or felling sacred trees," the "spirit ancestors" now demanded that everyone "heal the land, reforest the earth, and protect the water" (Gottlieb 2006: 119). For one member of the African Association of Earthkeeping Churches, Mwari has been revisioned as God and will be cared for by the congregants "'because in Jesus Christ you [Mwari] are one with us'" (cited in Daneel 2001: 185). This syncretism of Christian and Zimbabwean traditions signals the possibility for a similar rapprochement in Zagoué.

Notes

1 Historian E.S.D. Fomin concurs in saying that, although the "effect of Christianity on land use in [western Cameroon] has not been wholly negative[,] the disrespect for those cultural practices that they deemed inimical to Christian practice affected the sacred forests in the region negatively. Many ardent Christian converts today have very little sympathy for royal sacred forests" (2008: 404).

2 A Goh phrase that means "undesirable persons," alluding to Christians.

References

Anadón, M. 2006. "La recherche dite 'qualitative': de la dynamique de son évolution aux acquis indéniables et aux questionnements présents." *Recherches Qualitatives* 1, 5–31.

Berkes, F. 1999. *Sacred Ecology: Traditional Ecological Knowledge and Resource Management*. Philadelphia, PA: Taylor & Francis.

Charnay, J.P. 1980. "Expansion de l'islam en Afrique Occidentale." *Arabica* 2, 140–153.

Chauveau, J.P. and Dozon, J.P. 1981. "Histoires de riz, histoires d'igname: le cas de la Moyenne Côte d'Ivoire." *Africa: Journal of the International African Institute* 2, 621–658.

Christoplos, I., S. Anderson, M. Arnold, V. Galaz, M. Hedger, R.J.T. Klein and K. Le Goulven. 2009. "The Human Dimension of Climate Adaptation: The Importance of Local and Institutional Issues." Commission on Climate Change and Development. Available at http://www.ccdcommission.org/Filer/report/HUMAN_DIMENSIONS.pdf (accessed July 25, 2009).

CIA (Central Intelligence Agency). 2011. *The World Fact Book*. Available at https://www.cia.gov/libraryy/publications/the-world-factbook/geos/iv.html (accessed October 8, 2011).

Cormier-Salem, M. and T.F. Bassett. 2007. "Nature as Local Heritage in Africa: Longstanding Concerns, New Challenges." *Africa* 77, 1–17.

Daneel, M.L. 2001. *African Earthkeepers*. Maryknoll, NY: Orbis Press.

——. 2005. "African Earthkeeping Churches." In *Encyclopedia of Religion and Nature*, ed. B. Taylor, pp. 21–24. London: Continuum International.

——. 2006. "African Initiated Churches as Vehicles of Earth-Care in Africa." In *Oxford Handbook on Religion and Ecology*, ed. R.S. Gottlieb, pp. 535–567. Oxford, U.K.: Oxford University Press.

Del Rio-Roberts, M. 2009. "How I Learned to Conduct Focus Groups." *Weekly Qualitative Report*. Available at http://www.nova.edu/ssss/QR/WQR/wqr2_34.html (accessed August 31, 2009).

Denzin, N.K. and Y.S. Lincoln. 2000. *Handbook of Qualitative Research*. Thousand Oaks, CA: Sage.

Eckberg, D.L. and T.J. Blocker. 1996. "Christianity, Environmentalism, and the Theoretical Problem of Fundamentalism." *Journal for the Scientific Study of Religion* 35.4, 343–355.

Ehui, S.K. and T.W. Hertel. 1989. "Deforestation and Agricultural Productivity in the Côte d'Ivoire." *American Journal of Agricultural Economics* 71.3, 703–711.

FAO (Food and Agriculture Organization of the United Nations). 2009. "Food Security and Agricultural Mitigation in Developing Countries: Options for Capturing Synergies." FAO. Available at http://www.fao.org (accessed October 15, 2010).

Fomin, E.S.D. 2008. "Royal Residences and Sacred Forests in Western Cameroon: The Intersection of Secular and Spiritual Authority." *Journal for the Study of Religion, Nature and Culture* 2.3, 391–407.

Geertz, C. 1983. *Local Knowledge: Further Essays in Interpretive Anthropology*. New York: Basic Books.

Gottlieb, R.S. 2006. *A Greener Faith: Religious Environmentalism and Our Planet's Future*. Oxford, U.K.: Oxford University Press.

Handmer, J.W. and S. Dovers. 1999. "Societal Vulnerability to Climate Change and Variability." *Mitigation and Adaptation Strategies for Global Change* 4, 267–281.

Hoffmann, U. 2011. "Assuring Food Security in Developing Countries under the Challenges of Climate Change: Key Trade and Development Issues of a Fundamental

Transformation of Agriculture." United Nations CTAD Discussion Papers. Available at http://www.unctad.org/en/docs/osgdp20111_en.pdf (accessed October 9, 2011).

IPCC (Intergovernmental Panel on Climate Change). 2007. "Synthesis Report." IPCC. Available at http://www.ipcc.ch/pdf/assessment-report/ar4/syr/ar4_syr.pdf (accessed October 10, 2011).

Juhé-Beaulaton, D. 2008. "Sacred Forests and the Global Challenge of Biodiversity Conservation: The Case of Benin and Togo." *Journal for the Study of Religion, Nature and Culture* 2.3, 351–372.

Kay, J. 1989. "Human Dominion over Nature in the Hebrew Bible." *Annals of the Association of American Geographers* 79.2, 214–232.

Kouadio Ahou, V. 2008. "Émergence des femmes pasteurs dans les nouvelles églises de Côte d'Ivoire." Unpublished thesis, Université de Bouaké.

Kouamé, A., J. Schellekens and M. Rabenoro. 2002. "Rural Development and Attitudes towards Family Size in Côte d'Ivoire." *Population* 57.2, 269–299. Available at http://www.jstor.org/stable/3246610 (accessed September 28, 2011).

Kurukulasuriya, P. and R. Mendelsohn. 2008. "How Will Climate Change Shift Agro-Ecological Zones and Impact African Agriculture?" World Bank Policy Research Working Paper 4717. Available at http://www-wds.worldbank.org/external/default/WDSContentServer/WDSP/IB/2008/09/11/000158349_20080911163038/Rendered/PDF/WPS4717.pdf (accessed October 5, 2011).

Loucou, J.N. 1984. *Histoire de la Côte d'Ivoire*. Abidjan: CEDA.

Mongabay. 2006. "Côte d'Ivoire." Available at http://rainforests.mongabay.com/20cotedivoire.htm (accessed October 8, 2011).

Nyamweru, C. and M. Sheridan. 2008. "Introduction." *Journal for the Study of Religion, Nature and Culture* 2.3, 285–291.

Nzuma, J.M., M. Waithaka, R.M. Mulwa, M. Kyotalimye and G. Nelson. 2010. "Strategies for Adapting to Climate Change in Rural Sub-Saharan Africa: A Review of Data Sources, Poverty Reduction Strategy Programs (PRSPs) and National Adaptation Plans for Agriculture (NAPAs) in ASARECA Member Countries." International Food Policy Research Institute (IFPRI). IFPRI Discussion Paper 01013. Available at http://www.ifpri.org/publication/strategies-adapting-climate-change-rural-sub-saharan-africa (accessed October 10, 2011).

Poisson, Y. 1991. *La Recherche Qualitative en Éducation*. Québec: Presses de l'Université du Québec.

Reed, D.B. 2001. "Pop Goes the Sacred: Dan Mask Performance and Popular Culture in Postcolonial Côte d'Ivoire." *Africa Today* 4, 67–85.

Schütz, A. [1932] 2008. "The Phenomenology of the Social World." Reprinted in *Contemporary Sociological Theory*, ed. C. Calhoun, J. Gerteis, J. Moody, S. Pfaff and I. Virk, pp. 32–41. Oxford, U.K.: Blackwell.

Tarakeshwar, N., A.B. Swank, K.I. Pargament and A. Mahoney. 2001. "The Sanctification of Nature and Theological Conservatism: A Study of Opposing Religious Correlates of Environmentalism." *Review of Religious Research* 42.4, 387–404.

United Nations. 2010. "Millennium Development Goals." UN.org. Available at http://www.un.org/millenniumgoals/environ.shtml (accessed March 6, 2012).

Wolkomir, M., M. Futreal, E. Woodrum and T. Hoban. 1997. "Denominational Subcultures of Environmentalism." *Review of Religious Research* 38.4, 325–343.

10 Climate change and indigenous African religion

A case study of the transitional ecological zone of Ghana

Paul Sarfo-Mensah and Samuel Awuah-Nyamekye

Introduction

Indigenous knowledge systems (IKS) have been proven by research to have potential to make significant contributions to solving many of the contemporary development challenges, including climate change (Nyong, Adesina and Osman Elasha 2007; Odora Hoppers 2002). In this chapter, we explore the potential contributions that indigenous African religious beliefs and practices can make to the global fight against the menace of climate change.

Climate change, the phenomenon where the temperature of the earth is increasing as a result of increasing greenhouse gases, is causing a lot of damage and severe conditions on earth (Budiani 2009). Amongst the continents, Africa is considered to have the highest vulnerability to the effects and stresses of climate change (Dybas 2008; IPCC 2007). This vulnerability is exacerbated by existing developmental challenges, including: endemic poverty; complex governance and institutional dimensions; limited access to capital, including markets, infrastructure and technology; ecosystem degradation; and complex disasters and conflicts (IPCC 2007). Indeed, researchers have emphasized, for instance, that agricultural production and food security in many African countries and regions are likely to be severely compromised by climate change and increasing climate variability (Lisk 2009: 10).

Finding sustainable solutions to climate change and its impact has become a global challenge (IPCC 2007; Nyong, Adesina and Osman Elasha 2007). One area that researchers admit has received little attention but has great potential is indigenous knowledge systems (Macchi et al. 2008). Indigenous communities have deep knowledge and understanding of nature (Sponsel 2007) and have devised social-cultural and biological systems that have enabled them to live in harmony with nature (Hulme 2008). Studies have confirmed that indigenous religions, for instance, can contribute to very different relationships between humans and nature, one with positive impacts on biodiversity, ecosystems, ecological processes, and related phenomena (Sponsel 2007; Tanner and Mitchell 2002). Current studies have emphasized the centrality of nature and culture in climate change discourses (Hulme 2009: 1), but these studies have made little effort to explore religion, a key aspect of nature and culture, for its integration into the global fight against climate change (Macchi et al. 2008). This chapter explores the contribution indigenous African religious beliefs and practices can make to the global effort in finding lasting solutions to climate change.

After an initial, brief description of the study site and our research methodology, we describe Ghanaians' indigenous worldview and the linkages between traditional beliefs and nature conservation. We then describe how various aspects of the process of modernization have undermined traditional ways and traditional beliefs. Next we discuss climate change, its manifestations and impacts, and how Ghanaians understand and explain those impacts. At the end of the chapter we argue that, although traditional ways of thinking currently produce misleading interpretations of climate change, indigenous Ghanaian ecological knowledge *could*, instead, prove to be of use in coping with climate change impacts.

The study site

This study was carried out in four communities in the transitional agroecological zone of Ghana. The communities (Bofie, Dotobaa, Buabeng Fiema, and Nkyeraa) (see Figure 10.1) are all in the Brong Ahafo region. The transitional agroecological zone spreads across several regions in Ghana (Sarfo-Mensah 2001).

The zone receives mean annual rainfall of between 1,300 and 1,500 millimeters, an amount that supports the production of cash crops (cocoa, cashew, tobacco, yams, and maize) and food crops (cassava, plantain, and vegetables), the mainstay of the local economy. Rainfall distribution in the area has generally become very erratic over the last 50 years, causing the zone to face tremendous challenges with annual bush and wild land fires (Amanor 1993). These changes, thought to be due to climate change, are already causing serious havoc to the local biodiversity and environment.

Research methodology

The study benefited from an earlier one we carried out on sacred groves in the transitional agroecological zone (Sarfo-Mensah and Oduro 2010). For the current study we did informal interviews and in-depth discussions, over a period of six months (July to December 2010), with local people in the four communities. The focus of the in-depth discussions was local people's perceptions of changes in climate and how their religious beliefs interpret and how their practices respond to those changes.

We conducted interviews with a cross-section of the local people, including community leaders, the elderly, local religious leaders (indigenous African religion believers, Christians, and Muslims), youth, men and women and children, to seek their views on changes in environmental conditions and their interpretations of the situation. The respondents were selected after initial contacts had been established in the respective communities and preliminary discussions had been held with community leaders on changes in the local environment. The interviews

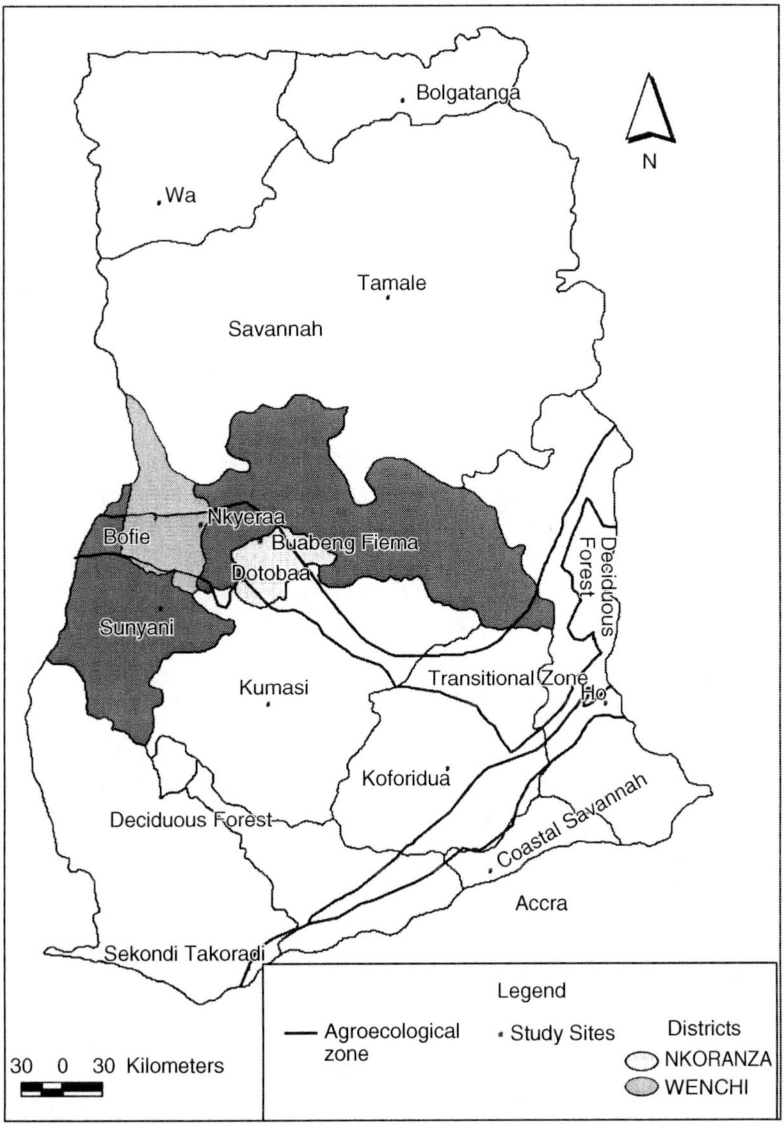

Figure 10.1 Map of the study area in Ghana

Source: Authors' construct.

and reviews with local people were complemented with field observations and extensive desk studies on the transitional ecological zone, its people, and its culture, including local religions, history of settlements, and their local ecological history.

The indigenous Ghanaian worldview

Like their counterparts in other parts of Africa, traditional Ghanaians are very religious (Mbiti 1989; Parrinder 1974). Religion's influence is evident in their socio-moral, political, and economic lives. Among the Akan of Ghana, an important group in our study area and by far the biggest ethnic group in Ghana, God is known as *Odomankoma* or *Oboadee* (creator), and is credited with the creation of the universe (Busia 1968; Danquah 1928; Opoku 1978).

According to the Akan cosmological myth, as narrated by Opoku (1978), there is no doubt that the universe is a religious entity. Their attitude toward it reflects the sacred nature of the universe and all that is therein. This has been the guiding principle in the Akan and in fact this reflects the general attitude of traditional Ghanaians towards creation (Opoku 1978). The traditional Ghanaians, like other Africans, see the natural environment as being intimately linked with integral human life, and without it human life may not be sustained (Nwosu 2010).

Ontologically, traditional African people believe that human lives are tied to their environment and that their survival is dependent on how they make use of their environment. This implies that what is now termed "environmental consciousness" has been part and parcel of the traditional African people (Awuah-Nyamekye 2009). Reichel-Dolmatoff (1976) made a similar claim in his analysis of the Tukano Indians of the Amazon rain forest. No wonder that many traditional African societies believe that human beings are socially and morally obligated to manage and conserve the environment for themselves and posterity (Abayie Boateng 1998). This socio-moral obligation constitutes the basis of traditional Ghanaian environmental ethics. The above theoretical framework thus forms the basis for interpreting or analyzing traditional Ghanaians' environmental conservation ideas. To ignore this deeply rooted ethical system is to misunderstand the people's attitude to nature, including how they handle issues of climate change.

Different studies across Africa indicate that indigenous African peoples encourage non-violence to the natural environment (Abayie Boateng 1998; Dopamu 2006; Ntiamoa-Baidu 2008; Nwosu 2010). Our interactions with the local people of the study area indicate that they have likewise, and for a very long time, put measures such as the institution of the sacred grove, reverence for water bodies, and belief in *sasa*[1] in place to check environmental degradation, thereby preventing or mitigating the effects of climate change.

We have argued that, among the Africans, acceptable environmental management has religio-cultural underpinnings, with belief in the ancestors playing a key role (Opoku-Ankomah, Amevenku and Obuobie 2010). In the traditional Ghanaian society, laws that have the backing of the ancestors and the gods usually endure, for their implementation does not cause a problem, owing to the awed reverence that people attach to such laws. This is in contradistinction to modern society, where people can break a law when they have reason to believe that no one is watching them. The people in our study area would abide by the laws, including environmental ones, enacted by the chief and his elders, because such laws were deemed to have spiritual backing. This is because the chief and his elders were believed to be the earthly representatives of the ancestors. Therefore, anyone who breaks or challenges any of the laws enacted by the chief and his elders is deemed to be daring the ancestors. Reichel-Dolmatoff (1976: 311) has made similar findings among the Tukano Indians of the Colombian Northwest Amazon.

Belief systems and nature conservation in Ghana

The above worldview of the Ghanaian in the past helped create a situation of a harmonious relationship between the people and the environment. In our study area, for example, we found sacred groves and other forest patches that have been preserved as a result of this relationship, driven mainly by traditional religious beliefs. We learned from the local ecological history that these sacred groves, especially those that were attached to sacred animals, such as the Buabeng-Fiema Sanctuary, were an integral part of the social lives of the people. In these communities, for instance, we were told that the animals and the forest were protected in the past and were still being protected because of the belief that the monkeys are the children of the local god, Daworo, and anybody who killed or maltreated any of the animals would be severely punished by the god. The people still associate the animals with the god and, therefore, the population of the animals has increased and the forest vegetation expanded over the years. In fact, the forest and the animals have become a tourist attraction. The people also say that they derive ecological benefits from the forests, including wind shelter, cooler temperatures, and rainfall.

In another study community, Bofie, we found that a forest patch which has been kept through similar beliefs had become a feed corridor where local people fed and watered their livestock during the dry season. In this community, we were told, the forest had been preserved because it was the abode of forest dwarfs, *gina*, which protected the local people against outside aggressors. However, local people admitted that the dwarfs had receded deep into the forest as a result of the use of modern equipment such as guns and chainsaws, which made too much noise and were, therefore, disliked by the dwarfs. But people were quick to add that, since they believed that the dwarfs still protected the local people, the forest had been preserved.

In the past, such beliefs helped maintain soil fertility and agricultural productivity in the study settlements. The people believed that Asase Yaa (Earth) was a goddess who needed to have days when she had to be accorded a serene atmosphere to rest. They also believed that during those days she would enhance the fertility of the soils. Thus, in all the study communities, we found a day in the week when people did not till the land or do any activity they considered a disturbance to Asase Yaa. The people believed that reducing human impacts in this way enhanced fertility, and that translated into improved agricultural productivity.

Our respondents confirmed that water catchment areas, wildlife, and entire ecosystems were in the past sustainably managed through local religious beliefs and practices. And, as we have mentioned above, our respondents emphasized the centrality of traditional chiefs as critical to the survivability of the belief systems.

Indeed, we found that the traditional authorities and the gerontocracy, custodians of the local culture, were the key pillars around which the belief systems revolved. But it became obvious from our interactions with the local people that modernity is working fast against tradition.

How modernity is undermining traditional systems

In the study area, the introduction of Christianity and Islam have had a major influence on traditional natural resources management practices. Local people considered the coming of these new faiths to be the most powerful force obliterating traditional society, including, importantly, traditional religion's environmental ethic. For instance, the central point of departure between Christianity and African indigenous religious beliefs in the study area is the strong belief of the latter group in *tumi*, or supernatural powers suffused in nature by Oyname (God), the Creator of the Universe (Sarfo-Mensah and Oduro 2007). Whilst indigenous religious believers have very elaborate taboos and sanctions to protect "sacred" or protected plants, animals, and water bodies, most Muslims and Christians are inclined to denigrate these taboos, and even interpret these as paganism and idol worshiping. Such beliefs make possible the overexploitation and degradation of natural resources.

Just as Lynn White (1967) blamed Christianity for the environmental woes of today (though this has been challenged—see Harrison 1999; Johnson and Butigan 1984), Christianity was noted to be largely responsible for this change in local perceptions and attitudes in the study area. Several studies on traditional people across Africa have revealed similar findings (Jehu-Beaulaton 2008; Nukunya 1986: 87; Opoku-Ankomah, Amevenku and Obuobie 2010; Parrinder 1961; Smith 1986: 86). Ecologists who blame Christians for current environmental problems in Ghana (e.g., Opoku-Ankomah, Amevenku and Obuobie 2010 and others) do not claim that the Christians are anti-environmental. Rather, these ecologists blamed Christians for the erosion of the indigenous religion, which acts as the live-wire of the traditional Ghanaian way of life. This is because, when the Christian missionaries arrived in Ghana, they launched an incessant attack on the indigenous religion. Indeed, this assault has not abated as of now; this is because some Christian pastors, particularly those of the Pentecostal strand, have sustained this attack. Today, many Ghanaians have become converted to Christianity. This has led to the erosion of the basis for the traditional way of conserving the environment.

Urbanization and the increasing presence of the modern market economy are two other forces that are undermining the belief in *tumi* and sustainable management. For instance, we found in the study communities that the attraction of urban life has caused many young people to drift from these indigenous communities to the cities. After acquiring money, many of these youngsters return home to set up farms or start other ventures. But, when they return, their worldviews have been changed, especially in terms of beliefs in local ways of life and management of natural resources. These returnees influence other youngsters, thereby building a critical mass of a new generation whose belief systems are completely alien to the local traditional beliefs of nature and creation. Generally, we found that traditional reverence and fear attached to nature are waning, and that youth who are more opened to the effects of urbanization and modernity are more inclined towards disbelief in indigenous environmental management systems.

The issue of the market economy and its attendant effects needs mentioning. The commercialization of timber, for example, has negatively impacted weather conditions in the study area. There is an unprecedented increase in the number of timber firms in the country, all of which export lumber abroad. Now some trees such as *onyina* (*Ceiba pentandra*), *otie* (*Tycnanthus angolensis*) and *akonkodie* (*Bombax buonopozense*), which were once considered non-timber trees, are being cut. This has left the land virtually bare. In turn, increased deforestation contributes to Ghana's rising temperatures.

In addition, we found that there is an increasing commercialization of crop production, especially maize, yams, cashew, cocoa, tobacco, and vegetables. This has changed the local economies and poses a threat to local environmental management. The trend of increasing commercialization leads to integration of the rural economies into national and international markets, which in turn affects traditional management practices (Fairhead and Leach 1998). Indeed, as a result of intense pressure on land, more marginal lands and traditionally protected areas—especially sacred groves and areas close to rivers and water bodies—have been brought under production and are being intensively cropped.

The pressure to exploit heretofore-protected land has driven some indigenous people, who fear retribution from local gods if they personally use such lands, to hire Christians and Muslims to farm those protected areas, believing that then the gods will not punish them. Such pressures on land can be horrendous for traditional management systems, both the micro-environments and the socio-cultural and economic arrangements those micro-environments sustain (Leff 1985).

Population increase has also contributed to changes in the environment and landscape in the transitional zone. As already noted, the population in the transitional zone has grown, especially because of substantial in-migration from the northern regions of Ghana. In the study communities we found that increase in population is not only putting excessive pressure on land but also impacting the agricultural practices of the local people. For example, we noted that the northern migrant farmers totally remove the tree stumps and burn the debris completely during land preparation. This farming practice compounds deforestation and land degradation in the zone.

We also observed that government's (district assemblies') inability to rigorously implement the 1994 Forest and Wildlife Policy is adversely affecting local environmental management in the transitional zone and throughout Ghana. The policy aims at conservation and sustainable development of the nation's forest and wildlife resources for maintenance of environmental quality and perpetual flow of optimum benefits to all segments of society (FC 1994). Among its specific objectives, which are very comprehensive, several explicitly aim to enhance environmental quality at the local level: to promote public awareness and involvement of rural people in forestry and wildlife conservation so as to maintain life-sustaining systems; to preserve scenic areas and enhance the potential of recreation, tourism, and income-generating opportunities; and to develop effective capability at national, regional, and district levels for sustainable management of forest and wildlife resources (FC 1994).

However, we discovered that, at the local level of the transitional agroecological zone, several local governments (district assemblies) did not have the full complement of decentralized agencies and resources to support the implementation of the Forest and Wildlife Policy objectives. Owing to lack of resources in terms both of material and of personnel, these assemblies are unable to educate their people and enforce their own bylaws on environmental management.

In some communities there were illegalities in the management of the environment, especially the utilization of tree resources. The government thus still faces serious challenges in implementing its policy objectives on illegal lumber production from both on- and off-forest reserves and other protected areas, including sacred groves. We observed that the transitional zone is one of the ecological zones that have suffered severely from the activities of illegal timber merchants and their local collaborators. The result of the gap in policy design and the reality in implementation is undermining local people's effort at combating deforestation and their contribution to the fight against climate change.

Climate change: impacts and explanations

In the course of the study we noticed that the people of the study area did not have indigenous terms for the greenhouse gases—carbon dioxide, methane, nitrous oxide, etc.—that are the major drivers of climate change. They did, however, notice that there had been a drastic change in the rainfall pattern in the area. The rain does not come at the time that it is expected and, when it comes, it does not last as long as it used to. The study area is experiencing prolonged dry seasons, erratic rainfall patterns, and increased widespread annual bush and land wildfires.

Our informants in the study area added that they had been experiencing excessive heat, which has badly affected food production. The heat, they said, was accompanied by various diseases. They further added that, as a result of the change in the weather, rivers in the area which used to be perennial are now seasonal. For instance, it was mentioned that *petra*, the practice of intensively growing vegetables within close proximity of perennial rivers during the dry season, can no longer be practiced because most streams in the area dry up completely. This is a clear indication that the people are fully aware that what is referred to as climate change is happening around them.

They are acutely aware of the manifestations of climate change, but their *explanations* of those manifestations are not anything like those produced by Western climate science. They say, for example, that such changes in weather conditions were never experienced when traditional authorities were fully in charge of affairs. According to them, traditional peoples have a unique way of protecting and conserving the environment and thereby maintaining beneficial weather patterns. They blamed the problem of changing weather on the activities of people today—especially their exploitation of the natural environment and disregard for traditional, environmentally friendly religious systems for social and economic gains—as well as the outright neglect by various governments of Ghana of indigenous ways of living with the environment. They were emphatic that traditional peoples, aided by their religion and culture, are able to live in harmony with their environment.

Indigenous Ghanaian ecological knowledge can combat climate change impacts

To sum up the argument so far: Even before the impacts of climate change grew to the point that they started to damage the local environment and threaten traditional rural livelihoods in Ghana, the processes of modernization had already long been undermining traditional Ghanaian religious beliefs and the ecologically wise practices those beliefs supported. It was in that deteriorating social context that climate change then has only further increased the damage. Rather than end on that despairing note, however, we wish finally to argue that, however formidable the above challenges are, indigenous ecological knowledge in Ghana can potentially contribute to fighting the menace of climate change.

Traditional Ghanaians make use of various beliefs and practices to protect and conserve the environment. For instance, traditional farming practices are friendly to the environment. One can cite shifting cultivation and bush fallowing practices, where local farmers make conscious efforts to allow a parcel of cropped land to rest for a period of years to regain its vegetative cover and to replenish its fertility through natural recycling. Even though the people could not explain the processes involved, they knew from their experience that the presence of forest can induce rain and can, as well, prevent water bodies from drying up. They said they learned this from careful observation and other experiences. The community leaders explained that this is one of the reasons why it is a taboo among them to farm close to river channels. We found this to be reasonable, for a number of the sacred groves in the study area serve as headwaters for many of the rivers in the study area. One can cite the Nyima River, which runs through the Asuoyima sacred grove in the Nkoranza traditional area of the Brong Ahafo region of Ghana.

The people of the study area see sacred groves as significant cultural artifacts, which they view as important for conserving nature. People are banned from entering these groves for any activity. Entry is granted only to the chief and his elders and the traditional priest or priestess of the community for religious purposes (this was emphasized by all our informants). This is also supported by studies carried out by other researchers elsewhere (Dorm-Adzobu, Ampadu-Agyei and Veit 1991; Falconer 1992; Gyasi 1996; Ntiamoa-Baidu 2008). These groves serve as a sanctuary for many animals and endangered species of trees (see Colding and Folke 1997). A case in point is the famous Buabeng-Fiema sacred grove, which serves as a sanctuary for monkeys and trees which are capable of adding fertility to the soil.

Although some scholars (Lebbie and Guries 2008) have claimed that these groves are primarily preserved for religio-cultural reasons and not for ecological reasons, our interaction with some chiefs and traditional priests in the study area showed that this claim is not entirely accurate. This is because we were told that the keeping of sacred groves is for both religio-cultural and ecological reasons. It was explained that, in addition to religious beliefs and practices, human beings are supposed to adapt to their environment by designing practical ways to ensure that there is always harmony between humans and nature so as to guarantee a continuous supply of human necessities for their survival—food, water, and shelter. It was explained that the attachment or involvement of religion is a sure way of making people respect and follow their ecological ethics. This is because religion is a powerful force in traditional Ghanaian society that makes compliance a matter of course as a result of the fear that punishment awaits those who disobey societal injunctions. This revelation lends credence to the advice of Sponsel (2007) that field researchers methodically and practically examine the environmental underpinnings of religious and spiritual beliefs, as well as people's attitudes and behaviors, in their case studies.

It is also important to stress that the practice of keeping sacred groves has been with the people for a very long time. This suggests that indigenous Ghanaian people (and indigenous peoples elsewhere) developed it before anthropogenic climate change became an issue. Still, we argue that the practice can contribute to checking or mitigating the effects of climate change, most especially in the areas where these groves are found.

Our interaction with the key informants showed that the people knew the old adage that, "When the last tree dies, the last man also dies." This is because we were told by many people that trees are key to their survival. They explained that trees give them food and shade, protect their roofs from being ripped off by storms, and serve as the source of medicinal herbs.

Belief in totems is another aspect of the study area which can be used to combat climate change-related problems. A totem is usually "an object such as an animal or plant that is believed to have a special (usually) sacred relationship with an individual, a tribe, a clan, a family or a group of people" (Awuah-Nyamekye 2009: 267). The Akan word for totem is *akyeneboa* (literally translated as "an animal that one leans upon or relies on for spiritual protection"). The reference to animal does not necessarily mean trees are not involved. There are traditional societies in Ghana which have plants as their totems. For instance, *atoa* (a kind of berry plant) is the totem for the people of Nsuta traditional area in the Ashanti region of Ghana. The people of Ekumfi traditional area have the *odum* (*Chlorophora excelsa*) as their totem. Other plants that are considered as totems include *homakyem* (*Spiropetalum heterophyllum*) and *odii* (*Okuobaka aubrevillei*). The protection of these plants through religious beliefs obviously has ecological implications which translate into climate change control measures. This is because the belief in totems helps in the preservation of certain trees, which in association with other trees can check excessive heat in their vicinity. The potential of totems and other beliefs such as taboos to conserve natural resources had long been recognized (e.g., Berkes 1997; Berkes, Folke and Gadgil 1995). The role of plants in mitigating the effects of climate change has also been recognized by Burley and

Haslam, who note that "trees are in the frontline of defense against climate change. . . . The United States, for instance, would be having an even greater impact on climate change if its forests weren't absorbing around a tenth of its carbon dioxide emission" (2008: 45). Although the study area's carbon dioxide emission is not as great as that of the United States of America, the belief in totems still undoubtedly helps preserve the area's forest cover and, by extension, reduce its contribution to climate change.

One other belief that has an implication for conservation is *sasa*. Like many traditional Ghanaians, the Akan consider certain trees and animals to be sacred because such plants and animals are believed to have spiritual power of some sort that can harm people (see Awuah-Nyamekye 2012; McCaskie 1995; Rattray, 1959; Warren 1986). Consequently, the Akan are very cautious in their attitude towards such plants and animals. The Akan word for such a belief is *sasa*. Examples of such plants include *odum* (*Chlorophora excelsa*), *esa* (*Celtis mildbraedii*), *tweneboa* (*Entant drophragma*), *homakyem* (*Spiropetalum heterophyllum*), *odii* (*Okuobaka aubrevillei*), and *Nyame dua* (God's tree) (*Alstonia boonei*) (personal communication, February 2010). The animals include *tromo* (*Tragelaphus euryceros isaacii*), *oyuo* (*cephalophus niger*), and *okoo* (buffalo) (personal communication, February 20, 2010). Most of the trees are of high medicinal value in addition to their ecological importance. The attitude of the people of the study area towards trees and the forest in general can also be said to reflect or emphasize the above statement from Burley and Haslam (2008: 45). It is on this basis that Schoffeleers (1978) has referred to African indigenous religions as "profoundly ecological."

Conclusion

We realized during the course of our research that in areas where the influence of the indigenous religious beliefs was strong the rate of environmental degradation was less than in those areas where the influence was weak. This confirms the findings of Byers, Cunliffe and Hudak (2001: 215) from their study in the Zimbabwean sacred forests. Such findings strengthen our conviction that traditional ecological knowledge is a potential resource which needs to be mainstreamed into modern ways of controlling climate change and environmental degradation in general in Ghana. Unfortunately, traditional peoples' views and interests are more often than not ignored by policy makers in Ghana when it comes to climate change. If traditional religion underpins the conservation practices of indigenous people, who in turn occupy the marginal and fringe ecosystems that have the highest vulnerability to climate change, it is imperative that a new positive attitude toward these beliefs be adopted.

The foregoing has identified several potential "arsenals" that could be positively deployed in the fight against climate change. Indigenous religious belief systems and their conservation practices most definitely fall within this category. Indigenous peoples' ways of addressing environmental problems are an important resource which can be tapped to handle some of the climatic problems of today. There is, therefore, the urgent need to study them in greater depth in order to find those that can be maintained and possibly integrated into modern ways of addressing the problem of climate change, which is threatening the very survival of humanity. We are, therefore, convinced that a comprehensive rethinking by policymakers in Africa and the rest of the world toward climate change, in which indigenous religions are brought into the mainstream policy guidelines to address climate change, is urgently required. But we are quick to point out that policy should focus not only on traditional knowledge—which certainly holds some promise of helping the people in the transitional ecological zone cope with climate change—but also on other social conditions (including population growth, urbanization, intrusion of market forces, and weak state institutions) that indirectly affect vulnerability and adaptive capacity to climate change.

Note

1 The belief that certain plants and animals have spirits which can be dangerous, for they can be vindictive if the right attitude is not used toward them.

References

Abayie Boateng, A. 1998. "Traditional Conservation Practices: Ghana's Example." *Institute of African Studies Research Review* 14.1, 42–51.

Amanor, K. 1993. "Wenchi Farm Training Project: Social/Environmental Baseline Study." Report prepared for the Overseas Development Administration.

Awuah-Nyamekye, S. 2009. "Salvaging Nature: The Akan Religio-Cultural Perspective." *World Views: Global Religions, Culture, and Ecology* 13.3, 251–282.

——. 2012. "Belief in Sasa: Its Implications for Flora and Fauna Conservation in Ghana." *Nature and Culture* 7.1, 1–15.

Berkes, F. 1997. "Indigenous Knowledge and Resource Management Systems in the Canadian Subarctic." In *Linking Social and Ecological Systems: Management Practices and Social Mechanisms for Building Resilience*, ed. F. Berkes and C. Folke, pp. 98–128. Cambridge, U.K.: Cambridge University Press.

Berkes, F., C. Folke and M. Gadgil. 1995. "Traditional Ecological Knowledge, Biodiversity, Resilience, and Sustainability." In *Biodiversity Conservation: Problems and Policies*, ed. C.A. Perrings, K.G. Mäler, C. Folke, C.S. Holling and B.O. Jansson, pp. 281–299. Dordrecht, Netherlands: Kluwer Academic.

Budiani, I. 2009. "Best Practices and Innovation of Biodiversity Conservation as Friendly Ways of Adapting to Climate Change in Forest, Coastal, and Agro Ecosystem in Indonesia." Report prepared for the Indonesian Biodiversity Foundation (Yayasan KEHATI).

Burley, H. and C. Haslam. 2008. *How Can I Stop Climate Change? What It Is, Why It Happens and What You Can Do to Help*. London: HarperCollins.

Busia, K.A. 1968 (1951). *The Position of the Chief in the Modern Political System of Ashanti*. London: Frank Cass & Co. Ltd.

Byers, Bruce A., Robert N. Cunliffe and Andrew T. Hudak. 2001. "Linking the Conservation of Culture and Nature: A Case Study of Sacred Forests in Zimbabwe." *Human Ecology* 29.2, 187–218.

Colding, J. and C. Folke. 1997. "The Relations among Threatened Species, Their Protection, and Taboos." *Conservation Ecology* 1.6. Available at http://www.consecol.org/vol1/iss1/art6/.

Danquah, J.B. 1928. *Akan Laws and Customs and the Akyim Abuakwa Constitution.* Oxford, U.K.: Oxford University Press.

Dopamu, A.T. 2006. "God, Nature and the Environment in Yoruba Perspective." In *Science and Religion in the Service of Humanity*, ed. P. Ade Dopamu, Olu Obafemi and O.B. Oloyede, pp. 163–179. Ilorin: Nigerian Association for the Study and Teaching of Religion and the Natural Sciences (NASTRENS) and Local Societies Initiative (LSI).

Dorm-Adzobu, C., O. Ampadu-Agyei and P.G. Veit. 1991. *Religious Beliefs and Environmental Protection: The Malshegu Sacred Grove in Northern Ghana.* Report edited by Centre for International Development and Environment and World Resources Institute, U.S.A. Washington, DC: World Resources Institute. Available at pdf.usaid.gov/pdf_docs/PNACA562.pdf.

Dybas, C.L. 2008. "Climate, Environment, and Infectious Diseases: A Report from the AIBS 2008 Annual Meeting." *Bioscience* 58.9, 792–797.

Fairhead, J. and M. Leach. 1998. *Reframing Deforestation: Global Analysis and Local Realities: Studies in West Africa.* London: Routledge.

Falconer, J. 1992. *Non-Timber Forest Products in Ghana.* Main report. London: Overseas Development Authority (ODA), London.

FC (Forestry Commission). 1994. *Ghana Forest and Wildlife Policy.* Accra: Forestry Commission of Ghana.

Gyasi, E.A. 1996. "Gyamfiase, Ghana: A Study in Threat and Counteracting Threat to Indigenous Forest Groves and Sustainable Forest Management Systems." Paper presented at the Workshop on Contested Terrain: West African Forestry Relations, Landscapes and Processes, April 12–13, University of Birmingham, Edgbaston, U.K.

Harrison, P. 1999. "Subduing the Earth: Genesis, Early Modern Science, and the Exploitation of Nature." *Journal of Religion* 79.1, 86.

Hulme, M. 2008. "Geographical Work at the Boundaries of Climate Change." *Transactions of the Institute of British Geographers* 33.1, 5–11.

——. 2009. *Why We Disagree about Climate Change: Understanding Controversy, Inaction and Opportunity.* Cambridge, U.K.: Cambridge University Press.

IPCC (Intergovernmental Panel on Climate Change). 2007. *Climate Change 2007: Synthesis Report. Contribution of Working Groups I, II and III to the Fourth Assessment Report of the Intergovernmental Panel on Climate Change* (Core Writing Team, R.K. Pachauri and A. Reisinger, eds.). Geneva, Switzerland: IPCC.

Jehu-Beaulaton, D. 2008. "Sacred Forests and the Global Challenge of Biodiversity Conservation: The Case of Benin and Togo." *Journal for the Study of Religion, Nature and Culture* 2.3, 351–372.

Johnson, P.N. and K. Butigan, eds. 1984. *Cry of the Environment: Rebuilding the Christian Creation Tradition.* Santa Fe, NM: Bear & Company.

Lebbie, A. and R.P. Guries. 2008. "The Role of Sacred Groves in Biodiversity in Sierra Leone." In *African Sacred Groves: Ecological Dynamics and Social Change*, ed. M.J. Sheridan and C. Nyamweru, pp. 42–61. Oxford, U.K. and Athens, OH: James Curry and Ohio University Press.

Leff, E. 1985. "Ethnobotany and Anthropology as Tools for Cultural Conservation Strategy." In *Culture and Conservation: The Human Dimension in Environmental Planning*, ed. J.A. McNeeley and D. Pitt, pp. 81–89. London: International Union for Conservation of Nature and Natural Resources.

Lisk, F. 2009. "Overview: The Current Climate Change Situation in Africa." In *Special Report: Climate Change in Africa: Adaptation, Mitigation and Governance Challenges*, ed. H. Besada and N. Sewankambo, pp. 8–14. Waterloo, Ontario: Centre for International

Governance Innovation (CIGI). Available at http://www.unicef.org/esaro/, Climate Change in Africa (accessed April 30, 2011).

Macchi, M., G. Oviedo, S. Gotheil, K. Cross, A. Boedhihartono, C. Wolfange and M. Howell. 2008. "Indigenous and Traditional Peoples and Climate Change." Issues paper. International Union for Conservation of Nature. Available at cmsdata.iucn.org/downloads/indigenous_peoples_climate_change.pdf.

Mbiti, J.S. 1989. *African Religions and Philosophy*, 2nd rev. and enlarged edn. Oxford, U.K.: Heinemann.

McCaskie, T.C. 1995. *State and Society in Pre-Colonial Ashanti*. Cambridge, U.K.: Cambridge University Press.

Ntiamoa-Baidu, Y. 2008. "Indigenous Beliefs and Biodiversity Conservation: The Effects of Sacred Groves, Taboos and Totems in Ghana for Habitat and Species Conservation." *Journal for the Study of Religion, Nature and Culture* 2.3, 309–326.

Nukunya, K. 1986. "Christianity, Western Education and Social Change in Africa: An Overview." In *Christian Missionarism and Alienation of the African Mind*, ed. R. Belhag and Y. El-Kabir, pp. 85–98. Tripoli, Libya: African Society of Social Sciences.

Nwosu, P.U. 2010. "The Role of Okonko Society in Preserving Igbo Environment." *Journal of Human Ecology* 31.1, 59–64.

Nyong, A., F. Adesina and B. Osman Elasha. 2007. "The Value of Indigenous Knowledge in Climate Change Mitigation and Adaptation Strategies in the African Sahel." *Mitigation and Adaptation Strategies for Global Change* 12.5, 787–797.

Odora Hoppers, C.A. 2002. "Indigenous Knowledge and the Integration of Knowledge Systems: Toward a Conceptual and Methodological Framework." In *Indigenous Knowledge and the Integration of Knowledge Systems: Towards a Philosophy of Articulation*, ed. C.A. Odora Hoppers, pp. 2–22. Claremont, South Africa: New Africa Books.

Opoku, Asare K. 1978. *West African Traditional Religion*. Accra: FEP International.

Opoku-Ankomah, Y., F.K.Y. Amevenku and E. Obuobie. 2010. "Research Findings in Water Bodies: The Human and Cultural Factors and Their Impact on the Environment." Report of the National Conference on Culture and Water organized by the National Commission on Culture, February 24–26, Akosombo, Ghana.

Parrinder, E.G. 1961. *West African Religions: A Study of the Beliefs and Practices of Akan, Ewe, Yoruba, Ibo and Kindred Peoples*. London: Epworth Press.

——. 1974. *African Traditional Religion*, 3rd edn. London: Sheldon Press.

Rattray, R.S. 1959. *Religion and Art in Ashanti*. London: Oxford University Press.

Reichel-Dolmatoff, G. 1976. "Cosmology as Ecological Analysis: A View from the Rain Forest." *Man* 11.3, 307–318.

Sarfo-Mensah, P. 2001. "The Spirituality of Forests and Conservation: The Dynamics of Change and Sustainability of Sacred Groves in the Transitional Zone of Ghana." Ph.D. dissertation, University of Greenwich, U.K.

Sarfo-Mensah, P. and W. Oduro. 2007. "Traditional Natural Resources Management Practices and Biodiversity Conservation in Ghana: A Review of Local Concepts and Issues on Change and Sustainability." *Nota di Lavoro* 90.2007, 1–14. Available at http://www.feem.it/Feem/Pub/Publications/WPapers/default.htm (accessed April 30, 2011).

——. 2010. *Changes and Perceptions about Environment in the Forest–Savanna Transitional Zone of Ghana: The Influence of Religion*. Fondazione Eni Enrico Mattei Working Papers. Available at http://www.bepress.com/feem/papers398 (accessed November 20, 2010).

Schoffeleers, J.M., ed. 1978. *Guardians of the Land: Essays on Central African Territorial Cults*. Harare, Zimbabwe: Mambo Press.

Smith, Noel. 1986. *The Presbyterian Church of Ghana, 1835–1960: A Young Church in a Changing Society*. Accra: Ghana Universities Press.

Sponsel, L.E. 2007. "Spiritual Ecology: One Anthropologist's Reflections." *Journal for the Study of Religion, Nature and Culture* 1.3, 340–350.

Tanner, R. and C. Mitchell. 2002. *Religion and the Environment*. New York: Palgrave.

Warren, D.M. 1986. *The Akan of Ghana: An Overview of Ethnographic Literature*. Accra, Ghana: Pointer.

White, L. 1967. "The Historical Roots of Our Ecologic Crisis." *Science* 155.3767, 1205–1207.

Part II

The Global North

11 Stepping up to the plate

Climate change, faith communities and effective environmental advocacy in Canada

Mishka Lysack

> [The] mainline Protestant denominations will have to step up to the plate. They long ago passed all the proper resolutions, decrying the destruction of creation, and certain congregations have launched some interesting initiatives. . . . But not many mainline Protestants have stepped far outside their comfort zones—in part because the denominations themselves are dwindling in number and beset by internal divisions.
>
> (Bill McKibben, 2008: 232)

As leaders from 20 Christian denominations quieted down, a hush of expectation filled the room at the Canadian Council of Churches. All eyes turned to Graham Saul, executive director of the Climate Action Network Canada, who described how faith communities could bring a crucial ethical frame and moral authority (Lysack 2008; Northcott 2007) to the public discussion about how Canadian society could best address climate change. As Saul underlined it, the key question is profoundly simple, yet deeply arresting: *What are the faith communities actually prepared to do to combat climate change?*

This question has yet to be adequately answered by the faith communities in Canada. As the gap between the severity and scale of climate change and the inadequacy of action taken by political and institutional leadership widens, actions taken by other social sectors such as faith communities acquire greater urgency. The impacts of climate change continue to intensify, outpacing even the most pessimistic projections and timelines of the UN's Intergovernmental Panel on Climate Change (Hamilton 2010; Hansen 2009). Despite the accelerating pace of climate change impacts, Canadian political leadership on climate change is weak, if not obstructionist. In its recent report, the National Round Table on Environment and the Economy assessed that actions taken by Canadian federal and provincial governments to reduce greenhouse gas (GHG) emissions will be insufficient to meet the federal government's targets (which climate scientists have argued are inadequate): reducing the country's emissions by 17 percent below 2005 levels by 2020 (Paris 2012). Drawing on Environment Canada data, the Canadian Environment and Sustainable Development Commissioner pointed out that, "in 2020, Canada's GHG emissions will be 7.4% above the 2005 level instead of 17%

below" (Vaughan 2012: 53).[1] Given the lack of political leadership, faith communities in Canada have the opportunity to enrich debate in the public square and to advocate for strong policies that would address climate change, thereby contributing to the creation of a political will to take action.

The faith community in Canada has great potential to exercise considerable political influence on political and institutional leadership, not only for the reasons outlined in Chapter 1, but also because of the sheer numbers of Canadians who are members of a faith community. The 2001 Canadian census lists the total number of Canadians who are participants or adherents in the 16 largest faith traditions at 23,720,160 (Bibby 2011). In this chapter, I provide a series of glimpses into the landscape of faith communities in Canada and their environmental advocacy regarding climate change. Many faith communities have position statements on climate change or, at least, on environmental decline. Having a climate change policy does not, however, equal effective environmental advocacy. As McKibben (2008) has observed, there is a troubling gap between *talking* about fighting climate change and actually *doing* something about it. This chapter outlines how the religious traditions in Canada are responding to climate change through advocacy in the public square with the goal of assessing their effectiveness, identifying barriers to success and proposing pathways for effective action.

The methodological approach

My position as a researcher seeking to investigate the landscape of faith-based climate activism in Canada is from the position of an insider to the processes I sought to observe and analyze. While working as an associate professor within a faculty of social work in a Canadian university, I am also an Anglican priest and facilitator in the nascent faith-based movement to advocate for the environment. More specifically, since 2008, I have worked as a public educator and organizational consultant with faith communities on both a local and a national level, assisting faith groups to build effective leadership capacity for faith-based ecological advocacy. I am therefore directly involved as a participant in climate change activities in faith communities in Canada, rather than a detached observer. While certainly affecting my perspective to some degree, my experience of facilitating educational and capacity-building workshops and conferences with faith communities across the country also served as a key source of data, granting me access that few outsiders would be able to attain.

As a participant in the events described and being "inside" what I was studying (see Ellis and Bochner 2000: 743), I chose to hybridize two qualitative methodologies: participant observation and autoethnography (Anderson 2006; Angrosino and Mays de Perez 2000; Ellis and Bochner 2000). Following the "reflexive turn" in ethnography, Angrosino and Mays de Perez (2000: 678) note that participant observation relies on a "conscious adoption of a situational identity." As such, the data is primarily the experience and observations of the subject embedded in the midst of the activity being studied. As I was both an active participant in events described and a subsequent researcher, the methodology of participant observation provided a disciplined inquiry into the experiences of research subjects, as well as my own.

As a form of inquiry, autoethnography is somewhat on the methodological margins (Anderson 2006; Bullough and Pinneager 2001; Ellis and Bochner 2000), although it is gaining more acceptance with the flourishing of qualitative inquiry. For Ellis and Bochner, autoethnography is an "autobiographical genre of writing and research . . . connecting the personal to the cultural" (2000: 739). Haluza-DeLay (2008) conducted a methodologically similar study of faith-based environmental education. Both participant observation and autoethnographic methodologies locate themselves in what Bruner (1990: 4–11) would call narrative knowledges centered on the "construction of meaning" rather than computational knowledges that focus on the "processing of information." My inquiry was not limited to my personal experiences, however, but also relied on hundreds of emails and written material gathered between 2008 and 2012.

In addition to studying my own experience as a facilitator, I conducted over 70 conversations (in person and via email) from 2008 to 2012 with leaders and environmental educators in eight mainstream Christian church denominations (United, Anglican, Presbyterian, Evangelical, Lutheran, Christian Reform, Mennonite, and Roman Catholic), as well as with eight leaders within the Evangelical organizations and churches in Canada and with leaders in three other faith traditions: Judaism, Islam, and the Baha'i community. Using my experience and observations as a facilitator as well as the written materials I gathered as a data source, I provide in this chapter a distillation of my own appreciative and critical inquiry into the emerging climate change leadership of the faith communities.

Participant observation and autoethnography were combined with research arising from another data source: a report completed by Citizens for Public Justice, a national faith justice organization, which reviewed the position documents on environment and climate change from 19 faith communities in Canada (Ljomov and Gunn 2010). I utilized the data in this report as a basis for further inquiry into the activities of faith communities in Canada regarding climate change.

The following four questions guided my research. First, which faith communities in Canada have taken a public stand on climate change as a moral issue by way of a position document? Second, what (if anything) have these faith communities done to take action on climate change through environmental advocacy? Third, what barriers and constraints exist for faith communities to develop leadership on global warming as a major public policy issue? Finally, how could these barriers be overcome?

Assessing the position statements and internal actions of faith communities

Based on my work with Canadian faith communities, I developed a rubric to assess the effectiveness of faith communities in working to address climate change. The rubric identified four ways in which a denomination could be ranked as a leader in faith-based environmental advocacy: 1) the existence of public policy positions on climate change or the environment; 2) methods of circulating policy positions and taking action internally within the faith community through education and worship; 3) methods of translating the positions on climate change externally into faith-based advocacy and engaging the government and public; and 4) the capacity for collaborating with allies and for contributing to strategic campaign initiatives to influence policy and mobilize deeper public support for reducing GHG emissions. Items 3 (advocacy) and 4 (collaborating with allies in campaigns) are covered in Table 11.3 later in the chapter. Table 11.1 summarizes items 1 and 2.

As Table 11.1 shows, while eight faith communities have either developed national documents or issued policy letters on issues which contain statements regarding climate change (United, Mennonite, Christian Reform, Anglican, Lutheran, Presbyterian, Roman Catholic, Baha'i), the United Church and the Baha'i community are the only faith communities in Canada that have intentionally and actively disseminated their document to their membership through a visible strategic communications plan (education, website, worship resources, webinars).

If the United Church and Baha'is have distinguished themselves through their intentional efforts to distribute their position documents within their faith community, the national leaderships of several other faith communities in Canada are conspicuous through their lack of a current public stance on climate change: Evangelical Christianity, Roman Catholicism, Judaism, and Islam. Since publishing its booklet "God's Earthkeepers" in 1995, the Evangelical Fellowship of Canada has not addressed climate change or the environment, despite several invitations from the Canadian Council of Churches and others to engage with these issues. As individual Roman Catholic bishops, Bishop Bouchard (2009) did write the exceptional pastoral letter on the oil sands, and Archbishop James Weisgerber signed a letter

Table 11.1 Public position of Canadian faith communities on environment and climate change, September 2011

	Document on environment	*Document on climate change*	*Climate change position actively disseminated within faith community*
Christian denominations:			
United Church of Canada	✓	✓	✓
Mennonite	✓	✓	
Christian Reform	✓	✓	
Anglican	✓	✓	
Lutheran	✓	✓	
Presbyterian	✓	✓	
Roman Catholic	✓	✓	
Evangelical	✓		
Other religious traditions:			
Muslim			
Baha'i	✓	✓	✓
Jewish			

to Prime Minister Harper expressing concerns about climate change impacts on the poor in the Global South. These actions notwithstanding, however, as a whole the Catholic bishops in Canada have been silent on climate change. In the case of both the Evangelical Christian Church and the Roman Catholic bishops, writer and journalist Dennis Gruending (2012) has speculated that their refusal to take a public stand on climate change may be due to these two organizations protecting their privileged access to the current Conservative federal government, and being preoccupied with right-wing "fundamentalist" issues, such as abortion. In the case of Judaism and Islam, these communities are likely hindered by their more decentralized networks with weak or non-existent national structures that could act as the Canadian voice for that faith community on any issue, including climate change.

The absence of these four faith communities is striking not only because of their significant number of members, but also because all three "Peoples of the Book"—Judaism, Christianity, and Islam (Abdul-Matin 2010)—have considerable spiritual and ethical resources in their sacred scriptures (Brueggemann 2002; Habel and Trudinger 2008; Murray 1992) that could inform a vibrant faith-centered advocacy for protecting the climate. Given the number of members—14,772,060 or 51 percent of the Canadian faith population in 2001 (Bibby 2011)—the engagement of any of these four faith communities in faith-based environmental advocacy would constitute a significant tipping point on the Canadian religious landscape.

With only two of the 11 faith communities surveyed having actively distributed their public position documents on climate change within their faith communities, and four large traditions (Roman Catholic, Evangelical Christian, Judaism, and Islam) having failed even to develop a unified public position on climate change, it seems that climate advocacy efforts within Canada's faith communities have so far been rather anemic. But in October 2011 new signs of faith involvement began to emerge, coalescing around developing a joint public position on climate change. Were the interfaith communities finally stepping up to the plate?

UN climate change talks 2011: Are interfaith communities stepping up to the plate?

A promising beginning: 2011

Two months before world leaders gathered to discuss the UN Framework Convention on Climate Change in South Africa, a series of events in Ottawa (the capital of Canada) seemed to signal a new chapter in the faith communities' engagement regarding climate change. On October 25, 2011, leaders from within five faith communities (Christian, Muslim, Hindu, Baha'i, and Buddhist) in Canada released a joint document entitled *Canadian Interfaith Call for Leadership and Action on Climate Change*. Arising from the Canadian Council of Churches and the nascent Interfaith Partnership, this position document was the first of its kind in Canada on faith and climate change to be developed and endorsed by a broad cross-section of faith leaders and communities (see Table 11.2).

Table 11.2 Faith communities in Canada that signed the *Canadian Interfaith Call for Leadership and Action on Climate Change*

	Signed Interfaith Call	*Comments*
Christian denominations:		
United Church of Canada	✓	
Mennonite	✓	
Christian Reform	✓	
Anglican	✓	
Lutheran	✓	
Presbyterian	✓	
Roman Catholic		
Evangelical		
Other religious traditions:		
Muslim	✓	
Baha'i	✓	
First Nations		1 independent leader
Jewish		2 local leaders
Hindu	✓	
Buddhist	✓	

The Call highlighted the spiritual roots and moral dimensions of the climate crisis, underlining the justice dimensions of global warming in its impacts on the poor, who are being impacted first and the most severely, even though they have contributed little to climate change. The interfaith communities proposed a framework of values for a low-carbon economy, and called upon the Canadian government to provide leadership at the upcoming climate change talks by enacting three policy goals: 1) a binding international agreement to set clear goals for emissions reduction to keep the global average temperature increase below a 2 °C increase; 2) a national plan to reduce emissions and encourage renewable energy; and 3) contribution to the UN Green Climate Fund to assist poor countries to mitigate and adapt to climate change impacts.

In order to launch the document, a series of events was organized in Ottawa, including: 1) a public evening event on faith and climate; 2) a day-long Interfaith Forum on Climate Change with over 50 faith leaders and a political panel with three members of parliament (MPs); and 3) an in-camera breakfast meeting with MPs from all five national political parties. The Canadian Council of Churches organized a press briefing about the Interfaith Call, but, because little effort was put into contacting journalists, media coverage of the events was minimal. While the Call's impact was therefore limited, the document itself and the publicizing activities surrounding it were nonetheless significant achievements for the Canadian faith communities, demonstrating that they were willing to move a critical public policy issue onto the public stage.

Building on this accomplishment, members of a number of local faith communities in several faith traditions (predominantly United Church, Anglican, and Roman Catholic) sent petitions to their MPs in four of the five federal political parties (the governing Conservative Party, New Democrats, Liberals, and Greens) to give public profile to the ethical concerns of the faith communities regarding global warming, underlining the urgency of the federal government taking action to reduce GHG emissions. These petitions were presented in the House of Commons by the MPs, thus directing attention to an issue that the governing Conservative Party would have preferred to avoid.

Sixty-one official representatives from faith communities and faith-based organizations from five different world faith traditions (Christian, Muslim, Baha'i, Hindu, Buddhist) signed the Interfaith Call. The Christian tradition was the most heavily represented among the signatories, with small numbers of leaders from the other world religions being included, Islam being the second-largest group. It is instructive to note which national faith leaders refused to endorse the Interfaith Call, when invited to do so by the Canadian Council of Churches. First, while 18 leaders from several different Catholic religious orders and Catholic organizations in Canada endorsed the Call, no bishop from the Canadian Conference of Catholic Bishops (CCCB) agreed to sign. Also curious was that only two local Jewish leaders in Canada signed the declaration, despite invitations being sent to the Centre for Israel and Jewish Affairs, the official voice of the Jewish community in Canada. Finally, while leadership from three Evangelical Christian denominations (i.e., Mennonite Church, Christian Reformed Church, and Salvation Army) signed the Call, the Evangelical Fellowship of Canada (the national organization representing Evangelical Christians) itself refused to sign.

Missed opportunities: 2012

Despite this promising step, engagement of faith leaders on climate change has dissipated. First, there were some significant missed opportunities at the events associated with launching the document. Although three faith leaders spoke at a press filming on October 25, 2011, no journalists appeared for the press conference to cover the events, owing to the lack of a communications plan and the staff to plan and implement communications. While organizers discussed engaging faith communities through worship to focus prayers and sacred readings on climate change, no such action was taken. Although organizers discussed getting the Ottawa events covered by two influential newspapers, in the end no action was taken. In addition, although faith leaders agreed to circulate the document in their own publications within their respective faith communities, very few leaders actually made certain that the document was so distributed. Thus the Interfaith Call had little circulation either publicly or within the faith communities.

Second, according to Willard Metzger (2012), leader of the Mennonite Church Canada, after October 2011 momentum around the issue of climate change disappeared. Although certain faith leaders such as the United Church moderator and Metzger continued to press for action on climate change, the majority of faith leaders have taken no further collective action or planned any follow-up. Indeed, some leaders promised to create a follow-up document highlighting a faith perspective on the international UN gathering Rio+20 for the Governing Council of the Canadian Council of Churches' 2012 meeting in Ottawa. But such a document was never completed.

Finally, the Call did not create momentum for greater organizational capacity building within the nascent faith-based climate movement. It did not result, for example, in a strategic action plan with realistic objectives, a communications strategy, or an infrastructure to support faith members to advocate with their political leaders. With little, if any, coordination of efforts, only fragmented and partial plans were implemented to more deeply engage faith communities through education and worship, and little was done to reach out to significant allies and potential partners. Critically, there was no plan to develop financial or organizational capacity to continue the movement forward, no body to coordinate or monitor progress to achieving targets and goals, and no backbone organization to coordinate and drive these actions forward in an effective and strategic manner. It therefore appears likely that little, if any, concrete action will emerge to enable faith-based advocacy on protecting the climate.

Is faith-based climate change action moving outside the religious institutions?

At the October 23, 2011 event at St. Andrew's Church in Ottawa, Lauryn Drainie, a young climate change organizer within the faith communities, warned that if faith communities fail to act seriously on climate change, religious institutions are in danger of becoming irrelevant. Further, she remarked that she and other young adults are already abandoning the churches, directing their energies elsewhere to develop solutions for climate change. As of August 2012, there are signs that Drainie's comments may have foreshadowed emerging trends in the faith communities.

Ron Ewart, the associate director of the Toronto United Church Council and executive director of the Green Awakening Network, has reported that many grassroots environmental leaders in the faith communities in Toronto believe that national faith leadership has "dropped the ball" on climate change, and are moving forward on their own, such as Anthony Ketchum and his leadership in "For Our Grandchildren," a non-profit organization devoted to "empower[ing] grandparents to take the urgent action needed to address . . . climate change" (http://forourgrandchildren.ca/). In May 2012, other grassroots faith leaders organized three events, including a public forum with former Toronto mayor David Miller (http://www.trinitystpauls.ca/content/cfja) and a climate change advocacy workshop with the Liberal Environment critic and scientist Kirsty Duncan. That faith-based activists are moving beyond the faith community further suggests the unwillingness of faith communities to nurture a climate advocacy movement.

Faith community leadership and advocacy on climate change

If Canada's faith communities have been slow to address climate change, it would be useful to attempt to identify barriers to engagement. Toward this end, I have created a rubric to evaluate the effectiveness of each faith community's climate advocacy activities thus far. Attempts by faith communities to influence public policy can be divided into three sub-types, under which I have listed specific activities as examples (the letters associated with the activities correspond to those listed in Table 11.3):

1 *Utilizing position documents for climate change advocacy in the public realm:*

   a) Have the faith communities strategically distributed their documents externally in the public square to government, the public, the media, and other decision makers?
   b) Have faith communities utilized the position document on climate change as a point of reference for advocacy with governments, advocating public policies addressing climate change?

2 *Developing relationships with allies for coalition building:*

   c) Has the faith community joined the Climate Action Network Canada (the Canadian network for coordinating climate change advocacy) as a member?

Table 11.3 Advocacy by faith communities on climate change

	(1) Utilizing position documents		*(2) Developing relationships*	*(3) Participating in campaigns*				*No. of actions taken*
	(a)	*(b)*	*(c)*	*(d)*	*(e)*	*(f)*	*(g)*	
Christian denominations:								
United Church of Canada	✓	✓	✓	✓	✓	✓	✓	7
Mennonite		✓		✓				2
Christian Reform								0
Anglican					✓		✓	2
Lutheran	✓							1
Presbyterian	✓	✓					✓	3
Catholic	✓							1
Evangelical								0
Other religious traditions:								
Muslim			✓	✓	✓			3
Jewish								0
Buddhist					✓			1
Baha'i			✓					1
Faith-based organizations:								
KAIROS	✓	✓	✓		✓	✓	✓	6
Citizens for Public Justice (CPJ)	✓	✓	✓			✓		4

3 *Participating in campaigns for climate change, renewable energy or fossil fuel policy:*

  d) Has the faith community acted on specific campaigns of the Climate Action Network?
  e) Did the faith community endorse the Tides Foundation Proposal for a Clean Energy Accord for Canada in 2012?
  f) Did the faith community protest the Canadian federal government's attack on environmental NGOs and its omnibus budget Bill C-38?
  g) Did the faith community take a position, opposing the Gateway pipeline by Enbridge?

Table 11.3 profiles the spectrum of engagement of faith communities in Canada regarding climate change, ranging from more intensive involvement (United) and possible emerging leaders (Mennonite, Muslim) to those faith communities that have not advocated for protecting the climate or environment at all (Evangelical Christian, Jewish, Roman Catholic bishops). A few faith communities have utilized their position documents on climate change strategically in the public domain (a and b). For instance, Mardi Tindal (United Church moderator) has consistently written op-eds, used webinars, written blogs, and made YouTube videos, using the church's position documents as pivot points. Similarly, the Mennonite Central Committee Canada (MCCC) is highlighting climate change in its international relief and development work by fostering a growing awareness between the excessive fossil fuel use in the Global North and impacts on the Global South by bringing "climate witnesses" who have already witnessed climate change impacts. Only sporadically have other faith leaders taken action through advocacy, such as Lutheran national bishop Susan Johnson decrying the Canadian federal government's withdrawal from the Kyoto Protocol in December 2011 in a press release.

Only the United Church and the Baha'i community have formally joined the Climate Action Network (CAN), which coordinates national campaigns and activities in Canada advocating for action on climate change (c). But, promisingly, other faith communities have joined CAN campaigns (d). The Canadian Council of Imams was a signatory on a letter with a number of NGO leaders to Peter Kent, which successfully halted the expansion of the Maxim coal plant. MCCC Canada also endorsed a letter from the Climate Action Network Canada to Prime Minister Harper and Finance Minister Flaherty to terminate subsidies to the fossil fuel industry in Canada. In 2011, the United Church of Canada also joined with the Climate Action Network and the Pembina Institute in an innovative educational tour of farming communities in southern Ontario, profiling the possibilities of renewable energy as a way of revitalizing rural communities and engaging key opinion leaders in several agricultural regions.

After consulting Canadians across Canada, the Tides Foundation developed a "Clean Energy Accord" (http://tidescanada.org/energy/accord/) with endorsements of over 700 companies and organizations across Canada to present to the provincial premiers at their meeting in Halifax in July 2012 (e). The Tides project is a strategic cross-sectoral initiative creating a frame for a sustainable energy policy that reduces GHG emissions. The United Church moderator and five Anglican bishops endorsed the Accord, as did the Canadian Council of Imams and the Noor Cultural Centre (Islam) and the Palyul Foundation (Buddhism).

The faith community was notably silent in the face of the federal Conservative government gutting federal environmental legislation and attacking environmental NGOs (f). In response to these actions, over 500 NGOs and organizations staged a national protest, blacking out their websites on June 4, 2012 in the "Black Out; Speak Out" campaign (http://www.blackoutspeakout.ca/index.php). Several key federal cabinet ministers cancelled their schedules to travel across Canada to present the administration's point of view, indicating the campaign's effectiveness. Yet not one national faith community participated in the "Black Out" protest (though two faith organizations, KAIROS and Citizens for Public Justice, did participate). Nor did any faith leaders (other than Moderator Tindal of the United Church) speak out publicly to criticize the government's assault on environmental NGOs, or to urge a more respectful and productive dialogue on environmental issues. Although some local faith leaders on the British Columbia coast have expressed concerns about the Gateway pipeline (Presbyterian; and six Anglican bishops regarding the fairness of the pipeline hearings), no national faith leadership other than the United Church has expressed its opposition to the pipeline (g).

Barriers to engagement

My consultations with faith-based organizations and conversations with leaders have led me to identify the following obstacles to the emergence of more robust environmental citizenship within the faith communities.

Shrinking membership and staff capacity

With recent demographic trends (Bibby 2011), many faith communities in Canada have suffered from both diminishing and aging membership. As a result, with decreasing financial assets, these faith communities have been forced to cut staff in their national offices in order to remain financially solvent. For the Anglican, United, Lutheran, and Presbyterian churches, these financial constraints have reduced their staff in the areas of social justice and the environment. Similar constraints of reduced staff capacity in the Canadian Conference of Catholic Bishops (Gunn 2010) and other faith communities, including those that are growing (Evangelical Christian), are becoming the norm. Consequently, traditional methods of utilizing full-time national staff to drive environmental agendas are no longer available.

Ignoring environmental advocacy by focusing on greening buildings and personal change

One of the greatest barriers to faith communities developing a capacity for leadership and advocacy on climate change is their tendency to conceptualize green initiatives as solely personal and local, rather than developing a more holistic model that would integrate personal or local change with large-scale political and economic policy change on a macro level (McKibben 2005; Turner 2011). As faith communities initially move into taking action, the "greening" of physical assets and religious buildings is frequently the only activity that local communities undertake, neglecting any climate change advocacy altogether. Greening one's lifestyle on a personal and local level is important, as it reduces the actual amount of GHG emissions and other pollution, providing valuable time for action and slowing the pace of climate change. Yet, as McKibben (2006) argues, while personal and local greening activities are psychological "calisthenics" that prepare individuals for the larger-scale action, personal greening activities are no substitute for climate change advocacy. The magnitude and speed of climate change and environmental decline are such that personal change alone is a grossly inadequate response and not the right scale for effectively mitigating climate change. Thus far, unfortunately, only a few faith leaders have recognized this.[2]

Apathy about environmental issues among the clergy

Several NGO staff have named the clergies' indifference or hostility to faith-based ecological initiatives as a consistent barrier to the development of leadership on climate change. The staff have suggested that this pattern of on-the-ground clerical disinterest is widespread through all the faith communities, although Roman Catholic clergy appear to be the most closed and resistant (despite the excellent documents on climate change that the bishops and the Vatican have produced). There is little point in writing position statements on climate change if the clergy in faith communities could not care less about integrating ecology and faith into their community activities.

Bureaucratic and hierarchical models of leadership within faith communities

In normal times when the issues are of small significance, what Burns (2003) terms *transactional* leadership can provide continuity and a functional governance structure where competing interests are brokered. At present, this bureaucratic model of leadership is characteristic of the faith communities and organizations in Canada. However, in times of crisis, Burns argues that *transformational* leadership facilitating significant systemic and durable social change can emerge, which can "define public values that embrace the supreme and enduring principles of a people . . . at testing times when people confront the possibilities—and threat—of great change" (2003: 29). This style of leadership may be necessary in order to confront the climate crisis. Yet, in my research, I found that more exclusive and transactional models of bureaucratic leadership appear to be favored by many faith institutions. These bureaucratic forms of leadership act as a "bottleneck" for any new initiatives, with the result that climate advocates are leaving institutional faith settings to create their own networks and organizations.

Not only do internal factors impede the emergence of leadership for climate change advocacy, but social forces external to faith communities also play a role in inhibiting the development of leadership. Such forces include the secularization of Canadian society, where faith has been relativized and diminished in importance, leading increasing numbers of individuals to be wary of the relationship between religion and intolerance.

Lessons learned

Faced with many barriers, Canada's faith communities have failed to ignite a powerful movement for effective climate change advocacy. How might these barriers be overcome?

Moving from bureaucratic to empowerment models of leadership

Chris Lind (2012), director of the Sorrento Centre (an Anglican conference centre in British Columbia), has suggested that faith communities would benefit from changing their bureaucratic models of leadership that dominate their organizational life. Not only are the faith communities in danger of losing their best up-and-coming leaders in faith and ecology, but the effectiveness of campaigns of environmental advocacy are adversely affected and diminished, as new grassroots leaders are exiled to the margins of decision making. Instead, Lind argues that, rather than inadvertently marginalizing grassroots leadership on the edges of bureaucratic structures of hierarchical authority, faith communities should foster more inclusive and empowerment-oriented models of leadership, integrating new transformational leadership into a covenantal community of heterarchical leadership (leadership that is distributed and with multiple sources of authority) and accountability.

Collaborating with strategic allies and partners

As we have seen, faith communities have made little effort to collaborate or partner with strategic allies or NGOs outside the faith community, although doing so would significantly magnify their influence. In particular, leveraging influence through networks, organizations, and coalitions would augment the faith community's ability to intervene in political decision-making processes. For example, more faith-based organizations could join campaigns, such as the Tides Foundation's Clean Energy Accord. The campaign already had a comprehensive media strategy in place (see, for example, Neumann and Smith 2012; Smith, Bondy and Rand, 2012), which the participation of faith communities could have amplified. Faith communities can also seek membership in strategic non-religious organizations such as the Climate Action Network Canada, or collaborate with NGOs with significant research and policy capacity regarding climate change, such as the Pembina Institute. Such collaborations will further foster effective climate change advocacy, while requiring minimal additional resources on the part of the faith community.

Enhancing coordination between faith communities

Social scientists Kania and Kramer (2011: n.p.) note that large-scale "social change requires broad cross-sector coordination, yet the social sector remains focused on the isolated intervention of individual organizations." If significant change is to be accomplished, they insist that collaboration and partnership between community organizations with shared strategic objectives is required. To this end, faith communities should collaboratively develop a national organizational structure or secretariat to effectively coordinate strategic advocacy campaigns to protect the climate.

For faith communities committed to taking action on climate change, there are several organizational models available. For several years, the Green Budget Coalition (GBC, http://www.greenbudget.ca/main_e.html), a collaborative effort undertaken by 20 Canadian environmental NGOs, has worked to influence the budget of the federal government. While it is a small-scale organization, it exercises an influence much greater than its size through interacting with MPs and staffers. Faith communities need to develop a similar organization, a new national organizational structure or secretariat, perhaps a "Green Faith Alliance," which would focus on coordinating faith-based advocacy campaigns to achieve effective government policies to protect climate and the environment. Such a secretariat could also coordinate a communications strategy through the media as well as function as a "backbone organization" (Kania and Kramer 2011) for communication between faith communities. As a support infrastructure, the secretariat would also link with faith communities for coordinating educational projects that build their capacity for advocacy in the public square. Several key environmental NGOs and scientists have a history of working with faith communities in Canada, and have already indicated that they would be willing to share educational support and research capacity.

Skills education in effective faith-based advocacy

In order to work effectively, faith communities also need to develop a program and a methodology for training members of faith communities in citizenship and advocacy skills. In the spring of 2012, a network emerged that centered on developing a faith-based education program for teaching the skill sets critical for a ministry of advocacy grounded in a resilient spirituality. This project is centered on fostering skills germane for effective advocacy in addressing both social justice and environmental justice concerns (Lysack 2009b, 2010, 2012a, 2012b, 2013). These skills sets include: 1) communication skills; 2) influencing and relationship skills with decision makers; 3) relationship-building skills with allies in networks; 4) developing personal spiritual resilience (Lysack 2009a); and 5) educating and mentoring others. A pilot project was completed in November 2012 in Toronto, with a follow-up event in the winter/spring of 2013.

A helpful organizational model for education would be that of Engineers without Borders (EWB, http://www.ewb.ca/), an NGO that combines high-level advocacy coordination by national staff with advocacy campaigns utilizing student volunteers. To educate volunteers to work in advocacy, national staff develop maps of the federal political parties through website research and interviewing MPs. After attending a three-hour training session organized by EWB staff, university students descend on the national Parliament to dialogue with and influence MPs. EWB emphasizes planning, preparation, and support for its volunteers to equip them to be effective advocates. Citizens Climate Lobby (CCL, http://citizensclimatelobby.org/) provides another useful model for faith communities. CCL affirms a two-fold goal: 1) "to create the political will for a stable climate"; and 2) "to empower individuals to have breakthroughs in exercising their personal and political power." CCL members focus on lobbying political leaders and shaping the public discourse through letters to the editor and op-eds in the media to influence politicians for a carbon fee or dividend and "an end of subsidies to fossil fuel companies." CCL embodies a hybrid of highly focused strategic objectives, a lean and focused educational and leadership structure, and a proven methodology of citizen engagement.

What are the faith communities actually prepared to do about climate change?

Lack of action on climate change still haunts faith communities. All faith communities have the potential to become effective leaders in advocating for creation and protecting the Earth's climate. They possess remarkable resources of covenantal community, ethical relationality, prophetic insight, and a transcendent joy in the gift of life and creation. Climate change represents a great opportunity to offer a unique vision of an alternative lifestyle that is more sustainable and just. Yet, at the same time, like a fig tree that fails to bear fruit, some faith communities will fail to fulfill their calling to protect creation and abdicate their moral responsibility, just as some faith communities failed to join the abolition of slavery movement (Lysack 2012a). Then, as now:

> Good intentionality and heroic personal practices are important but they are not adequate. What is necessary now is political will to reverse the politics of self-indulgence and the economics of commoditization that places our world in jeopardy. . . . [T]he hard issues concern political will. It is precisely the task of religious communities to summon political will so that there can be freedom and courage to do the right thing.
>
> (Brueggemann 2011: 172)

Acknowledgments

The author wishes to express his appreciation to both Joe Gunn, executive director of Citizens for Public Justice, and Julia Kilpatrick, communications manager of the Pembina Institute, for their kind assistance in reviewing this chapter.

Notes

1 For comparison, Canada signed the Kyoto Protocol in 2002, committing to the reduction of emissions to 6 percent below 1990 levels of 461 megatonnes of GHG. The Kyoto target was 433 megatonnes, considerably lower than the 2007 commitment to reduce to 607 megatonnes.

2 I am thinking here of Ron Ewart (director, Green Awakening Network) and Ted Reeve (director, Greening Sacred Spaces), as well as Willard Metzger (executive director, Mennonite Church Canada) and Mardi Tindal (moderator, United Church).

References

Abdul-Matin, I. 2010. *Green Deen: What Islam Teaches about Protecting the Planet*. San Francisco, CA: Berrett-Koehler.

Anderson, L. 2006. "Analytic Autoethnography." *Journal of Contemporary Ethnography* 35.4, 373–395.

Angrosino, M. and K. Mays de Perez. 2000. "Rethinking Observation: From Method to Context." In *Handbook of Qualitative Research*, 2nd edn., ed. N.K. Denzin and Y.S. Lincoln, pp. 673–702. Thousand Oaks, CA: Sage.

Bibby, R. 2011. *Beyond the Gods and Back*. Lethbridge, AB: Project Canada Books.

Bouchard, Bishop Luc. 2009. *A Pastoral Letter on the Integrity of Creation and the Athabasca Oil Sands*. St. Paul, AB: Diocese of St. Paul.

Brueggemann, W. 2002. *The Land*, 2nd edn. Minneapolis, MN: Fortress Press.

——. 2011. *Disruptive Grace*. Minneapolis, MN: Fortress Press.

Bruner, J. 1990. *Acts of Meaning*. Cambridge, MA: Harvard University Press.

Bullough, R.V. and S. Pinneager. 2001. "Guidelines for Quality in Autobiographical Forms of Self-Study Research." *Educational Researcher* 30.3, 13–21.

Burns, J.M. 2003. *Transforming Leadership*. New York: Grove Press.

Ellis, C. and A.P. Bochner. 2000. "Autoethnography, Personal Narrative, Reflexivity: Researcher as Subject." In *Handbook of Qualitative Research*, 2nd edn., ed. N.K. Denzin and Y.S. Lincoln, pp. 733–768. Thousand Oaks, CA: Sage.

Gruending, D. 2012. Presentation on faith and climate change, January 27, University of Calgary.

Gunn, J. 2010. "Muted and Maligned Voices: Public Justice and the Canadian Churches." *Ecumenist* 47.4, 1–6.

Habel, N. and P. Trudinger. 2008. *Exploring Ecological Hermeneutics*. Atlanta, GA: Society of Biblical Literature.

Haluza-DeLay, R.B. 2008. "Churches Engaging the Environment: An Autoethnography of Obstacles and Opportunities." *Human Ecology Review* 15.1, 71–81.

Hamilton, C. 2010. *Requiem for a Species: Why We Resist the Truth about Climate Change*. London: Earthscan.

Hansen, J. 2009. *Storms of My Grandchildren*. New York: Bloomsbury.

Kania, J. and M. Kramer. 2011. "Collective Impact." *Stanford Social Innovation Review*, Winter. Available at http://www.ssireview.org/articles/entry/collective_impact (accessed June 20, 2012).

Lind, C. 2012. Personal conversation.

Ljomov, D. and J. Gunn. 2010. "Faith Communities and Public Stances on Creation Care and Climate Change." Unpublished research monograph. Citizens for Public Justice.

Lysack, M. 2008. "Global Warming as a Moral Issue: Ethics and Economics of Reducing Carbon Emissions." *Interdisciplinary Environmental Review* 10.1 and 2, 95–109.

——. 2009a. "From Environmental Despair to the Ecological Self: Mindfulness and Community Action." In *Mindfulness and Social Work*, ed. S. Hick, pp. 212–218. Chicago, IL: Lyceum Publications.

——. 2009b. "Practices and Skills for Building Social and Ecological Resiliency with Individuals and Communities." In *Structural Social Work in Action: Examples from Practice*, ed. S. Hick, H. Peters, T. Corner and T. London, pp. 211–228. Toronto, ON: Canadian Scholars Press.

——. 2010. "Environmental Decline, Loss, and Biophilia: Fostering Commitment in Environmental Citizenship." *Critical Social Work* 11.3. Available at http://www.uwindsor.ca/criticalsocialwork/2010-volume-11-no-3.

——. 2012a. "The Abolition of Slavery Movement as a Moral Movement: Ethical Resources, Spiritual Roots, and Strategies for Social Change." *Journal of Religion and Spirituality in Social Work* 31.1 and 2, 150–171.

——. 2012b. "Building Capacity for Environmental Engagement and Leadership: An Ecosocial Work Perspective." *International Journal of Social Welfare* 21, 260–269.

——. 2013. "Emotion, Ethics, and Fostering Committed Environmental Advocacy." In *Environmental Social Work*, ed. M. Gray, J. Coates and T. Hetherington, pp. 231–245. London: Routledge.

McKibben, B. 2005. *The Comforting Whirlwind: God, Job, and the Scale of Creation*. Cambridge, MA: Cowley Publications.

——. 2006. *The End of Nature*, 2nd edn. New York: Random House.

——. 2008. *The Bill McKibben Reader: Pieces from an Active Life*. New York: Henry Holt and Company.

Metzger, W. 2012. Personal communication, January 7.

Murray, R. 1992. *The Cosmic Covenant: Biblical Themes of Justice, Peace, and the Integrity of Creation*. London: Sheed & Ward.

Neumann, K. and R. Smith. 2012. "Energy Strategy: Let's Look at Things Differently, Move Forward Together." *Halifax Chronicle Herald*. Available at http://thechronicleherald.ca/opinion/120601-energy-strategy-let-s-look-at-things-differently-move-forward-together (accessed August 13, 2012).

Northcott, M. 2007. *A Moral Climate: The Ethics of Global Warming*. Maryknoll, NY: Orbis Books.

Paris, M. 2012. "Emissions Target Won't Be Met without Action, Panel Argues." *CBC News*. Available at http://www.cbc.ca/news/canada/calgary/story/2012/06/12/pol-nrtee-round-table-environment-economy-report.html (accessed June 22, 2012).

Smith, M., K. Bondy and T. Rand. 2012. "A Bold Canadian Energy Strategy for a Stronger Canada." *Toronto Star*, July 25. Available at http://www.thestar.com/opinion/editorialopinion/article/1231817--a-bold-canadian-energy-strategy-for-a-stronger-canada (accessed August 14, 2012).

Turner, C. 2011. *The Leap: How to Survive and Thrive in the Sustainable Economy*. Toronto, ON: Random House of Canada.

Vaughan, S. 2012. *Report of the Commissioner of the Environment and Sustainable Development to the House of Commons*. Ottawa, ON: Office of Auditor General of Canada.

12 The U.S. Catholic response to climate change

Michael Agliardo

Introduction

When the Vatican announced plans in 2007 to become the world's first carbon-neutral state (CNA Staff 2007), it challenged Catholic and non-Catholic leaders alike to acknowledge the seriousness of anthropogenic climate change. In recent decades, beginning with Pope John Paul II naming Saint Francis of Assisi patron of ecology in 1979, he and his successor, Benedict XVI, increasingly underscored the moral and theological import of environmental issues, and of anthropogenic climate change in particular (Benedict XVI and Koenig-Bricker 2009; Keenan 2002). In his 2010 World Day of Peace address, typically reserved to highlight a theme of significance for the whole Church, Benedict XVI (2010) affirmed the theological grounds for Catholic concern over climate change, warning that peace between peoples and justice for the poor depended on respect for God's creation. The next year, prior to the Durban climate negotiation meetings, he again spoke out forcefully, urging "a responsible, credible and supportive response to this worrisome and complex phenomenon, keeping in mind the needs of the poorest populations and of future generations" (Thavis 2011). At its highest levels the Roman Catholic Church has promoted an environmental ethic that is combined with an understanding of justice, and underscored the urgency of climate change. In order to shed light on Catholic mobilization in response to climate change, this chapter examines how one local Church, that of the United States, has taken up the challenge of climate change.

With an estimated worldwide membership of 1.1 billion, the Catholic Church is a sizable presence in most countries and the majority faith in 67 (Hackett and Grim 2011: 23–25). In recent history, from the Philippines (Moreno 2006) to Poland (Casanova 1994) it has mobilized the forces of civil society to stare down state power and usher in real social change. At the same time, the Catholic Church is not the unified corporate actor that many imagine. Rather, it is a vast, segmented institutional conglomerate, and that makes mobilization a daunting prospect. Moreover, how the Church operates in different societies can vary considerably. The scope of the present account is limited to the U.S. Catholic response to anthropogenic climate change. Nonetheless, the hope is that it will shed light on factors that affect the broader Catholic response to this important environmental issue;

given the size, resources, and influence of the U.S. Church, it is also noteworthy in its own right.

U.S. Catholic responses to climate change largely correspond to two dominant, relatively distinct ways of framing the human relationship to nature: an "environmental justice" frame, which consciously grounds itself in the Catholic tradition and in Catholic Social Teaching,[1] and a "creation spirituality" frame, which critiques anthropocentrism and calls for a deep transformation of the human relationship to nature.[2] These frames support different but overlapping environmental ethics and are maintained by people who travel in relatively distinct institutional circles. A number of other, subsidiary frames also structure Catholic engagement with environmental issues, intersecting with the two dominant frames in complex ways. Of note are the traditions associated with religious orders (such as the Franciscans and the Benedictines) and the work of academic theologians. This chapter offers a preliminary assessment of the strengths and limitations faced by proponents of these various frames in responding to climate change.

This account of Catholic climate action draws on the ongoing research on religious environmentalism and public life that I have conducted since 2004 (Agliardo 2008), including on-site research at nine Catholic environmental centers across the United States in 2011 funded by a grant from the Association for the Sociology of Religion. Site visits and 15 open-ended interviews with national and local key informants conducted early in 2012 provided further insight. Interviewees cited by name gave permission for doing so. Document analysis and web scans of American Catholic engagement with climate change helped to triangulate interview data. Given the size and complexity of the U.S. Catholic Church, the present discussion makes strategic use of institutional exemplars and key informants in the Catholic climate change movement. Insofar as these informants are leaders in this effort, their accounts are not always representative of the larger Church. I should also note that, as an ordained Jesuit priest teaching at Loyola University in Chicago, I enjoy a certain insider status; nonetheless, being a newcomer to Loyola, as well as a priest often still sorting through the complexities of his own Church, I acknowledge in advance my limitations as an "insider."

Making sense of the Catholic world

The transnational size and complexity of the Roman Catholic Church warrants description of its structure. In terms of the Roman Catholic Church worldwide, the Holy See[3] is in a unique position to articulate Catholic doctrine with the highest level of authority. At the same time, in the normal course of affairs the Pope is Bishop of Rome and, while the Holy See may commend an issue to the attention of Catholics worldwide, the other bishops, as leaders of the local Church in their own right, have considerable autonomy in deciding the priorities of their respective dioceses.

In this light, the U.S. Church is best conceived of as a series of relatively autonomous dioceses. It does support a coordinating mechanism known as the United States Conference of Catholic Bishops (USCCB). The USCCB maintains a staff at its headquarters in Washington, DC, but its designation as a "conference" specifically underscores its consultative and collaborative nature; it is not a governing structure. Within a diocese, most Catholics live out their faith lives in a parish headed by a priest who serves as its pastor. In keeping with their role as pastors, priests themselves have considerable independence and discretion when it comes to coordinating parish life.

Beyond the regular structure of dioceses and parishes, numerous other institutional entities populate the Catholic universe. These include religious orders: communities of sisters, brothers, or priests who have made a vowed commitment to their common life. Such communities are formally recognized in canon law, have their own constitutions, and operate on the basis of their own norms, traditions, and governing structures. As a result, they, too, maintain a certain relative autonomy within the Catholic world. In Catholic parlance, a "religious" is a member of such a community, and the adjective "religious" may refer to someone or something pertaining to such a group. Religious orders run a wide range of institutions in the United States, including colleges, retreat and spirituality centers, hospitals, publishing houses, and other works, and as a result they often play a crucial role in advancing a Catholic response to climate change. Finally, since the Second Vatican Council (1962–65) many Catholic-affiliated institutions have become separately incorporated, so that each follows its own path in articulating its Catholic identity.

Overall, U.S. Church membership stands at between 66 million and 75 million, depending on the measure used (Center for Applied Research in the Apostolate 2012; Hackett and Grim 2011: 23). Catholics are approximately one-fifth of the population of the United States. *The Official [U.S.] Catholic Directory* (2012), an institutional directory of the parishes, dioceses, religious orders, and Catholic-affiliated institutions of the U.S. Church, runs for some 2,100 pages of minuscule print. On the one hand, given its vast membership and institutional heft, the U.S. Catholic Church has considerable resources and potential impact. On the other, given its highly segmented structure, the Church finds that mobilization around any given issue, even one backed by the Pope, is no simple matter.

Framing climate change as environmental justice

In the U.S., Catholic religious environmentalism had begun to pick up momentum only by the late 1980s. Prior to Vatican II, certain sectors of the U.S. Church did foster an environmental ethic. For example, the National Catholic Rural Life Conference (founded in 1923) regularly published prayer and liturgical materials that promoted a spirituality of the land (Witte 1948). Historians have identified important pre-Vatican II precedents for the contemporary environmental justice frame (Hamlin and McGreevy 2006). However, even in the post-Vatican II era, social justice advocates in the U.S. Church were most concerned with questions of poverty, peace, and racial equality, and at first they typically cast environmentalism as a distraction from these more "real" issues. Meanwhile, a number of issues, including concerns over human fertility and population control, led to estrangement between the Catholic community and the environmental movement (Agliardo 2011: 39–44; Allitt 1998).

In certain respects, the U.S. Catholic community lagged behind a number of other nations where local Catholic bishops' conferences were publishing position papers on the environment, faith, and the common good (Catholic Bishops' Conference of the Philippines 1988; Catholic Bishops of Northern Italy [1988] 1996; Guatemalan Bishops' Conference [1988] 1996). At around that time, Los Angeles Archbishop Roger Cardinal Mahony, then chair of the USCCB Domestic Policy Committee, raised the question of the U.S. bishops following suit (Christiansen 2002). A number of Catholics had already participated in the 1987 inaugural gathering of a Christian environmental alliance known as the North American Conference on Christianity and Ecology (NACCE), including representatives of the National Catholic Rural Life Conference (Lucas 1995: 190–191, 253; Muratore 1987). When John Paul II brought environmental issues to the fore in his 1990 World Day of Peace address, he provided further impetus for the U.S. bishops to act. The next year they issued their first statement on the environment, *Renewing the Earth* (NCCB 1991).

Renewing the Earth framed its environmental ethic, as its subtitle indicates, "in light of Catholic social teaching." The document identified a need to link "natural ecology and social ecology" and "to build bridges among the peace, justice, and environmental agendas and constituencies." This way of framing environmentalism reflected both the standing commitment to justice issues of the progressive wing of the U.S. Church and the desire of the more traditional wing not to abandon concerns for human society. It also reflected the international context of the Catholic Church. In most regions of the world, where resources and land have long been the objects of intense social competition, environmental issues are rarely viewed as anything but justice issues.[4] *Renewing the Earth* stressed that "it is the poor and powerless who most directly bear the burden of current environmental carelessness" (NCCB 1991). Most mainstream Catholic teaching on the environment has since remained firmly tied to questions of concern for the poor and powerless, integral human development, and the common good. Accordingly, the USCCB Environmental Justice Program lists "Care for God's Creation" as one of the seven integral "Themes of Catholic Social Teaching" which the Church endorses (USCCB 2012).

The Episcopal Conference and the Catholic Climate Covenant

In 2001 the USCCB brought its environmental justice frame to bear on climate change, publishing a 24-page statement entitled *Global Climate Change: A Plea for Dialogue, Prudence, and the Common Good* (2001). In keeping with the environmental justice frame, the document sought to balance "respect for nature" and "social well-being." It proceeded to underscore the authoritative nature of IPCC reports. And it concluded with a call for action: "In facing climate change, what we already know requires a response. . . . Significant levels of scientific consensus . . . justifies [*sic*], indeed can obligate, our taking action intended to avert potential dangers." *Global Climate Change* forcefully pointed out that, if wealthier nations take climate change seriously enough to provide for their own adaptation, they must not neglect the plight of the poor: "At its core, global climate change is not about economic theory or political platforms. . . . It is about the future of God's creation and the one human family" (2001: n.p.).

Despite its forcefulness, during the half-decade after its publication *Global Climate Change* had little impact. Mark Potosnak currently works with the Catholic Coalition on Climate Change. Despite being a climate scientist and a faithful Catholic, he confides, "I had no idea that back in 2001 they [the U.S. bishops] had such a strong statement about climate change" (Potosnak 2012). In an article for the *Journal of Catholic Social Thought* Braun, Hellwig and Byrnes (2007: 373–374) make a similar observation. Overall, the document did not generate significant attention or activism, even among the bishops who had adopted it.

Finally, in 2006 two departments of the USCCB and 11 other national-level Catholic organizations collaborated to found the Catholic Coalition on Climate Change (CCCC).[5] Dan Misleh (2011), the Coalition's executive director, explains that this form of sponsorship lent the CCCC institutional legitimacy, while also affording it the freedom to engage in the kind of activism that the USCCB normally eschews. The major campaign of the Coalition is the Catholic Climate Covenant, launched on Earth Day 2009. Taking the Covenant's "Saint Francis Pledge" entails committing to a five-fold program of reflection and action tailored to various constituencies (families, dioceses, parishes, high schools, universities, and so forth).[6] "Covenant partners" disseminate the Coalition's resources throughout various networks of the broader Church. They also provide a mechanism for drawing on the institutional and symbolic resources of their various constituencies. By fall of 2012, partners included 31 national Catholic organizations, six dioceses and 15 Catholic universities or colleges.[7]

The Catholic Climate Covenant is firmly rooted in the stewardship and environmental justice frame of the institutional Church. Its logo features co-joined slogans, "Care for creation" and "Care for the poor." Its website foregrounds a shoeprint with the challenge "Who's under your carbon footprint?" This theme brings out the degree to which consumption by the wealthy has an impact on the poor, a theme that resonates with the standing commitment to human service and social justice of many of the members of the Covenant. Covenant leaders develop this theme in a way that points U.S. Catholics beyond their own society. For example, in October 2012, the Covenant distributed to network members the Academy Award-nominated film *Sun Come Up*, along with background resources and discussion guides; the film documents the environmental plight of the residents of the Carteret Islands in the South Pacific.

To provide educational programming, the CCCC partners with JustFaith, an organization founded in 1989 to make the tenets of Catholic Social Teaching accessible to parish groups. The Coalition has also trained some two dozen Catholic climate ambassadors across the country to serve as locally based advocates. Some are teachers or college professors, others are involved in ministerial formation, while still others are diocesan officers or parish ministers. These climate ambassadors have introduced the campaign into a wide range of institutional settings.

Insofar as the goals of the campaign range from prayer to transformation of lifestyle to civic engagement, they are diffuse, especially when compared to the more targeted activities and campaigns of secular environmental organizations. In part, this is because religious organizations take up environmental frames and activities in view of larger pastoral considerations.[8] Walter Grazer (2007), former director of the USCCB's Environmental Justice Program, confirms that that program operated with a view to fostering the larger pastoral mission of the Church. He adds that, when it comes to public policy initiatives, a certain political realism necessarily enters in on several levels. The bishops (and most Catholic institutional actors) are aware that their political capital is limited and that their involvement in American public life is a sustained one. For that reason, they may lag behind other groups in advancing an issue until they are convinced that it is one they can back over the long term and one that accords with their standing commitments.

Dioceses and parishes

Despite the consistent and often forceful collective positions on environmental issues and climate change taken by the U.S. Bishops' Conference, by the fall of 2012 only six of the 210 American dioceses had signed up as Covenant partners. That list does include such heavyweights as Los Angeles. And, even if a diocese is not a Covenant partner, that does not mean it is not engaged. The Archdiocese of Chicago has not yet signed the Saint Francis Pledge, but nonetheless it features the Covenant on the webpage of its Office of Peace and Justice. Tom Howard (2012), the program director for Justice Education for the archdiocese, has integrated environmental concern into the work of his office. He ties this directly to the USCCB prioritizing of "care for creation" as one of the seven endorsed "Themes of Catholic Social Teaching." Mark Potosnak, the local climate ambassador, has been active in the archdiocese. Howard believes combining the issues of justice and the environment will play an important part in the efforts of the archdiocese to engage young people in the work of the Church. He highlights a number of ways that Catholic teaching on climate change dovetails with local pastoral priorities. Similarly, the Archdiocese of Seattle has yet to make the Saint Francis Pledge. Nonetheless, Catholic climate ambassador Patty Bowman (2012) introduced JustFaith's eight-week module on climate change at the Cathedral parish with great success. She attributes this success to the way it links the issue to matters of faith. In Stockton, one of the dioceses that has become a Covenant partner, Betsy Reifsnider serves as a Catholic climate ambassador and is also the environmental justice coordinator at Catholic Charities. In a region where one in five children suffers the effects of smog and ozone pollution, Catholic Social Teaching's emphasis on environmental justice resonates deeply (Diocese of Stockton 2012; Wulff 2011).

Not all dioceses have been as receptive as Chicago, Seattle, or Stockton. One climate ambassador I interviewed described working on environmental issues for some time and had hoped that receiving certification from the USCCB would promote the legitimacy of this work among local diocesan officials. However, one particular official who did not take the scientific consensus on climate change seriously—

despite *Renewing the Earth*—did not facilitate access to local Catholic schools. Nonetheless, this ambassador still received calls for presentations from schools and parishes in the area. Feedback from these events underscored the importance of making the connection between science, Catholic faith, and social justice.

In dioceses that have become Covenant partners, the receptivity and leadership of the bishop have been instrumental. Bishops can provide key resources and staff to advance an issue. Nonetheless, there are limits to what a bishop might do, even in his own diocese. Chicago's Tom Howard (2012) stressed that the archdiocese maintains a collaborative culture, adding that it might even be counterproductive for the cardinal to push an issue from above. Be that as it may, when compared to other issues many bishops have championed, their climate activism often seems demure.[9]

At this stage, Catholic climate mobilization at the parish level is modest. In this respect the Chicago archdiocese may be typical. Clare Butterfield is executive director of Faith in Place, a community organization linking faith, justice, care for creation, and green energy. She relates that at any one time perhaps a half-dozen or fewer Catholic parishes have gotten involved in her organization (Butterfield 2012). One that has is Saint Gertrude's. There Sister Terri MacKenzie's (2012) efforts to help green the parish have found warm welcome. She reports that a range of activities from prayer for creation to gardening to increasing energy efficiency have been integrated into the parish culture over time.

Patty Bowman (2012) observes that, when it comes to promoting particular initiatives such as climate concern in the parish of Saint James Cathedral, one criterion of selection was what is "in keeping with the mission of the parish." Whether the referent is parish culture or parish mission or pastoral considerations, key actors who seek to mobilize parish concern about climate change do so with an eye to a larger set of pastoral priorities. In this process the role of the pastor is often pivotal. According to parishioners, the current and former pastors at Saint Gertrude's have played key roles in shaping that parish's culture (Sharma 2012). Sister Terri MacKenzie also described another parish where the pastor was not receptive to environmental concerns, and as a result people who raised those concerns went elsewhere. While few pastors take up initiatives such as climate change,[10] they serve as gatekeepers when it comes to shaping parish culture and mission.

Assessment of environmental justice and the hierarchical Church

In sum, actors at the highest levels of the Catholic Church have made bold statements urging action on climate change. The United States Conference of Catholic Bishops developed a framework for addressing environmental issues in 1991, and one for climate change in 2001. In the case of both sustainability and climate change, Church authorities at many levels have so consistently articulated various prudential, ethical, and spiritual concerns that Catholic conservatives, in particular, will have difficulty maintaining that *as Catholics* they can readily dismiss these issues.[11]

At the same time, the segmented structure of the Catholic Church means that whether and how climate change is taken up will depend upon the initiative of local actors: a concerned bishop, an engaged pastor, a motivated parish committee, or some fortuitous combination of these. Some U.S. dioceses have formally endorsed the Catholic Climate Covenant, and others have taken important steps advancing the issue. However, few U.S. bishops have exercised vigorous leadership on climate change. And, while some parishes have taken impressive steps to integrate sustainability and climate concern into their culture, this process is often shaped by a larger set of "pastoral considerations." Indeed, the diffuseness of the goals of the Catholic Climate Covenant explicitly respects the prerogative of local actors and the reality that climate activism takes place within some larger institutional frame.

"Pastoral considerations" can mean a range of things, but the implication is that they signal the degree to which climate activism must align with existing institutional priorities. Walter Grazer pointed out that bishops had to balance their backing of one issue against other commitments and priorities. Tom Howard indicated that addressing climate change fit within an existing commitment to Catholic Social Teaching and certain pastoral initiatives of the Chicago archdiocese, such as youth outreach. Patty Bowman and Sister Terri MacKenzie noted that, where they were active, environmental concerns fit within an existing parish culture and mission. As one looks to the future, this then raises the question of the extent to which pivotal Catholic leaders operating out of the environmental justice frame will make climate change a genuine priority at the expense of other concerns.

Creation spirituality and climate change

Broadly speaking, creation spirituality is a religious approach that emphasizes the immanence of the divine, directing one to find God at work within the self, within culture, and within the natural order. The advocates of creation spirituality see their approach as the necessary ground of a renewed environmental ethic, correcting those strands of Western tradition that treat nature merely as an object for exploitation. In the contemporary Christian context, creation spirituality is associated with such figures as Matthew Fox, Thomas Berry, and John Shelby Spong.

Thomas Berry and the creation spirituality frame

U.S. Catholics who operate out of a creation spirituality frame consistently cite the work of Thomas Berry, a Catholic priest who belonged to the religious order known as the Passionists. Berry emphasized that creation itself is the primary revelation of the Divine, and therefore nature is a privileged locus of sacred encounter. Berry viewed nature not so much as a "collection of objects" but as a "communion of subjects" with whom we are called to exercise solidarity (Berry 1988). Because he believed that the standard motifs that shape the Christian imagination were less than helpful for sustaining an adequate ecological ethic, he proposed a "New Story" (1978, 1987). Teaming up with mathematical cosmologist Brian Swimme, Berry authored a narrative that celebrates cosmic evolution, locating the rise of humanity as one event within this larger process (Berry and Swimme 1992).[12]

In Berry's view climate change was the product of human arrogance, a pathology of the soul which led human beings to objectify and despoil nature (Berry and Clarke 1991: 5–6). It was as if the human species were suffering a kind of autism, an exaggerated sense of isolation from the rest of creation. At the same time, Berry looked forward to what he termed the "Ecozoic Era," when human beings would exist in a mutually sustaining way with all the Earth's diverse life forms. However, he argued that, to reach that end, technology and science alone would not suffice. Instead, humanity would have to undergo a radical transformation of consciousness.

These themes typify the loose movement associated with the Catholic creation spirituality frame. Such themes appear to appeal to Catholics (and other Christians) who find traditional Christian categories confining. They also form a standpoint for critiquing structures of authority in the Church and Christian tradition itself. In this respect, the creation spirituality frame resonates with similar themes in liberation theology and feminist thought. Many Catholic women and female religious have found that Berry's insights and those of creation spirituality speak to their own experience.

Collaborators of Berry have established numerous retreat and study centers with a creation spirituality orientation. Perhaps the most well known or significant is Genesis Farm in Blairstown, New Jersey, founded by Sister Miriam MacGillis, OP. Beginning with a community-supported organic garden in 1986, it had developed into an ecological learning center by 1990. Genesis Farm has introduced hundreds of people from around the country, including many women religious, to a living experiment in how to re-inhabit the land. Its "Earth literacy" courses have taken a range of formats, often combining elements of a seminar and elements of a retreat. In the process they integrate Berry's "New Story" with natural history, bioregionalism, sustainable living, organic farming, spirituality, liberation and feminist theology, personal empowerment, and a commitment to social justice. After this experience, numerous religious have returned to their sponsoring communities to initiate their own centers. Sarah McFarland Taylor (2007: 289–292) documented some 50 such sites. Many of these centers have expended considerable effort to adopt specific carbon-friendly practices, in the process serving as models for the larger community.

In terms of the dedication, innovation, and institutional resources they bring, American women religious contribute perhaps the most important constituency of Catholic supporters of the creation spirituality frame. Moreover, many have been instrumental in bringing its insights to the parish activities, educational programs, healthcare facilities, community services, and retreat houses that they staff in every U.S. diocese. Saint Gertrude's own Sister Terri MacKenzie is an example. She had studied Berry's work early in her long career of service. Largely because of her efforts, ecology is not simply one more program at Saint Gertrude's; it is integrated into the activities and prayer of the parish. This impact reflects the emphasis of the creation spirituality frame on personal and communal transformation.

Separate institutional circles

At the same time, while the themes of creation spirituality have found friendly reception among some Catholics, others have held the frame at arm's length. Matthew Fox was originally a priest in the Catholic Church, but facing censure he eventually severed ties and was received into the Episcopal Church in 1994 (see his own account in Fox 1996). While no Catholic officials have ever officially censured Berry, his writing has provoked criticism from more traditional Christians of many affiliations. When NACCE sponsored its first gathering in 1987, Berry's proposals sparked acrimonious debate. In the end, a faction coalesced around Berry to form a separate organization (Kearns 1996: 61–62). Detractors argued that his emphasis on the natural order as the primary revelation of the Divine largely bypassed the Bible, the centrality of Christ, the doctrine of redemption, and traditional teaching concerning the role of the Church (Muratore 1988).[13]

One USCCB staffer I interviewed reports that his office had invited Berry to participate in a consultation on environmental matters on one occasion, probably in 1989. Apart from that, generally speaking, they did not find Berry's approach helpful for what they were trying to accomplish. When I asked the head of a key center associated with creation spirituality whether he ever collaborated with the local diocese, he responded, "They're busy. We're busy." More generally, the leading voices of the creation spirituality and environmental justice frames tend to inhabit separate institutional circles.

In sum, proponents of the creation spirituality frame model an ethic that supports both personal and cultural transformation. By fostering changes in both attitude and lifestyle, they integrate a range of carbon-friendly practices and a comprehensive spirituality into a greener way of life. Through the centers they themselves maintain and the pastoral role many play in numerous institutions of the larger Church, they have introduced a broad spectrum of Catholics to an environmental ethic that is theologically grounded. The geocentric vision they advance is one that would reorder the priorities of Church and society. Thomas Berry has repeatedly argued that our ecological woes require more than a technical fix. Creation spirituality is a "prophetic" alternative, one that offers itself as a paradigm for a new era.

At the same time, in part owing to the radicality of its critique, some of proponents of creation spirituality like Matthew Fox have found their way out of the Catholic Church, while others remain but travel largely in their own circles. Recently, relations have become more strained. When the Vatican's Congregation for the Doctrine of the Faith (2012) released the results of a doctrinal assessment of the Leadership Conference of Women Religious (LCWR), one bishop had this to say: "LCWR speakers also explore themes like global spirituality, the new cosmology, earth-justice and eco-feminism in ways that are frequently ambiguous, dubious or even erroneous with respect to Christian faith" (Blair 2012). Unless bridges between creation spirituality and the institutional Church are repaired, further penetration of this frame into Catholic circles is likely to be limited.

Another crucial consideration is the numerical decline of American women religious. To briefly encapsulate a complex phenomenon: In 1965 the number of women religious in the United States topped 181,000; by 2011 that number had dropped to some 56,000. Projections indicate that this process of numerical decline will continue. As a result not only are women religious themselves not in a position to expand their involvements, but they will have to close or hand over operations of some of the ecological learning centers they have already started. This trend, too, raises questions regarding the long-term viability of creation spirituality as a force within the Catholic Church.

Religious orders and Catholic universities

Catholic religious orders

Religious orders occupy a distinctive position within the Catholic community. While many of their members serve in various capacities throughout the Catholic community and the larger society, as noted above they also have founded a wide range of their own institutions, from high schools and retreat centers to publishing houses and institutions of higher education. In addition, they maintain their own structures of internal governance, and recent Church documents on religious life affirm the importance of their distinctive traditions. The autonomy these arrangements lend has allowed some religious orders to articulate creation spirituality as a frame independent of the more mainstream environmental justice frame; it has also meant that members of religious orders have developed approaches to environmentalism drawing on symbolic and institutional resources of their own.

In the 1970s a number of scholars already had pointed out the roots for distinctive environmental ethics in the traditions of various religious orders (for example Dubos 1974). More recently, religious operating out of the framework of their own traditions have wrestled with contemporary environmental issues. Ilia Delio of the Order of Saint Francis (2003, 2008) draws on his sense of kinship with all creation to reflect upon the human relationship with Sister Earth. The Carmelites are members of an order known for contemplative prayer. In a position paper on climate change developed for the Carmelite NGO, the author treats rampant consumerism and the climate change it gives rise to as the result of disordered affections of the soul (Scarel 2012). The Jesuits have also been quite intentional about developing an ecological ethic, drawing on the insights of their founder, Ignatius of Loyola, a mystic who could "find God in all things." In their recent general congregations (worldwide assemblies), they have increasingly highlighted environmentalism as a priority. The Jesuit general curia now maintains a Social Justice and Ecology Secretariat, and it has just published a 68-page document, "Healing a Broken World" (Curia Societas Iesu and Alvarez 2011–12), to guide ecological reflection in Jesuit institutions worldwide. The document aligns environmentalism with an existing commitment to social justice. This framework supplies the backdrop for discussion of a number of environmental issues, including climate change and its impact on the poor.

Catholic colleges and universities

There are 251 Catholic degree-granting institutions in the United States (according to the most recent count of the Association of Catholic Colleges and Universities), the vast majority of which are sponsored by religious orders. By the early 2000s a number of Catholic colleges and universities had begun to develop significant eco-friendly momentum, often stemming from the environmental commitments of their sponsoring religious orders. In 2009 Notre Dame sponsored "Renewing the Campus: Sustainability and the Catholic University," a conference to promote networking around issues of sustainability. Some 45 institutions participated (University of Notre Dame 2009). The conference also highlighted the ethical and religious implications of climate change, and Dan Misleh spoke on behalf of the CCCC. To further promote the Catholic Climate Covenant among colleges and universities, the CCCC developed "Sustainability and Catholic Higher Education: A Toolkit for Mission Integration" (Catholic Coalition 2011). By highlighting institutional mission, the toolkit supported efforts to frame a response to climate change in terms of both Catholic Social Teaching and the distinctive heritage of a given institution.

Climate ambassador Mark Potosnak and several leaders at DePaul University attended the Notre Dame conference; at the time DePaul had begun working through its own path to sustainability. Though DePaul has not yet signed the Catholic Climate Covenant's Saint Francis Pledge, it has had a Sustainability Initiatives Task Force since May of 2010. Sustainability has been integrated into the curriculum, and all new buildings are being built according to LEED specifications. Moreover, since one of the leaders of the Task Force is also assistant vice president in the Office of Mission and Values, the effort to green the campus and respond to climate change is pursued "in light of our mission as a Catholic, Vincentian,[14] urban school" (DePaul University 2012; Potosnak 2012).

Loyola University Chicago, sponsored by the Jesuits, made the Saint Francis Pledge in the fall of 2012 (Loyola University Chicago 2012). That decision built on an existing commitment to sustainability institutionally located in its Center for Urban Environmental Research and Policy, in particular. Under its current president, Michael Garanzini, SJ, and vice provost Nancy Tuchman, Loyola has made sustainability and carbon reduction an institutional mandate. It aspires to be a "net-zero" energy campus. Among other activities, Loyola supports a bike shop to promote green commuting, uses native landscaping to provide habitat for indigenous species, operates green roofs throughout the campus, and runs a farmer's market to support local produce. In addition, Loyola established a hundred-acre retreat and ecology campus with an organic garden and a native habit restoration project. With an ecologist on staff who also happens to be a Jesuit priest, the campus combines ecological awareness with spiritual renewal.

In addition to developing these initiatives on its own campus, Loyola is part of various networks of Jesuit institutions—those of the "Heartland Delta" region (the central U.S.), the Association of Jesuit Colleges and Universities (a partner of the Catholic Climate Covenant), and the more informal worldwide network of Jesuit secondary educational institutions. In February of 2012, Loyola hosted a conference for the Heartland Delta schools entitled "Jesuit, Catholic, and Green." One focus of the conference was the Jesuit Social Justice and Ecology Secretariat's document "Healing a Broken World" (Curia Societas Iesu and Alvarez 2011–12), which, incidentally, Loyola's own Nancy Tuchman helped to draft. Currently, President Garanzini serves as assistant for higher education to the superior general of the Jesuits in Rome. Tuchman's and Garanzini's efforts have placed Loyola in a position to contribute to the greening of Jesuit education worldwide.

Academics working in Catholic colleges and universities have engaged both the environmental justice and the creation spirituality frames, and they have done so from a range of other perspectives, as well. Some in the U.S. or the wider English-speaking world also have worked to supply Berry's cosmological ethic with a more robust Christological dimension (for example, Delio 2008, 2011; Edwards 1991, 1995, 2001; Galiardi 2008; Wessels 2000, 2003). By pushing the boundaries these scholars also made connections between the environmental justice and creation spirituality frames, opening possibilities for dialogue between their respective proponents. Catholic scholars writing specifically on climate change have produced a number of works elucidating the moral and theological dimensions of climate change, drawn on nearly 20 centuries of reflection on the relations between humanity, Creator, and the rest of creation (Miller 2010; Schaefer 2011).

In sum, Catholic religious orders and the institutions they sponsor in the United States have generated some of the most vigorous responses to climate change. They draw on symbolic resources from the creation spirituality and environmental justice frames, as well as their own traditions. Catholic higher education also serves as a site for building bridges between various frames, and its considerable institutional resources enable the integration of sustainability into the curriculum and campus life. Moreover, because of their connection with the larger Church and their national and international networks, Catholic religious orders and universities act as potentially significant catalysts for promoting a climate-friendly ethic in the wider Catholic community.

Concluding observations

When John Paul II highlighted environmental issues in his 1990 World Day of Peace address, the vast majority of U.S. Catholics, both traditionalist and progressive, did not view environmental issues as religious issues. In the two decades since, leaders in many sectors of the Church have worked hard to frame environmental concerns, including climate change, as a Catholic priority.

Even before the Pope spoke out in 1990, proponents of creation spirituality were summoning Catholics to an ecological conversion, and Catholic women religious in particular responded. The creation spirituality frame is highly prophetic in the way it calls for a reordering of priorities. Its insights resonate with many, Catholic and non-Catholic alike. However, at present it is not well integrated into the life of the mainstream Church.

More central to the Catholic ecological imaginary is a stewardship and environmental justice frame. It has achieved deeper institutional penetration and symbolic integration with standing Catholic commitments. In developing this frame, proponents have laid out a multifaceted response to Catholic naysayers who deny the reality of global climate change or its religious salience. Nonetheless, to date the bishops and priests who lead the many dioceses and parishes of the United States have not taken up the issue of climate change in a systematic and vigorous way. The limited sampling undertaken here suggests that in many Catholic dioceses and parishes "pastoral considerations" that defer to standing institutional priorities may modulate the efforts of proponents of the environmental justice frame to meet the challenge of climate change.

At the same time, the work of religious orders and the activities taking place in some of the universities they sponsor have been innovative and far-reaching. In some cases these developments have arisen quite rapidly, despite the often considerable investment of personnel and resources involved. Insofar as they bypass the dichotomy of prophetic mobilization versus institutional penetration, these developments suggest that Catholic universities may play an essential role in catalyzing a more adequate response to climate change in the U.S. Church.

Finally, the environmental justice frame, the creation spirituality frame, and the witness of Catholic religious orders all envision a response to climate change that goes beyond a technical quick fix. The very "diffuseness" of religious environmentalism links this response to personal and communal transformation. While U.S. Catholics often strongly identify as Americans, the countercultural norms of Catholic Social Teaching and the transnational ties of the Catholic Church also lend a degree of critical distance from the dominant culture. While current mobilization is modest, as more and more Americans search for an adequate response to climate change American Catholics have in a short time developed robust cultural and institutional resources to support this response.

Notes

1 Catholic Social Teaching is the body of Catholic doctrine that addresses social justice and the good of society.

2 In an early discussion of U.S. Christian environmentalism, Kearns (1996) argued that various U.S. Christian constituencies (mainly Protestant and Catholic) drew on three distinct frames, "Christian stewardship," "eco-justice," and "creation spirituality." As Shibley and Wiggins (1997) explain, in the early 1990s leaders within the U.S. Catholic Church developed a position that they termed simply "environmental justice," but which combined both stewardship and social justice perspectives. I use that designation here, but not in such a way as to imply that the creation spirituality frame does not include significant justice implications. In fact, creation spirituality is closely allied with liberation theology and maintains its own very significant justice concerns.

3 This term refers to the episcopal jurisdiction of the Catholic Church in Rome and the seat of the Pope. It is often called The Vatican, although this should really refer only

to Vatican City, a sovereign state within the boundaries of Rome that is geographical home to the Holy See.

4 See similar discussions in the World Council of Churches, of which the Roman Catholic Church is only an observer and not a member (Ch. 18 by Kerber, this volume).

5 For a list of Coalition members see "Who We Are," http://catholicclimatecovenant.org/about-us/coalition-members/.

6 The five parts are Pray, Learn, Assess, Act, and Advocate. "Toolkits" for implementing the Saint Francis Pledge may be found on the CCCC's website.

7 The full list of Covenant partners may be found at http://catholicclimatecovenant.org/about-us/covenant-partners/.

8 A recent analysis of religious environmental movement organizations (REMOs) brings out the extent to which religious affiliation and theological frames, and not simply achieving a particular environmental goal, shape their environmental activism (Ellingson, Woodley and Paik 2012). See also Chapter 6 by Amri (this volume) for a similar observation relative to Muslim pastoral organizations.

9 For example, just before election day 2012, Bishop Jenky of Peoria ordered his clergy to read from the pulpit a letter concerning Federal health policy. Sean Michael Winters (2012), a columnist for the *National Catholic Reporter*, offers some commentary.

10 When a pastor does take leadership in an area, he can have a significant impact. An example is the Rev. Charles Morris, a priest of the Archdiocese of Detroit. When pastor of St. Elizabeth's in Wyandotte, Michigan, Father Morris led efforts to make the parish energy efficient. In 2002 he founded Michigan Interfaith Power and Light, the Michigan affiliate of the same network to which Faith in Place belongs. He also has his own green blog on the *National Catholic Reporter* (http://ncronline.org/blog/8807).

11 Such dissent is not impossible. One famous Catholic climate dissenter was 2012 Republican presidential hopeful Rick Santorum. While not directly taking on the Pope or the bishops, neither did he share their conviction that mainstream climate science is credible (Lieberman 2012). Other, more explicit Catholics add that dissent on this issue is legitimate, because the bishops' teaching on climate change is not an obligatory article of the faith. Such dissent begs comparison with William F. Buckley, Jr., who in the mid-1980s similarly critiqued the U.S. bishops' pastoral letter on the economy.

12 Hope and Young (1999) provide a fuller discussion of Berry's thought and its context.

13 For an example of a brief critique of Berry from a traditional Catholic perspective, see the website of the Saint Kateri Tekakwitha Conservation Center, http://conservation.catholic.org/creation_spirituality.htm.

14 The Vincentians, more formally the Congregation of the Mission, are a Catholic order founded by Saint Vincent DePaul.

References

Agliardo, Michael. 2008. "Public Catholicism and Religious Pluralism in America: The Adaptation of a Religious Culture to the Circumstance of Diversity, and Its Implications." Ph.D. dissertation, Department of Sociology, University of California, San Diego.

——. 2011. "Restoring Creation and Redressing the Public Square: Religious Environmentalism and the Columbia River Pastoral." In *Green Discipleship: Catholic Theological Ethics and the Environment*, ed. Tobias L. Winright, pp. 37–59. Winona, MN: Anselm Academic.

Allitt, Patrick. 1998. "American Catholics and the Environment, 1960–1995." *Catholic Historical Review* 84, 263–280.

Benedict XVI. 2010. *If You Want to Cultivate Peace, Protect Creation.* January 1. Vatican City: Holy See. Available at http://www.vatican.va/holy_father/benedict_xvi/messages/

peace/documents/hf_ben-xvi_mes_20091208_xliii-world-day-peace_en.html (accessed March 7, 2012).

Benedict XVI and Woodeene Koenig-Bricker. 2009. *Ten Commandments for the Environment: Pope Benedict XVI Speaks Out for Creation and Justice*. Notre Dame, IN: Ave Maria Press.

Berry, Thomas. 1978. "The New Story." *Teilhard Studies* 1, 1–13.

——. 1987. "The New Story." *CrossCurrents* 37, 187–194.

——. 1988. *The Dream of the Earth*. San Francisco, CA: Sierra Club Books.

Berry, Thomas and Thomas Clarke. 1991. *Befriending the Earth: A Theology of Reconciliation between Humans and the Earth*. Mystic, CT: Twenty-Third Publications.

Berry, Thomas and Brian Swimme. 1992. *The Universe Story: From the Primordial Flaring Forth to the Ecozoic Era: A Celebration of the Unfolding of the Cosmos*. San Francisco, CA: Harper San Francisco.

Blair, Leonard P. 2012. "The LCWR, CDF and the Doctrinal Assessment." *Catholic Chronicle*, 8 June. Available at http://www.catholicchronicle.org/index.php/Statements/reality-check-the-lcwr-cdf-and-the-doctrinal-assessment.html (accessed October 11, 2012).

Bowman, Patty. 2012. Personal interview, June 28.

Braun, Gerald W., Monika K. Hellwig and W. Malcolm Byrnes. 2007. "Global Climate Change and Catholic Responsibility: Facts and Faith Response." *Journal of Catholic Social Thought* 4, 373–401.

Butterfield, Clare. 2012. Personal interview, June 26.

Casanova, José. 1994. *Public Religions in the Modern World*. Chicago, IL: University of Chicago Press.

Catholic Bishops' Conference of the Philippines. 1988. *What Is Happening to Our Beautiful Land?* January 29. Tagaytay: Catholic Bishops Conference of the Philippines. Available at http://www.aenet.org/haribon/bishops.htm (accessed January 22, 2008).

Catholic Bishops of Northern Italy. [1988] 1996. "Ecology: The Bishops of Lombardy Address the Community." In *And God Saw That It Was Good*, ed. Drew Christiansen and Walter Grazer, pp. 295–307. Washington, DC: United States Catholic Conference (USCC).

Catholic Coalition on Climate Change and Association of Catholic Colleges and Universities. 2011. "Sustainability and Catholic Higher Education: A Toolkit for Mission Integration." Available at http://catholicclimatecovenant.org/wp-content/uploads/2011/08/Sustainability-and-Catholic-Higher-Education-A-Toolkit-for-Mission-Integration.pdf (accessed October 11, 2012).

Center for Applied Research in the Apostolate. 2012. "Frequently Requested Church Statistics." Available at http://cara.georgetown.edu/caraservices/requestedchurchstats.html (accessed October 30, 2012).

Christiansen, Drew. 2002. Personal interview, July 10.

CNA Staff. 2007. "Vatican Announces Plans to Become First 'Carbon Neutral State' in the World." *Catholic News Agency*, July 13. Available at http://www.catholicnewsagency.com/news/vatican_announces_plans_to_become_first_carbon_neutral_state_in_the_world (accessed July 5, 2012).

Congregation for the Doctrine of the Faith (CDF). 2012. *Doctrinal Assessment of the Leadership Conference of Women Religious*. April 12. Available at http://www.usccb.org/loader.cfm?csModule=security/getfile&pageid=55544 (accessed October 11, 2012).

Curia Societas Iesu Task Force on Ecology and Patxi Alvarez. 2011–12. "Healing a Broken World." *Promotio Iustitiae* 106.

Delio, Ilia. 2003. *A Franciscan View of Creation: Learning to Live in a Sacramental World*. St. Bonaventure, NY: Franciscan Institute, St. Bonaventure University.

——. 2008. *Christ in Evolution*. Maryknoll, NY: Orbis Books.

——. 2011. *The Emergent Christ: Exploring the Meaning of Catholic in an Evolutionary Universe*. Maryknoll, NY: Orbis Books.

DePaul University. 2012. "Sustainability at DePaul." Available at http://distinctions.depaul.edu/columns/Pages/SustainabilityatDePaul.aspx (accessed July 16, 2012).

Diocese of Stockton. 2012. "Environmental Justice." Available at http://www.stockton-diocese.org/Pastoral-Ministries/Social-Ministry/Social-Issues/Environmental-Justice (accessed July 10, 2012).

Dubos, René. 1974. "Franciscan Conservation versus Benedictine Stewardship." In *Ecology and Religion in History*, ed. David Spring and Eileen Spring, pp. 114–136. New York: Harper & Row.

Edwards, Denis. 1991. *Jesus and the Cosmos*. New York: Paulist Press.

——. 1995. *Jesus the Wisdom of God: An Ecological Theology*. Maryknoll, NY: Orbis.

——. 2001. *Earth Revealing—Earth Healing: Ecology and Christian Theology*. Collegeville, MN: Liturgical Press.

Ellingson, Stephen, Vernon A. Woodley and Anthony Paik. 2012. "The Structure of Religious Environmentalism: Movement Organizations, Interorganizational Networks, and Collective Action." *Journal for the Scientific Study of Religion* 51, 266–285.

Fox, Matthew. 1996. *Confessions: The Making of a Postdenominational Priest*. San Francisco, CA: Harper San Francisco.

Galiardi, Margaret. 2008. *Where the Pure Water Flows: The New Story of the Universe and Christian Faith*. San Antonio, TX: Sor Juana Press.

Grazer, Walter. 2007. "Catholic Social Teaching and the Environment: Pastoral Challenge and Strategy." *Journal of Catholic Social Thought* 4, 211–225.

Guatemalan Bishops' Conference. [1988] 1996. "Cry of the Land." In *And God Saw That It Was Good*, ed. Drew Christiansen and Walter Grazer, pp. 275–293. Washington, DC: United States Catholic Conference (USCC).

Hackett, Conrad and Brian Grim. 2011. "Global Christianity: A Report on the Size and Distribution of the World's Christian Population." Washington, DC: Pew Research Center. Available at http://pewforum.org/Christian/Global-Christianity-worlds-christian-population.aspx (accessed November 9, 2012).

Hamlin, Christopher and John T. McGreevy. 2006. "The Greening of America, Catholic Style, 1930–1950." *Environmental History* 11, 464–499.

Hope, Marjorie and James Young. 1999. *Voices of Hope in the Struggle to Save the Planet*. New York: Apex Press.

Howard, Tom. 2012. Personal interview, June 22.

John Paul II. 1990. *Peace with God the Creator, Peace with All of Creation*. January 1. Rome: Holy See. Available at http://www.vatican.va/holy_father/john_paul_ii/messages/peace/documents/hf_jp-ii_mes_19891208_xxiii-world-day-for-peace_en.html (accessed July 22, 2007).

Kearns, Laurel. 1996. "Saving the Creation: Christian Environmentalism in the United States." *Sociology of Religion* 57, 55–70.

Keenan, Marjorie. 2002. *From Stockholm to Johannesburg: An Historical Overview of the Concern of the Holy See for the Environment*. Vatican City: Vatican Press.

Lieberman, Bruce. 2012. "The Catholic Church and Climate Change." In *Climate and Major Religions: Yale Forum on Climate Change and the Media*. Available at http://www.yaleclimatemediaforum.org/2012/02/the-catholic-church-and-climate-change/ (accessed July 5, 2012).

Loyola University Chicago, University Newsroom. 2012. "Loyola University Chicago

Signs International, National, State, and Faith-Based Commitments to Sustainability." Available at http://blogs.luc.edu/pressreleases/2012/09/04/loyola-university-chicago-signs-international-national-state-and-faith-based-commitments-to-sustainability/ (accessed October 11, 2012).

Lucas, Phillip Charles. 1995. *The Odyssey of a New Religion: The Holy Order of Mans from New Age to Orthodoxy*. Bloomington: Indiana University Press.

MacKenzie, Terri. 2012. Personal interview, June 27.

Miller, Richard W., ed. 2010. *God, Creation, and Climate Change: A Catholic Response to the Environmental Crisis*. Maryknoll, NY: Orbis Books.

Misleh, Dan. 2011. Personal interview, August 5.

Moreno, Antonio F. 2006. *Church, State, and Civil Society in Postauthoritarian Philippines: Narratives of Engaged Citizenship*. Quezon City: Ateneo de Manila University Press.

Muratore, Stephen. 1987. "Ecumenical Ecology: The North American Conference on Christianity and Ecology." *Epiphany* 7, 66–69.

——. 1988. "The New 'Teilhard' at the Nacce." *Epiphany* 8, 6–14.

NCCB (National Conference of Catholic Bishops). 1991. *Renewing the Earth: An Invitation to Reflection and Action on the Environment in Light of Catholic Social Teaching*. November 14. Washington, DC: United States Catholic Conference (USCC). Available at http://www.usccb.org/sdwp/ejp/bishopsstatement.shtml (accessed September 7, 2006).

Official Catholic Directory, The. 2012. New York: P.J. Kennedy & Sons.

Potosnak, Mark. 2012. Personal interview, July 5.

Scarel, Eduardo Agosta. 2012. "A Position of the Carmelite NGO on Climate Change." Carmelite Order. Available at http://ocarm.org/en/content/ocarm/position-carmelite-ngo-climate-change (accessed July 5, 2012).

Schaefer, Jame, ed. 2011. *Confronting the Climate Crisis: Catholic Theological Perspectives*. Milwaukee, WI: Marquette University Press.

Sharma, Lucas. 2012. "From Individual Conscience to Parish Culture: A Study of Two Parishes in the Archdiocese of Chicago." MA thesis, Department of Sociology, Loyola University Chicago, Chicago.

Shibley, Mark A. and Jonathon L. Wiggins. 1997. "The Greening of Mainline American Religion: A Sociological Analysis of the Environmental Ethics of the National Religious Partnership for the Environment." *Social Compass* 44, 333–348.

Taylor, Sarah McFarland. 2007. *Green Sisters: A Spiritual Ecology*. Cambridge, MA: Harvard University Press.

Thavis, John. 2011. "Pope Urges International Agreement on Climate Change." *Catholic News Service*, November 28. Available at http://www.catholicnews.com/data/stories/cns/1104646.htm (accessed July 6, 2012).

USCCB (United States Conference of Catholic Bishops). 2001. *Global Climate Change: A Plea for Dialogue, Prudence, and the Common Good*. Washington, DC: USCCB. Available at http://www.usccb.org/issues-and-action/human-life-and-dignity/environment/global-climate-change-a-plea-for-dialogue-prudence-and-the-common-good.cfm (accessed March 3, 2012).

——. 2012. "Themes of Catholic Social Teaching." Available at http://www.usccb.org/beliefs-and-teachings/what-we-believe/catholic-social-teaching/themes-of-catholic-social-teaching.cfm (accessed July 2, 2012).

University of Notre Dame, Office of Sustainability. 2009. "Renewing the Campus." Available at http://green.nd.edu/education/renewing-the-campus/ (accessed July 5, 2012).

Wessels, Cletus. 2000. *The Holy Web: Church and the New Universe Story*. Maryknoll, NY: Orbis.

——. 2003. *Jesus in the New Universe Story*. Maryknoll, NY: Orbis.

Winters, Michael Sean. 2012. "Bp. Jenky Goes Off the Reservation." *National Catholic Reporter*, November 1. Available at http://ncronline.org/print/blogs/distinctly-catholic/bp-jenky-goes-reservation (accessed November 7, 2012).

Witte, Raymond Philip. 1948. *Twenty-Five Years of Crusading: A History of the National Catholic Rural Life Conference*. Des Moines, IA: National Catholic Rural Life Conference.

Wulff, Deanna Lynn. 2011. "God, the Earth and Ozone." *Bilingual Weekly*, December 11. Available at http://bwnews.us/2011/12/01/god-the-earth-and-ozone-stocktons-diocese-pushes-for-clean-air-pairing-environmental-advocacy-with-faith/ (accessed July 10, 2012).

13 "How many Presbyterians does it take to change a light bulb?"

Confronting global climate change in the Presbyterian Church, U.S.A.

Patricia K. Townsend

Among the many light bulb jokes that satirize common stereotypes of ethnic, cultural, and occupational groups, a subset aims at the foibles of religious denominations. The Presbyterian variant of the light bulb joke ran something like this:

> "How many Presbyterians does it take to change a light bulb?"
>
> "Presbyterians? Change?"

Alternatively, the web page of the Texas state affiliate of Interfaith Power and Light attributed the unwillingness to change to Lutherans. They suggested that for Presbyterians to change a light bulb it should require about five committees to review the idea first. If each is staffed with half a dozen members, that's what . . . 30?[1]

A review of the development of environmental and energy policies of the Presbyterian Church (U.S.A.) over the past half-century suggests that the second version of the joke was closer to accurate: The PC(USA)[2] stood at the cutting edge of developing policies related to climate change, despite the denomination's unpromising demographics and political leanings. In a 2008 survey, nearly half of the laity reported their political preference as Republican, 20 percent as independent, and 30 percent as Democrat.[3] The first question posed in this chapter is: How did this leadership role emerge?

Secondly, of what significance are such policy statements of a religious denomination anyway? Presbyterians themselves commonly denigrate their potential contribution when they whine, "Who cares what a declining mainline denomination thinks?" A later section of the chapter will explore the ways in which such policies are translated into behavioral changes that may have a cascading influence throughout local communities.

Who are the Presbyterians?

An ecumenical denomination

The mainline denominations are sometimes referred to as "ecumenical" churches, with reference to their participation in the National Council of Churches (NCC), the World Council of Churches (WCC), and various interfaith councils. With 2.7 million members, the PC(USA) is third largest of the ecumenical denominations behind the United Methodist Church (UMC, 7.7 million) members and the Evangelical Lutheran Church in America (ELCA, 4.3 million) (Lindner 2012). The ecumenical denominations have all been declining in membership over the past several decades, and the average worshipper is therefore older. Presbyterians are predominantly white and non-Hispanic and are above average in income. They are highly educated: 66 percent of members and elders have a college or graduate degree (more than twice the percentage in the US population), and 100 percent of clergy have a graduate degree (Research Services, PC(USA) 2009).[4]

A denomination in the Reformed tradition

To make sense of the differing positions on social issues that are taken by churches, it is necessary to understand some of the ideological diversity within Christianity. The PC(USA) is best described as theologically moderate. Scripture is important to Presbyterians, but few would argue for a literal interpretation of the Bible.

The first part of the PC(USA) Constitution is the Book of Confessions (PC(USA) 1996), containing 11 historic creeds, confessions, and catechisms ranging from the fourth through the twentieth centuries, including six from the Reformation era of sixteenth- and seventeenth-century Europe. Upon ordination, ministers and elders promise to be "guided by" the confessions, rather than subscribing to every detail. The Constitution identifies certain theological themes as being distinctive to the Reformed tradition. These are an emphasis on the sovereignty of God, the election of God's people and their call to work for the transformation of society, covenant life in the church, and "a faithful stewardship that shuns ostentation and seeks proper use of the gifts of God's creation" (PC(USA) 2004: G-2.0500).

A denomination with a representative form of government

Presbyterians are named for their form of polity, "presbyter" being derived from the word in biblical Greek for an elder. The ruling elders are elected by members of the congregation to serve on the Session, which then makes almost all decisions on behalf of the congregation. While the elder is elected to the Session for a limited term as an active ruling elder, he or (since 1930) she is ordained to this office for life. Ruling elders and teaching elders (ministers) together serve on other ruling bodies of the church: the regional presbyteries and synods, and the national general assemblies of the PC(USA) that met annually from 1789 through 2004 and since then biennially. This bottom-up polity of Reformed churches is distinct from episcopal polity (a hierarchy of bishops) and from congregational polity. Reformed polity can give lay leaders a strong voice on issues about which they are well informed and passionate.

The General Assembly (GA) elects its various standing committees, including the Advisory Committee on Social Witness Policy (ACSWP), a committee of 12 that is responsible for the process of developing and recommending social witness policy to the GA for its vote. The term "social witness policy" refers to issues in the public order. The recommendations in a policy statement may include a call to specific actions on the part of individual members, local congregations, seminaries, other institutions, and the denomination's staff.

The history of energy policy in the church

Church bodies, like governments, attend to energy policy in cycles initiated by crises in the supply of oil. There were two such crises in the 1970s—the OPEC oil embargo of 1973 and the Iranian revolution of 1979. These events bracket the period during which the National Council of Churches of Christ in the USA (NCC)[5] conducted its deliberations on energy policy.

In 1974 the NCC invited Margaret Mead and René Dubos to co-chair a special committee on the use of plutonium as nuclear fuel. Dieter Hessel, a Presbyterian minister, chaired the committee that received their report.[6] Before it could vote on the proposed policy, which condemned the commercial use of plutonium, the Council heard an unprecedented outpouring of opinion on both sides of the issue. Because the strip mining of coal was another controversial issue of the time, the NCC mandated a broader study of the ethical implications of energy production and use.

Following prolonged controversy within the NCC, in 1979 the governing board adopted an 11-page document entitled "The Ethical Implications of Energy Production and Use" (cited in Hessel 1979), and sent it to the churches for their study. It developed the concept of "an ecologically just society, one based on the values of sustainability, equity, and participation." Despite the opposition the NCC had heard from the laity employed in the energy industries, the study made clear that coal and nuclear power were high-risk technologies. Long before climate change had become an important public policy issue, the document also recognized the risk it posed: "Burning more coal could also result in less certain but irreversible climatic damage from increased carbon dioxide in the atmosphere. Changes in climate can turn agricultural land into deserts" (Hessel 1979: 123).

The 1981 Presbyterian energy policy

The Presbyterian energy policy that was developed in response to the NCC call for study was titled "The Power to Speak Truth to Power." The consultant employed to draft the background paper was Robert L. Stivers, a Presbyterian minister teaching ethics at Pacific Lutheran University.[7] The document declared, "The era of cheap and abundant energy is over," and called for Christians to "live frugal lives of energy sufficiency," focusing especially on economic justice and the need to transfer resources to poor countries (Presbyterian Church and United Presbyterian Church 1981). There was no mention of global climate change as one of the risks of energy choices. Like the NCC report, the Presbyterian document asserted the educational role of the churches in communicating a theological and ethical framework for evaluating energy issues.

The 1990 PC(USA) environmental policy

The next major environmental action of the denomination occurred after the 199th General Assembly in 1987 received overtures that led to the formation of the Task Force on Eco-Justice. The final report of the Task Force, *Restoring Creation for Ecology and Justice*, was adopted by the 202nd General Assembly (PC(USA) 1990). The Task Force on Eco-Justice was chaired by Robert Stivers, author of the background paper for the 1981 energy policy document. Another member of the Task Force on Eco-Justice was Holmes Rolston III, whose article "Is there an Ecological Ethic?" helped launch the discipline of environmental ethics (1975). The consultant and writer for the Task Force was William Gibson (2004), a campus minister at Cornell University. The broad ethical norms that underlay the environmental policy were sustainability, participation, sufficiency, and solidarity. The final report to the General Assembly made detailed policy recommendations in five general areas of concern:

- sustainable agriculture;
- water quality;
- wildlife and wildlands;
- reducing and managing solid and hazardous wastes;
- global warming and ozone depletion.

Of the issues dealt with in the 1990 policy, concerns about climate change had a leading voice in Gibson, and he continued to advocate in this policy area until his death in 2006. The 1990 document made ten specific recommendations about climate policy, which ranged from encouraging U.S. participation in firm international agreements for reducing greenhouse gases, to stronger fuel economy standards for vehicles, to technology transfer for developing countries to achieve energy sufficiency. In the following years, other denominations produced their own environmental policies, though the Presbyterians led the way in concern for climate change. The issue was kept before the denomination, as PC(USA) General Assembly meetings in 1998, 1999, and 2003 voted to support U.S. ratification of the Kyoto Treaty.

Revisiting PC(USA) energy policy in the twenty-first century

In 2002, the Presbytery of Susquehanna Valley (Gibson's presbytery) presented an overture to the 214th General Assembly that asked for review of the denomination's 1981 energy policy. Within the denomination, awareness of global climate change was increasing. In 1997, global warming had barely registered on the list, the top two being air pollution and destruction of the rainforest, followed by destruction of natural areas and destruction of the ozone layer. But, by May 2009, members, elders, and clergy ranked oil and gas consumption and global climate change as the two most serious environmental problems of the time.

Pamela McVety was chosen to chair the Energy Resolution Team formed to work on the revision. She was a biologist and ecologist who had held high-level positions in the Florida Department of Environmental Conservation and served as restoring creation enabler (see below) for the Presbytery of Florida. Another member of the Energy Team with strong environmental credentials in state government was Jananne Sharpless, who had chaired the California Air Resources Board from 1985 through 1993, pressing for cleaner fuels and electric cars. This work led to her being prominently interviewed in the film *Who Killed the Electric Car?* (Paine 2006). Another influential member of the Energy Team was John Topping, director of the Office of Air and Radiation of the United States Environmental Protection Agency under the Reagan administration and founder in 1986 of the Climate Institute in Washington, DC, the first environmental organization dedicated to climate issues.

The ten-member Energy Resolution Team differed from the task forces that produced the 1981 and 1990 policies in ways that ultimately proved significant for the acceptance of its report. The team had strong lay leaders who were informed, idealistic, and committed to the urgency of action, but it had no members who were ministers or theologians. It also lacked strong support from denominational staff, coming at a time of financial stringency and after a controversial ACSWP trip to the Middle East had resulted in the firing of relevant staff.

The team met three times to work together for several days during 2004 and 2005. As committee members felt that existing Presbyterian policies were adequate but that Presbyterians, like other Americans, had not implemented serious changes in energy use, they determined to try a different approach. The team decided to present a bold challenge to carbon neutrality.

The Energy Team's 2005 report to the ACSWP, "A Christian Witness on Energy," was deliberately brief, at 2,700 words, omitting the extensive biblical, theological, and ethical background of earlier policy documents and cutting quickly to recommendations for changes in public policy and individual behavior. This witness would speak powerfully to one's neighbors about one's personal commitment to preventing devastating climate change. In addition to breaking with the tradition of lengthy scholarly study papers, the report also marked a shift away from advocating structural solutions in favor of more personal ones. The leaders of the Energy Resolution Team had decided to focus on personal responsibility rather than to engage with issues that were potentially divisive among Presbyterians in the pews, such as the energy impact of the Iraq war or the failings of the Bush administration with respect to energy.

After the chair presented the report, it was rejected by the ACSWP.[8] The ACSWP members and staff I interviewed indicated that the committee could not entertain a policy paper that made no reference to the vast energy costs expended by the military, against which the energy conservation efforts of a tiny minority of American households seemed to pale. It was also clear from interviews that the ACSWP's own energies at the time were depleted by their committee's grappling with Middle East issues.

The 2006 Carbon Neutral Resolution

In order to save something from her team's work, the elected GA moderator encouraged the chair to present a commissioner's resolution to the 2006 General

Assembly making the case for the urgency for Presbyterians to become carbon neutral. Concerned that commissioners might not know the meaning of carbon neutrality, the team arranged to hand out fans (anticipating the summer heat of the Birmingham meeting site) with a concise definition: "Becoming 'carbon neutral' means that we have neutralized the effect of our personal greenhouse gas (carbon) emissions, so that our personal and household activities no longer contribute to the dangers of global warming." The message on the fan suggested that one could become carbon neutral by taking three steps:

1 We calculate our carbon emissions.
2 We reduce our carbon emissions to the world average of 4 tons per person per year by saving energy.
3 We aspire to offset these remaining carbon emissions to zero by purchasing carbon offsets.

Radical as it was in environmental terms and in its expectation of individual sacrifice, the Carbon Neutral Resolution received almost no publicity beyond its brief announcement by the denomination. The website of the Climate Institute, Topping's organization, did give extensive coverage to the GA action—but it was almost the only website to do so.[9]

Not until the following year was there a mention of the Presbyterian action in the media, and it came in the form of a sarcastic article about carbon offsets in *USA Today* (Schweizer 2007). "Everyone from Al Gore to the Presbyterian Church" is pushing carbon offsets to escape the moral quandary of not wanting to change one's lifestyle despite the belief that it is having dire effects on the climate. Parodying the concept, the author suggested "adultery offsets" and opportunities to become "fat neutral" without actually dieting or exercising, failing to notice that the Presbyterian resolution asked individuals to reduce their own carbon emissions as much as possible and purchase offsets for what remained, not to purchase indulgences but to increase investment in renewable energy and carbon sinks. Newspaper coverage of the mainstream denominations in those years was almost entirely directed toward the gay ordination and gay marriage controversies.

The 2008 energy policy

The ACSWP's new staff engaged James Martin-Schramm as a consultant to write the much longer report that it sent to the 2008 General Assembly. Martin-Schramm, a professor of religion at Luther College in Iowa, had studied ethics with Stivers as an undergraduate before doing his Ph.D. in social ethics at Union Seminary. Though Martin-Schramm was a Lutheran, the ELCA and PC(USA) had a long history of collaboration on environmental concerns.[10]

The revised energy study paper bore the title "The Power to Change," echoing the 1981 report title. Its understanding of climate change issues and recommendations did not differ greatly from those the Energy Team had proposed in 2005, but the report was more polished and echoed the ethical foundation that had been the hallmark of Presbyterian study papers. It also included a few paragraphs about the Iraq war, carefully worded to sidestep political minefields. The new report passed easily in the 2008 General Assembly (PC(USA) 2008). Despite the long and bumpy road from the 2001 request for revision to final adoption of the policy and the discouragement of participants, from an environmentalist perspective the outcome was favorable: The urgency of action on climate change was put before Presbyterians in 2006 and the polished materials for study by congregations and for advocacy came along in the election year of 2008, as the culmination of more than 20 years of denominational leadership on climate change. This leads us to examine what in Presbyterian culture underlies the history of forming policy on this issue.

The greening of Presbyterians in the late twentieth century

Lynn White, Jr., a Presbyterian elder and historian of technology, accelerated the "greening" of American theology by arguing in an influential paper that Christianity had set in motion the forces that led to today's environmental problems (1967). The bold (and over-simplified) thesis became widely popular among environmentalists, despite the extensive critique it drew from historians and theologians alike. Stimulating this debate was surely White's intention, for he once remarked, "It is better for a historian to be wrong than to be timid" (Whitney 1993: 168).

In the post-White flourishing of green theology across the spectrum of denominations, the historical, theological, and biblical response to White's argument came mostly from non-Presbyterians. Among the Presbyterians, the leading environmental thinkers among the clergy of the late twentieth century were not primarily biblical scholars or systematic theologians; instead they were ethicists. This network included Dieter Hessel, William Gibson, Robert Stivers, Holmes Rolston, Richard Cartwright Austin, and others. They drew on the social ethics of Reinhold Niebuhr (1892–1970), who had taught at Union Seminary in New York (Dorrien 2009). They found that Niebuhr's critique of the injustice spawned by early-twentieth-century capitalism was newly relevant to the injustices and environmental sins of late-twentieth-century capitalism, damaging the health of humans and whole ecosystems through strip mining, nuclear and industrial pollution, and limitless greed that stole from future generations and the world's poor.

In the same year that White's essay was published, Presbyterians adopted a new confession rooted in social ethics. In the *Confession of 1967* (PC(USA) 2004, Part I: 259–270) they held up the theme of reconciliation, which reflected their deep engagement with the civil rights movement of the era. The final section of the document, "The Fulfillment of Reconciliation," looked forward to God's redemption of the whole of human life, which in their view "includes the natural environment as exploited and despoiled by sin."

If Presbyterians had not been environmentalists even before Lynn White's 1967 lecture and Earth Day 1970, it could be argued that they were culturally pre-conditioned by their distinctive theology and worship to develop "a greener faith" (Gottlieb 2006). Presbyterian doctrines, if adhered to, contradict certain American cultural themes that are hostile to environmentalism. An emphasis on the sovereignty of God may counter an anthropocentrism that justifies the exploitation of the rest of God's creation. The theme of covenant life counters American individualism. The exhortation to stewardship that "shuns ostentation" points congregants toward frugal living. The belief that God's people are elected for service and to work for the transformation of society encourages active engagement in public life. All these Reformed doctrines are friendly to acceptance of the denomination's environmental and energy policies.[11]

Ritual, which both reflects and shapes theology, may also play a role in Presbyterianism's friendliness to environmental concern. Only two sacraments are recognized: the Lord's Supper and baptism. The liturgies for both emphasize the theology of covenant, a recognition of community that counters American individualism. This is explicitly recognized in infant baptism, in which the congregation promises to nurture the (helpless) child, and in the concept of communion as heavenly banquet. This emphasis on community may encourage environmental concern and activism, since environmental problems by their nature often require collective responses and may require individuals to make sacrifices for the common good.

Worship in the Reformed tradition differs only slightly from worship in other Christian denominations in that Reformed worship centers on the reading and preaching of the Scriptures.[12] The strong focus on preaching reflects the high value placed on an intellectual approach to the faith, which in turn is related to the high level of education that is observed in the pews. Within the New Testament, the preaching of the Gospels is central, but Presbyterians give more attention than do most Protestants to the Old Testament. In particular, they emphasize reading the Old Testament in worship and study and reading and singing psalms. This form of worship gives Reformed Christians a lifelong immersion in the themes of creation and land that run through the Hebrew Scriptures. These Old Testament passages, familiar to Presbyterians, are often cited in both popular and scholarly writings on environmental theology and spirituality (e.g. Hessel 1996; Hoezee 1998).

In addition to this Old Testament heritage, Reformed theology has a view of the end times that stands in contradiction to Rapture theology, also known as premillennial dispensationalism. In this view, which developed in nineteenth-century England and was popularized by American television preachers and the "Left Behind" series of books and movies (e.g. LaHaye and Jenkins 1995), Christians will be taken off the earth prior to its destruction, which many premillennialists argue is imminent. Scholars have suggested that with this theology there is little need for environmental stewardship (Curry 2008; Curry-Roper 1990).

In contrast, Reformed theology affirms that God's people are called to work for the transformation of society. In doing so, they share with Christ in establishing the kingdom that Jesus inaugurated—God's rule on earth. This reading of the Scriptures affirms that God intends to redeem and heal the whole of creation rather than to snatch believers away in the Rapture. Finally, heaven will come down to a renewed earth, according to a Reformed reading of the last chapters of the Bible (Davis 2011; Johnston 1997; Rossing 2004). This reading of the biblical texts informs both the content of the denomination's policies and its activism in implementation of those policies.

Implementing Presbyterian policies

Implementation at national level

The most direct channel for influencing Congress is the advocacy effort of the Presbyterian Washington office on Capitol Hill, which is supported by the letter-writing campaigns and Congressional office visits by church members. Staff are directed to advocate only for policies as directed by the General Assembly. They keep in touch with interested Presbyterians through newsletters and e-mails and are arguably more effective than secular environmental organizations in mobilizing letter-writing campaigns (Moody 2002).

It seems unlikely that the staff lobbying or citizen advocacy effects of denominational policies can constitute anything like a counterweight to the massive political contributions, lobbying, and media presence of petrochemical industry-financed climate skepticism. Yet sociologists have concluded that the mainline denominations have been "quietly influential" in Washington (Wuthnow and Evans 2002). The fact that the denominations have collaborated ecumenically helps them to maintain their outreach even as staff and finances shrink.

The small PC(USA) Washington staff has included persons with expertise and interest in climate change legislation. Presbyterians participate actively in annual Ecumenical Advocacy Days for Global Peace with Justice. These conferences offer a long weekend of training for several hundred attendees, culminating in group visits to members of Congress. The overall theme was climate change in the March 13–16, 2009 conference, but some attention has been given to climate in workshops in other years as well.

Presbyterian policies may have their strongest national impact through organizations that promote socially responsible investment. PC(USA) endowment and pension funds are substantial, and the denomination's participation in the movement for socially responsible investment has been significant. The office of Mission Responsibility through Investment (MRTI) has been active since the late 1960s and played a key role in the divestment movement in South Africa in the 1980s, a movement widely credited as a major factor in the successful end of apartheid.

Some of MRTI's proposals are bound to stir controversy both within and outside the denomination, as did the proposal to divest from several companies listed as contributing to Middle Eastern conflict. In 2004, when phased divestment in Caterpillar (manufacturer of equipment used by Israel to wall out Palestinians) was first proposed to the GA, PC(USA)'s total investment portfolio of approximately $8 billion in retirement and endowment funds provided it with "substantial leverage" for negotiation with companies (Clarke 2005: 47).

Presbyterians participate in both the Interfaith Center for Corporate Responsibility (ICCR) and Ceres. An organization dedicated to fostering sustainable investment, Ceres was formed in 1989 after the *Exxon Valdez* disaster to enlist the voluntary compliance of corporations in a set of principles for corporate environmental responsibility. Bill Somplatsky-Jarman of MRTI was the first clergyperson on the Ceres board of directors. He represented the PC(USA) at the 1992 Rio Conference and has attended nearly all of the United Nations climate conferences since then, including COP17, the 2011 Conference in Durban. At COP8, held in Delhi, India in 2002, he delivered the WCC statement to the plenary, "A Call to Action in Solidarity with Victims of Climate Change" (WCC 2010). He has successfully negotiated with owners of power plants for more environmentally responsible policies through shareholder resolutions and discussion without having to resort to divestment.

Restoring Creation for Ecology and Justice, the environmental policy adopted in 1990, called for an infusion of eco-justice concerns throughout the church's existing programs. The report also resulted in the establishment of new staff positions dedicated to environmental stewardship. This meant an expansion of the Washington office (for advocacy) and in the national office, in Louisville (for a program to increase the environmental engagement of congregations).

The Environmental Justice office at the Louisville headquarters developed, trained, supported, and coordinated a national network of restoring creation enablers (RCEs). By 2000 there was at least one RCE serving in 66 of the PC(USA)'s 173 regional presbyteries, the average presbytery containing 60-some congregations.[13]

The irony behind the enthusiastic work of the Energy Team in 2005 was that, just as they were completing their report, the denomination was struggling with finances and the dramatic downsizing of staff. In the spring of 2006, cuts in the proposed budgets for 2007 and 2008 completely eliminated staff in Environmental Justice, the office that was responsible for implementing a new energy policy. The grassroots outcry that resulted from the cut led the 2008 General Assembly to vote to reestablish the office in 2009, but momentum for implementation of energy issues had been lost. Nonetheless, staff paid through other funding streams were able to continue work related to climate change. For example, funding for Presbyterian work on hunger continued to flow in from special offerings, enabling a staff person to prepare materials on the impact of global climate change on agriculture.

Another vehicle for implementation of environmental policy was the grassroots organization Presbyterians for Restoring Creation (PRC), which was organized in 1995.[14] PRC held national conferences every two years, published a newsletter, and employed a part-time staff member. From 2000 until 2007, this staff member was Rebecca Barnes-Davies, whose PRC work led her into writing a book, *50 Ways to Help Save the Earth: How You and Your Church Can Make a Difference* (2009), which gracefully combined education on global climate change, advocacy, spirituality, and lifestyle change.

Over a period of several years, the educational focus of the PRC conferences showed an increasing emphasis on climate change. The PRC conference "Earth's

Energy, God's Light," held in June 2002 at Linfield College in Oregon, was almost entirely dedicated to renewable energy. The conferences were also the occasion of recruiting and training events for RCEs; hence they had educational impact on the wider church beyond the few hundred attending each conference.

Implementation at congregational level

The broadest impact of Presbyterian policy statements is on the people in the pews. Adult education in various forms, whether in Sunday morning classes or weekday and evening discussion groups, makes use of materials developed by the denomination or its ecumenical partners to discuss current issues in the light of faith.

"Ethics is most productive when it is conducted by a community of moral deliberation," writes Martin-Schramm (2010: xvii), in a book based, in part, on his work on the Presbyterian energy policy statement. The committees, boards, Sunday school classes, and men's and women's fellowship groups in their religious congregation are the "community of moral deliberation" in which discussions of social ethics take place for many Americans. Hence, the light bulb joke with which I began this chapter was not far from the mark, highlighting the propensity of Presbyterians to form study committees. The many hundreds of local, regional, and national Presbyterian committees and classes that have worked through their understandings of a Christian social ethic of environment and energy are at the heart of this case study of a mainline denomination.

Discussions of energy arise not only in an intellectual context but in the practical context of buildings and grounds committees struggling with how to manage church properties responsibly. All of the Presbyterian congregations with which I am familiar have volunteers attempting to improve energy efficiency in order to reduce expenses, making it only a short step for them to consider other reasons for energy stewardship.

After McVety completed her work as chair of the Energy Team in 2006, her home congregation, First Presbyterian of Tallahassee, became a model for action on the carbon neutrality resolution. They implemented carbon neutrality through the installation of photovoltaic panels, while engaging in outreach to their community by installing energy-saving home improvements for poor families. A cursory internet search conducted in 2012 indicates that Presbyterian churches in at least 12 states have installed photovoltaic systems. Still others have installed solar water heating or geothermal heating and cooling systems or purchased green power.

In another move toward implementation, the denomination began in 2006 to purchase renewable energy credits from NativeEnergy to offset the carbon emissions of the hotel accommodations and meeting spaces for its General Assembly meetings and urged attendees to do the same. The idea of buying carbon offsets from an outside company in order to become carbon neutral has not proven attractive to most congregations, struggling as they are simply to pay the utility bills on aging buildings. The suggestion that a congregation help needy people attain energy sufficiency is another matter, however; it is a charitable outreach ministry that can be embraced even by congregations whose members do not consider themselves environmentalists.

Many Presbyterian congregations have volunteers retrofitting homes or building new energy-efficient houses through Habitat for Humanity. Providing solar power for an isolated clinic or school in Haiti or Kenya is a mission project that appeals to congregations who might not consider installing solar panels for their own building. A project to install small photovoltaic systems on volunteer mission trips was started in 2009 by the Synod of the Sun, consisting of all the PC(USA) churches in the region that includes Texas, Oklahoma, Louisiana, and Arkansas (the heart of the petrochemical industry). It was patterned after an existing project of the neighboring Synod of Living Waters, which installs water systems on mission trips. A solar school trains volunteers in the appropriate technology.[15]

Conclusions

By the mid-2000s, the PC(USA), other ecumenical denominations, and even many non-denominational evangelicals were prepared to break with conservative politicians on climate change and torture, two issues in which Christian social ethics are profoundly at stake.[16] Climate and energy issues were relatively conflict-free among Presbyterians, especially in comparison to two highly charged issues (divestment from certain American firms doing business with Israel's government and the ordination of gay men and lesbians) that were being debated in the same decade. Congregations were far more comfortable discussing environmental ethics by comparison.

How did it come to be that the PC(USA), despite its politically conservative membership, became an early advocate concerning human-caused climate change? This chapter outlined several possible reasons. We identified a small circle of respected social ethicists who supported the declaration of human-caused climate change as environmental sin very early and stayed with the issue over long careers. Their persistence would not have paid off in acceptance in pulpit and pew had it not resonated with a Reformed theology that was hospitable to an ethic of creation care. This theology was well able to accommodate the idea of climate justice, the obligation owed to the poor and vulnerable and to later generations by those who have produced and benefited from the burning of greenhouse gases. Church members who retained doubts about climate science could accept energy conservation and renewable energy on the basis of reduced pollution and the frugal use of finite resources.

Perhaps more importantly, how effective has this framing been at inspiring change at the grassroots level? As we have seen, with the Presbyterian form of government, it is only possible to formulate national denominational policies when they resonate with grassroots concerns among ministers and lay leaders, who vote on the policies. In such cases, the study materials provided by the national denomination (both during and after the policy adoption process) can encourage continuing and deepening the conversations that take place at the local level.

These conversations can help other participants come to greater clarity about their ethical basis for difficult decisions. It may persuade them to take action on matters on which they were previously neutral or undecided. Indeed, I have observed Presbyterians who had been on the fence about climate change as a political issue express relief when given the opportunity to develop an ethical basis for taking a position on energy issues. Yet I do remain skeptical that policies formulated at the level of the national denomination have much effect in changing beliefs that are already strongly formulated.

Still, Presbyterians may be contributing something vital to the conversation about climate. It has become increasingly clear that the framing of climate change as an issue of science and politics has been both polarizing and inadequate in the United States. In the past few years new campaigns of climate change activism have begun to use the language of ethics and values. As this chapter has shown, the PC(USA) is far ahead of secular environmentalists when it comes to the use of this framing. Implementation of climate policy through energy conservation and renewable energy also fit into the practical realities of congregational life. Where, outside of religious congregations, are Americans more likely to gather in local voluntary groups to manage a building as they do their houses of worship? The principles by which they decide to do this may spread along their social networks into other congregations, homes, and businesses.

Acknowledgments

In the interest of full disclosure, I am an elder at North Presbyterian Church, Williamsville, NY, and a board member of New York State Interfaith Power and Light. An ecological anthropologist, I had no interest in doing research on religion until in the 1990s I discovered the activist role that mainstream Christians played at the mining and Superfund sites where I was then doing research. I am grateful for the generous telephone conversations and e-mail correspondence with present and past Presbyterian committee and staff members contacted for this research.

Notes

1 Interfaith Power and Light is a religious response to climate change with affiliates in 37 states (http://interfaithpowerandlight.org/about/mission-history/, accessed November 24, 2011). The reference for the light bulb jokes is http://txipl.org/lightbulbjokes (accessed August 13, 2011).

2 In addition to the PC(USA), there are several smaller, theologically more conservative Presbyterian denominations in the United States, including the Presbyterian Church in America, the Orthodox Presbyterian Church, and the Evangelical Presbyterian Church. The PC(USA) was formed from the union in 1983 of the northern (UPCUSA) and southern (PCUS) churches, which had been split since events surrounding the Civil War.

3 Presbyterian clergy are much more likely to think of themselves as Democrats, according to the Presbyterian Panel 2008 survey, though open discussion of politics is uncommon: pastors—50 percent Democrat, 25 percent independent, 23 percent Republican; specialized clergy such as counselors and educators—65 percent Democrat, 20 percent

independent, 13 percent Republican. Since clergy and elders have equal votes in the General Assembly, it is not surprising that policy votes tend to be seen as skewed leftward by many members.

4 The PC(USA) is extraordinarily transparent in the collection and publication of data by its Research Services. The PC(USA) website shows detailed membership and financial trends. Results of the survey research of the Presbyterian Panel are presented online at the denominational website and as printed reports. The Panel consists of representative samples of ordained ministers, elders, and members drawn every three years. The surveys of May 1997 and May 2009 addressed environmental issues, and the August 2004 survey dealt with energy.

5 The NCC is the main ecumenical body in the United States. It includes Orthodox and African American denominations, but it does not include the Roman Catholic Church or many of the evangelical and Pentecostal denominations.

6 Hessel went on to be one of the founding staff of the NCC Eco-Justice Working Group. For further details on the Eco-Justice Working Group see http://nccecojustice.org/ (accessed July 13, 2012). The Working Group was organized after a period of intense ecumenical collaboration among the denominations in communities at toxic waste sites, initiated at the grassroots level by Presbyterians at Love Canal in 1979 (see Townsend 2003), the United Church of Christ in Warren County in 1982, and Lutherans at Times Beach in 1983.

7 See Stivers (1976).

8 Copies of October 2005 document and PowerPoint in files of the author.

9 The Institute's web page called the Presbyterian action a "stunning development" (http://www.climate.org/topics/national-action/presbyterians-climate-neutral.html, accessed December 15, 2011).

10 See Hessel (1992). The denominations shared program ideas; for example, Presbyterians adopted sustainably harvested "eco-palms" for Palm Sunday from the Lutherans.

11 The theology outlined here is shared by other, smaller Reformed denominations that have produced important work in environmental theology. For example, Granberg-Michaelson wrote a call to Christian environmental stewardship that expressed concern about global warming (1984), before joining the staff of the World Council of Churches, leading its initial work on global climate change. He left to become general secretary of the Reformed Church in America.

12 On average, Presbyterian worship is less formally liturgical than Episcopalian or Lutheran worship but more so than Methodist or Baptist. The norms for Presbyterian worship are described, though details are not narrowly prescribed, in the *Directory for Worship* (PC(USA) 2004, Part II) and the *Book of Common Worship* (PC(USA) 1993). In addition to such published material, these conclusions are based on participation in and observation of church services in a wide variety of settings and informal sampling of online sermons, newsletters, and bulletins.

13 The title was later revised to "stewardship of creation enabler." In 2011 the program was replaced by the EMAN, a network of volunteer environmental advocates.

14 In 2009 the group changed its name to Presbyterians for Earth Care (PEC) to avoid confusion with "creationism," an anti-evolutionary stance that few, if any, of its members hold (http://www.presbyearthcare.org/, accessed October 20, 2011).

15 Further details are provided at http://www.solarunderthesun.org (accessed November 21, 2011).

16 Sir John Houghton, the Oxford atmospheric physicist who co-chaired the Intergovernmental Panel on Climate Change (IPCC) and edited the first three reports of the IPCC, is an elder in the conservative Presbyterian Church of Wales. It was Houghton who persuaded Richard Cizik, vice president for governmental affairs of the National Association of Evangelicals, of the importance of the issue (Cizik 2007; Heltzel and Safina 2009: 288).

References

Barnes-Davies, R. 2009. *50 Ways to Help Save the Earth: How You and Your Church Can Make a Difference*. Louisville, KY: Westminster John Knox Press.

Cizik, R. 2007. "New Moral Awakening, or How I Changed My Mind." *Reflections* 94.1, 58–60.

Clarke, D.L. 2005. "Mainline Protestants Begin to Divest from Israel: A Moral Imperative or 'Effective' Anti-Semitism." *Journal of Palestine Studies* 35.1, 44–59.

Curry, J. 2008, "Christians and Climate Change: A Social Framework of Analysis." *Perspectives on Science and Christian Faith* 60.3, 156–164.

Curry-Roper, J.M. 1990. "Contemporary Christian Eschatologies and Their Relation to Environmental Stewardship." *Professional Geographer* 42.2, 158–169.

Davis, D.M. 2011. *Left Behind and Loving It: A Cheeky Look at the End Times*. Eugene, OR: Cascade Books.

Dorrien, G. 2009. *Social Ethics in the Making: Interpreting an American Tradition*. Walden, MA: Wiley Blackwell.

Gibson, W.E., ed. 2004. *Eco-Justice: The Unfinished Journey*. Albany: State University of New York Press.

Gottlieb, R.S. 2006. *A Greener Faith: Religious Environmentalism and Our Planet's Future*. Oxford, U.K.: Oxford University Press.

Granberg-Michaelson, W. 1984. *A Worldly Spirituality: The Call to Redeem Life on Earth*. San Francisco, CA: Harper & Row.

Heltzel, P.G. and C. Safina. 2009. "Evangelicals for Environmental Responsibility: Uniting to Care for Creation." In *Journeys into Justice: Religious Collaboratives Working for Social Transformation*, ed. Nile Harper, pp. 285–304. Minneapolis, MN: Bascom Hill Publishing Group.

Hessel, D.T., ed. 1979. *Energy Ethics: A Christian Response*. New York: Friendship Press.

——. 1992. *After Nature's Revolt: Eco-Justice and Theology*. Minneapolis, MN: Fortress Press.

——. 1996. *Theology for Earth Community: A Field Guide*. Maryknoll, NY: Orbis Press.

Hoezee, S. 1998. *Remember Creation: God's World of Wonder and Delight*. Grand Rapids, MI: William B. Eerdmans Publishing Company.

Johnston, C. 1997. *And the Leaves of the Tree Are for the Healing of the Nations: Biblical and Theological Foundations for Eco-Justice*. Louisville, KY: Office of Environmental Justice, PC (USA).

LaHaye, T. and J.B. Jenkins. 1995. *Left Behind: A Novel of the World's Last Days*. Carol Stream, IL: Tyndale House Publishers.

Lindner, E., ed. 2012. *Yearbook of American and Canadian Churches*. National Council of Churches of Christ. Nashville, TN: Abingdon Press.

Martin-Schramm, J.B. 2010. *Climate Justice: Energy, Ethics, and Public Policy*. Minneapolis, MN: Fortress Press.

Moody, M. 2002. "Caring for Creation: Environmental Advocacy by Mainline Protestant Organizations." In *The Quiet Hand of God: Faith-Based Activism and the Public Role of Mainline Protestantism*, ed. R. Wuthnow and J.H. Evans, pp. 237–264. Berkeley: University of California Press.

Paine, C. 2006. *Who Killed the Electric Car?*Documentary film. Sony Pictures DVD.

PC(USA) (Presbyterian Church [U.S.A.]). 1990. *Restoring Creation for Ecology and Justice: A Report Adopted by the 202nd General Assembly (1990)*. Louisville, KY: Office of the General Assembly.

——. 1993. *Book of Common Worship*. Louisville, KY: Westminster John Knox Press.

——. 1996. "Part I, Book of Confessions." *The Constitution of the Presbyterian Church (U. S. A.)*. Louisville, KY: Office of the General Assembly.

——. 2004. *The Constitution of the Presbyterian Church (U.S.A.)*, Parts I and II, 2004–2005. Louisville, KY: Office of the General Assembly.

——. 2008. *The Power to Change: U.S. Energy Policy and Global Warming*. Louisville, KY: Office of the General Assembly, PC(USA). Available at http://www.pcusa.org/media/uploads/acswp/pdf/energyreport.pdf (accessed November 2, 2011).

Presbyterian Church in the U.S. and United Presbyterian Church in the U.S.A. 1981. "The Power to Speak Truth to Power." Available at http://www.pcusa.org/media/uploads/acswp/pdf/power_to_speak_truth_to_power.pdf (accessed July 13, 2012).

Research Services, PC(USA). 2009. *Religious and Demographic Profile of Presbyterians, 2008: Findings from the Initial Survey of the 2009–2011 Presbyterian Panel*. Louisville, KY: PC(USA).

Rolston, H. 1975. "Is There an Ecological Ethic?" *Ethics* 85.2, 93–109.

Rossing, B.R. 2004. *The Rapture Exposed: The Message of Hope in the Book of Revelation*. New York: Basic Books.

Schweizer, P. 2007. "Offset Away Our Guilt; If We Can Buy 'Carbon Offsets' for Our Environmental Mishaps, Why Not for Our Other Sins?" *USA Today*, September 25, 13A.

Stivers, R.L. 1976. *The Sustainable Society: Ethics and Economic Growth*. Philadelphia, PA: Westminster Press.

Townsend, P.K. 2003. "Congregations, Coalitions, and Chemicals: A Study in Applied Anthropology." In *Languages and Life: Essays in Memory of Kenneth L. Pike*, ed. M.R. Wise, T.N. Headland and R. Brend, pp. 597–610. Dallas: SIL International and the University of Texas at Arlington.

WCC (World Council of Churches). 2010. *Climate Change and the World Council of Churches: Background Information and Recent Statements.* Available at http://climatejusticeonline.org/wp-content/uploads/WCC-climate-change-booklet.pdf (accessed November 29, 2011).

White, L., Jr. 1967. "The Historical Roots of Our Ecologic Crisis." *Science* 155.3767, March 10, 1203–1207.

Whitney, E. 1993. "Lynn White, Ecotheology, and History." *Environmental Ethics* 15.2, 151–169.

Wuthnow, R. and J.H. Evans, eds. 2002. *The Quiet Hand of God: Faith-Based Activism and the Public Role of Mainline Protestantism*. Berkeley: University of California Press.

14 Keep Christianity brown!

Climate denial on the Christian Right in the United States

Bernard Daley Zaleha and Andrew Szasz

Introduction

Climate change is now considered the most important, most threatening, most dire aspect of our contemporary environmental crisis. As scientists and activists spoke out, ever more insistently, about what rising concentrations of carbon dioxide and other greenhouse gases in the atmosphere will do to the Earth's climate, and what impacts that will have for societies all across the planet, many of America's religions began to express concern and call for action: the Roman Catholic Church; all mainline Protestant churches; conservative and Reform Judaism; the Greek Orthodox Church; and others. Several chapters in this book describe how some of these faiths have engaged with the climate threat.

Even faith communities that we normally associate with conservative political causes were swept up in this process. The Evangelical Environmental Network (EEN) launched its Evangelical Climate Initiative in 2006. In a short document titled "Climate Change: An Evangelical Call to Action," it laid out four "claims": 1) "Human-Induced Climate Change is Real and increasing international instability"; 2) "The Consequences of Climate Change Will Be Significant, and Will Hit the Poor the Hardest"; 3) "Christian Moral Convictions Demand Our Response to the Climate Change Problem"; and 4) "The need to act now is urgent" (Evangelical Climate Initiative 2006). In 2008, the Southern Baptist Declaration on the Environment and Climate Change—an independent project spearheaded by an environmentally concerned Baptist seminary student—went live on the web and attracted hundreds of endorsements (Southern Baptist Environment and Climate Initiative 2008; Van Biema 2008). In both instances, though, these efforts provoked powerful—and for now largely successful—counter-organizing. Within evangelicalism at large, that counter-effort was led by Calvin Beisner and the Cornwall Alliance. Among Southern Baptists in particular, it was led by Richard Land and the Southern Baptist Convention's political arm, the Ethics and Religious Liberty Commission (ERLC). These efforts to resist the greening of evangelicalism and the Southern Baptist denomination are the subject of this chapter.

Textual foundations of conservative Christian ideas about the environment

For those not already familiar with Christian, and especially conservative Christian, metaphysical claims about the nature of God, some historical background will help illuminate contemporary environmental debates within conservative Christianity. According to Hebrew Bible scholar Richard Elliott Friedman, the ancient Israelites produced the first enduring monotheism. The "difference between Israelite monotheism and pagan religion[s]" of that era was not simply a "matter of arithmetic: one God rather than many. . . . Pagan religion personified [nature's] forces, ascribed a will to them, and called them: gods." In contrast, ancient "Israel's monotheism, for the first time" in recorded human history, "conceived of a God who was outside of nature" and through omnipotent power controlled the now desacralized forces of nature (Friedman 1995: 87–88).

This divorce of God from Nature had other ramifications. It allowed biblical writers to imagine that humans occupied a more exalted position in the natural order than the nature-based pagan religions conceived. One example is a passage from the Hebrew Bible's book of Psalms, which declares that God made humanity "a little lower than God," with "dominion" over God's creation, putting "all things under their feet, all sheep and oxen, and also the beasts of the field, the birds of the air, and the fish of the sea, whatever passes along the paths of the seas" (Ps. 8:6–8). Another psalm declares: "The heavens are the Lord's heavens, but the earth he has given to human beings" (Ps. 115:16). As historian Clarence Glacken noted, the theme that humans, "sinful" though they are, occupy "a position on earth comparable to that of God in the universe . . . has been one of the key ideas in the religious and philosophical thought of Western civilization regarding [humanity's] place in nature" (Glacken 1967: 155).

This concept of human superiority to and dominion over the rest of non-human nature was decisively carried forward into early Christianity by the church father, missionary, and theologian Paul of Tarsus. In his Epistle to the Romans, a book included in the Christian New Testament, Paul declared that God had abandoned pagans because they "worshiped and served the creature rather than the Creator" (Rom. 1:25, NRSV), thereby incorporating into early Christianity the Jewish idea that there is an absolute separation between God and the natural created order, and that to venerate nature was to engage in idolatry. Germane to our project here, this passage is one of the most cited whenever conservative Protestant Christians discuss human responsibility for the environment, and it is why the famed Protestant theologian Paul Tillich declared that within Christianity the term *pantheist* is "a 'heresy' label of the worst kind" (Tillich 1951: 233).

Evolution of the conservative Protestant discourse on the environment: 1970 to the present

Early conservative Protestant reaction to environmentalism

A key development in the discussion of the role of Christianity in American attitudes toward the environment was the publication in 1967 of Lynn White's still influential and widely cited essay "The Historical Roots of Our Ecologic Crisis." There, recounting the separation of God from Nature within Judaism and Christianity, White declared that, "in its Western form, Christianity is the most anthropocentric religion the world has seen," and opined that much—if not most—environmental degradation is directly traceable to Christianity's radical anthropocentrism (White 1967: 1205). White's indictment provoked an explosion of theological and philosophical reflection, both within and outside Christianity (Fowler 1995: 20). Poet and environmentalist Wendell Berry has gone so far as to assert that, because of White's argument, "the culpability of Christianity in the destruction of the natural world, and the uselessness of Christianity to any effort to correct that destruction, are now established clichés of the conservation movement" (Berry 1993: 93).

The first major Christian response to White was from the then prominent and a still widely cited thinker within evangelical Christianity, Francis Schaeffer. In *Pollution and the Death of Man* (1970), Schaeffer acknowledged the intensifying problems of environmental pollution, but denied that these problems stemmed from Christian arrogance and anthropocentrism. He confirmed that, yes, unique in all nature, only humans were created in "the image of God" and were therefore uniquely exalted by God. According to Schaeffer, however, humans only held nature "in stewardship for the true Owner. When we have dominion over nature, *it is not ours*, either. It belongs to God, and we are to exercise our dominion over these things not as though entitled to exploit them, but as things borrowed or held in trust, which we are to use realizing that they are not ours intrinsically. [Humanity's] dominion is under God's dominion and under God's Domain" (Schaeffer 1970: 51, 70, emphasis in the original).

Schaeffer was no doubt aware that more pantheistic reinterpretations of Christian doctrine were already well underway within mainline Christian denominations under the influence of the process philosophy of Alfred North Whitehead and the Christian process theology of Charles Hartshorne (Cooper 2006). Thus, though he argued that human authority over God's creation was not unlimited, Schaeffer also cautioned throughout his book, faithful to Paul of Tarsus's creator–creation distinction and consistent with White's thesis, that humans must not let worries over environmental pollution tempt them into a more pantheistic understanding of nature. Faithful Christians must instead, according to Schaeffer, stay within the framework of humanity as the proper and rightful stewards of God's creation. While other evangelical theorists and theologians such as Loren Wilkinson (De Vos, Wilkinson and Calvin Center for Christian Scholarship 1980) and Calvin B. DeWitt (DeWitt 1994) have further developed Schaeffer's foundational ideas, the basic framework of the stewardship model remains in place to the present, and stewardship is the main trope that is used by all evangelicals and literalistic fundamentalists who discuss environmental issues.

The next significant development within evangelical environmentalism was its move into policy advocacy. In 1993, the Evangelical Environmental Network (EEN) formed as a subsidiary ministry of Evangelicals for Social Action (Larsen 2005a: 624), a social justice advocacy organization founded in 1973. It formed in order to provide evangelical participation in the then emerging National Religious Partnership for the Environment, a Jewish and Christian umbrella organization whose purpose was to "promote environmental causes in church and temple teaching" (Womersley 2005: 1158). Though primarily an educational organization formed to distribute environmental information to evangelical congregations, the EEN also led a political campaign in 1996 to help fend off Republican efforts in Congress to weaken the federal Endangered Species Act (Larsen 2005a: 625). In the course of that campaign, evangelical lay-theologian and environmental studies professor Calvin B. DeWitt declared that the federal Endangered Species Act was "the Noah's ark of our day" and that "Congress and special interests are trying to sink it." As part of the campaign to protect the Act, DeWitt and other evangelical leaders organized a network of evangelical "Noah" congregations who pledged to oppose the congressional efforts to weaken federal species protection (Steinfels 1996). The campaign was successful, and evangelical participation in the campaign was widely credited as a substantial factor in this political victory.

The beginning of conservative Protestant backlash to creation care

Even before the EEN's Noah campaign in support of the Endangered Species Act, some Christian thinkers were beginning to push back on what they regarded as an unfortunate entanglement of some Christians in advocating for and promoting environmental regulation by government. An early leader in this backlash and countermovement was Catholic priest Robert Sirico. In 1990, shortly after his ordination, he founded a libertarian think tank and advocacy organization, the Acton Institute for the Study of Religion and Liberty, in order "to promote a society that embraces civil liberties and free-market economics" (Larsen 2005b: 849).

Within evangelical Protestantism, E. Calvin Beisner—in the early 1990s still a little known Orthodox Presbyterian theologian and interdisciplinary studies professor at Covenant College, a small conservative Christian college in Georgia—was an early critic and opponent of the EEN. Shortly after the formation of the EEN, he penned a critical article in the Christian news magazine *World*, in which he challenged the allegations of environmental problems laid out in the EEN's foundational document, "The Evangelical Declaration on the Care of Creation" ("Creation Care Declaration") (Beisner 1993). There, Beisner pioneered his basic manner of attack. First, while acknowledging that caring about the environment is important and expressing appreciation to the drafters for firmly asserting "the distinction between Creator and creature," Beisner asserts that the problems alleged either do not exist at all or are much less severe than suggested in the Creation Care Declaration. Second, Beisner says the only scientist involved with drafting the Declaration was Calvin DeWitt, and that the drafters therefore lacked sufficient scientific expertise. Third, Beisner associates the EEN with figures considered questionable within conservative Christian circles. For instance, Beisner suggests the EEN and the Declaration drafters had fallen under the influence of secularists like "Carl Sagan, a Marxist professor of astronomy" and the "well-known theological liberal Rev. James Morton, dean of the Episcopal Cathedral of St. John the Divine," which is code within conservative Christianity for being a center of illegitimate forms of "New Age" Christianity (Marrs 1990: 283–284). Finally, Beisner accused EEN of theological error for failing to acknowledge, pursuant to Psalms 115:16, that God specifically gave the Earth to humans, and that God specifically expects humans to transform "cursed earth—the wilderness—into a garden" such as existed before Adam and Eve's sin (Beisner 1993).

Perhaps in response to the EEN's successful 1996 Noah campaign, Sirico's Acton Institute published a book about a year later by Beisner. In *Where Garden Meets Wilderness: Evangelical Entry into the Environmental Debate*, Beisner expressed appreciation for evangelicals' entry into the discourse about environmental issues, but then expands his 1993 article into a book-length critique of what he regards as theological mistakes being committed by his fellow evangelicals (Beisner 1997b). In particular, Beisner seems interested in criticizing Calvin DeWitt—a scientist who became an important leader of the evangelical Creation Care movement. He targets DeWitt's view that the Earth is properly analogized to the Garden of Eden in all its perfection and fruitfulness, and that the divine command to humans to "subdue" and exercise "dominion over" the Earth found in Genesis 1:26–28 is constrained and modified by the subsequent command to Adam to "till" and "keep" the "garden of Eden" found in Genesis 2:15. According to DeWitt, "Creation's garden abundantly yields blessed fruits, sustainably supporting us and all life in its God-declared goodness. We 'disciples of the first Adam' have made the choice to extract more and yet more of the fruits of creation—even at the expense of destroying creation's protective provisions and blessed fruitfulness." Yet, "even from the time of Adam and Eve, humanity has not been satisfied with the fruitfulness and grace of the Garden—the productive and beautiful creation that God has provided us" (DeWitt 1994: 35, 45).

Beisner flatly rejects this view, arguing in a literalistic fundamentalist manner that "The Garden of Eden ceased to exist—and essentially the whole earth became wilderness—with [Noah's] Flood." Because God himself put a curse upon the Earth because of Adam and Eve's sin, "creation by itself simply does not abundantly yield blessed fruits . . . but it becomes abundantly fruitful only under the wise and resolute hand of man," especially by Christians who must now "seek to transform cursed ground back into blessed ground." Noting the "forcefulness" that God must have used when separating "light from darkness" or "the waters from dry land," Beisner concludes that, "While tender cultivation is suited to a garden, forceful subduing is suited to all of the earth that has not yet been transformed into the garden." Only a "strong, forceful subjugation" of the Earth by humans will accomplish the redemption of the fallen and cursed Earth to being again blessed and fruitful. Citing the Psalms 115:16 passage noted above, Beisner chastises and laments the "tendency among some environmentalists to blur or erase the metaphysical and ethical hierarchy of man over other living creatures," thereby failing to recognize "the Biblical teaching that God gave man a mandate to rule the earth" (Beisner 1997b: 16, 20, 103, 78).

ExxonMobil, the Koch brothers, and other extractive industries subsequently provided funding to spread Beisner's message. Given Beisner's articulation of what he has elsewhere simply called "forceful rule" (Beisner 1997a: 10), it is perhaps easy to see that extractive industries might indeed want to see his message spread as a counter-force to evangelical creation care.

The Cornwall Declaration: packaging Christian anti-environmentalism for the public

Beisner's book articulating humanity's Christian duty of "forceful rule" over the Earth was an innovative conservative Christian polemic in favor of neoliberal industrial capitalism. But of course few people would ever read or even hear about Beisner's book. How might Beisner's new gospel be spread? Robert Sirico's next move was to have the Acton Institute found in 2000 the Interfaith Council for Environmental Stewardship (ICES) as a new challenge to the National Religious Partnership for the Environment.

The ICES brought together theologically and politically conservative Jewish, Catholic, and Protestant leaders to promote a document called the Cornwall Declaration on Environmental Stewardship (Interfaith Council for Environmental Stewardship 2000). The Cornwall Declaration, so named because it was drafted at a conference center in West Cornwall, Connecticut, in October 1999, is primarily a two-page distillation of the main arguments of Beisner's book. The Declaration acknowledges ongoing harmful pollution in developing countries while asserting that environmental problems have mostly been solved in the developed world as a result of the application of free market capitalism and its resulting development of natural resources. The key to solving problems in developing countries will be to ensure that the "relationships between stewardship and private property are fully appreciated, allowing people's natural incentive to care for their own property to reduce the need for collective ownership and control of resources and enterprises, and in which collective action, when deemed necessary, takes place at the most local level possible." Curiously, this preference for "private property" and lack of collective control receives no theological justification, but is merely asserted. The Declaration also asserts that "fears of destructive manmade global warming, overpopulation, and rampant species loss" are "unfounded" or "undue," and that such concerns are at most "speculative," calling for "solutions" that "are unjustifiably costly and of dubious benefit." Predictably, the Declaration reminds its readers that they must not "deify nature or oppose human dominion over creation," but must instead remember that, as the only "bearers of God's image" on Earth, humans alone are capable of adding "to the earth's abundance." In a passage echoing the late Cornucopian economist Julian Simon (with whom Beisner had collaborated and whom he extolled in his book), the Declaration proclaims, in a robust display of human exemptionalism (Catton and Dunlap 1980), that "the human person is the most valuable resource on earth" (Interfaith Council for Environmental Stewardship 2000).

The Declaration attracted the support of a number of well-known conservative Protestant leaders, including Bill Bright of Campus Crusade for Christ, James Dobson, and Charles Colson. It also attracted support from outside Protestantism, including figures such as conservative Catholic priest John Neuhaus and Rabbi

Daniel Lapin. The ICES sent the Declaration out to 37,000 religious leaders, along with an explanation of the dangers being posed to traditional biblical faiths by the National Religious Partnership for the Environment, which they accused of attempting to "redefine traditional Judeo-Christian teachings on stewardship." The wholesale attack forced EEN to defend itself to its evangelical constituents in "a series of semi-public letters with ICES leaders" (Larsen 2005b: 850).

Evangelical climate care and its backlash: 2004 to the present

In 2004, the National Association of Evangelicals (NAE) published "For the Health of the Nation," adding "creation care," "peace," and "human rights" to more traditional evangelical concerns, such as the "sanctity of life," thereby creating a foundation for potential engagement by evangelicals in the issue of climate change (National Association of Evangelicals 2004). Two years later, in 2006, the recently organized Evangelical Climate Initiative (ECI) issued the "Evangelical Call to Action" summarized in our introduction.

To oppose this, Beisner renamed the ICES as the Interfaith Stewardship Alliance (ISA) and in its name drafted and delivered an open letter to the leadership of the NAE urging that they decline to sign or endorse the ECI's "Call to Action" (Barton et al. 2006). The letter was signed by 22 politically powerful evangelical leaders (Cooperman 2006), including Richard Land, head of the ERLC of the Southern Baptist Convention (SBC), America's largest Protestant denomination. This high-profile intervention received coverage from both the *Washington Post* (Cooperman 2006) and the *New York Times* (Goodstein 2006).

Ted Haggard, the NAE's president at the time, had spoken positively about the environment being "a values issue" (Cooperman 2006). The NAE's vice president for governmental affairs, the NAE's top lobbyist in Washington, Richard Cizik, said he had "had a conversion experience on the climate issue not unlike conversion to Christ" (Little 2005). The NAE might well have endorsed the Call. The conservatives' intervention had the desired effect, however; neither Haggard nor Cizik signed the Call, and the NAE refrained from endorsing it. Of interest to us in this chapter, the conservative evangelicals' letter was signed both by Beisner and by Land, each the leading figure in one of the organizations we have identified as leading institutional centers of American conservative Christian climate denial. This would be the first of several occasions in which these two seemingly independent institutions would join in a common, concerted effort to oppose the greening of conservative Christianity in the U.S.

This clash within the evangelical community received considerable attention, not only in the media, but from scholars as well (McCammack 2007). Indeed, famed journalist Bill Moyers produced a PBS documentary airing in late 2006, "Is God Green?," which explored the conflicts within evangelical Christianity over climate change and environmental policy generally. When seeking interviews with prominent evangelical leaders like James Dobson and Charles Colson who had signed the ISA's letter opposing the ECI, he was instead directed to interview, in Moyers's words, "Calvin Beisner, Ph.D." Beisner had by this time earned a doctorate in Scottish religious history at the University of St. Andrews in 2003, conferring a new level of professional prestige. In Moyers's documentary, he is identified as a professor of historical theology and social ethics at Knox Theological Seminary in Fort Lauderdale, Florida (Casciato 2006).

In the course of his interview, the full, unabridged text of which is published on Moyers's website,[1] Moyers queried Beisner about his doctrine of "forceful rule." Beisner explained that, "Well, quite frankly, if you are going to mine for precious metals, for fossil fuels, for anything else, you don't do that with a feather brush, you know. You don't do it by quietly blowing a breath on something as you would to blow out a candle." When asked if humans have "a moral obligation to clean up the consequences of that force," Beisner said, "As far as we are able to and, always, we also have a moral obligation to do rational, cost benefit analysis," implying that the cost of cleanup might not survive a cost–benefit analysis. Beisner also acknowledged that his work has received funding from ExxonMobil, but saw nothing scandalous in this, saying that it is only to be expected that ExxonMobil would want voices such as his to be heard.

One more insight into Beisner's thinking emerged from Moyers's interview that does not surface either from the Cornwall Declaration or from Beisner's book. When pressed by Moyers as to whether he worried that he might be wrong about global warming, Beisner said that global warming, even if it turns out to be real and harmful, is of little ultimate importance, because whether one is "going to live [in either heaven or hell for] eternity" should be any reasonable human's primary concern. This is a telling detail. While this kind of dismissal of the Earth's importance is not usually expressly articulated by conservative Christians in their polemics against environmental regulation, this exchange suggests the possibility that a fatalistic eschatology that envisions imminent divine intervention in earthly affairs may form at least part of the foundation of conservative Christian anti-environmentalism. Beisner himself acknowledged the influence upon him of his belief that what should most matter to any rational actor is where he will spend eternity, not the fate of his or her temporary abode, this planet, a sentiment captured in the lyrics of a popular Protestant folk hymn, "This world is not my home. I'm just a passing through. If heaven's not my home, then Lord what will I do" (Brumley 1938). According to a Pew study in 2006, 20 percent of Americans believe Jesus will return in their lifetimes (Pew Forum on Religion and Public Life 2006). The influence of this apocalyptic strain within conservative Protestant Christianity is evidenced by a bumper sticker produced by a Baptist church near Boise, Idaho ("Forget 'Save the Earth'; What About Your Soul? The Earth is Going to Burn; What About You?") and a bumper sticker and T-shirt available from ChristianBook.com ("Global Warming is Nothing Next to Eternal Burning"). The role of such apocalyptic thought in anti-environmental ideology is still little understood (Barker and Bearce 2012).

In May 2007, Beisner renamed his organization one more time, this time to the "Cornwall Alliance for the Stewardship of Creation," with the stated reason being to highlight the organization's connection to "its flagship document" (www.cornwallalliance.org/about). Cornwall's most significant recent project was its launch in 2010 of its "Resisting the Green Dragon" campaign, the primary tool of which was 12 half-hour lectures distributed through a box set of four DVDs. The promotional two-minute video at the campaign's website is striking for it scary music, flashy graphics, and strident rhetoric. In the introduction to the lecture series, Christian radio host Janet Parshall states that:

> In what has become one of the greatest deceptions of our day, radical environmentalism is striving to put America and the world under its destructive control. This so-called "Green Dragon" is seducing your children in our classrooms and popular culture. Its lust for political power now extends to the highest global levels. And its twisted view of the world elevates nature above the needs of people, of even the poorest and the most helpless. With millions falling prey to its spiritual deception, the time is now to stand and resist. Around the world, environmentalism has become a radical movement, something we call the Green Dragon, and it is deadly. Deadly to human prosperity, deadly to human life, deadly to human freedom, and deadly to the gospel of Jesus Christ. Make no mistake about it, environmentalism is no longer your friend; it is your enemy. And the battle is not primarily political or material; it is spiritual.[2]

While the promotional video is flashy, the rest of the content is more blandly presented, consisting of video recordings of 12 lectures recorded in an unidentified meeting room before a relatively small audience. Each of the lectures, two of them by Beisner, is preceded with short clips from various speakers, some of them prominent conservative evangelicals like David Barton, Bryan Fischer, Richard Land, and Tony Perkins. Of the 12 lectures, only one is by a credentialed climate scientist, David Legates, who is well known in climate denial circles. The rest of the speakers serve to put into video format the intellectual program Beisner laid out in the Cornwall Declaration and *Where Garden Meets Wilderness*. It seems clear that the series' intended audience is conservative evangelical congregations, not the general public. No real effort is made to critique mainstream climate science or explore it in any detail. The goal appears to be maintaining environmental skepticism in general and climate denialism in particular among a constituency that is already inclined to such skepticism, and to attempt to neutralize the influence of progressive evangelicals such as Calvin DeWitt and the EEN.

Southern Baptists on environment and climate, 1970–2011

The SBC defines stewardship

Years before the issue of climate change became a hot issue for Southern Baptists, the SBC sought, through a series of resolutions passed at its annual meetings, to articulate its understanding of environmental "stewardship."

The "Resolution on the Environment," passed in 1970, reads like a straightforward, sympathetic appropriation of 1970-era Earth Day environmentalism with a

Christian overlay. "God has created man to be a creature who needs clean air, pure water, and an environment which contributes to his general health," but "man has created a crisis by polluting the air, poisoning the streams, and ravaging the soil." Pollution is the newest manifestation of humans' fallenness: we are so sinful that we are destroying the conditions upon which we depend for our very survival. The Resolution ends on a positive note, urging "Christians everywhere to practice stewardship of the environment and to work with government, industry and others to correct the ravaging of the earth" (Southern Baptist Convention 1970).

The Resolution passed by the SBC four years later is quite similar. "God . . . is the author of the universe [and He] views his creation as being very good. . . . The Scripture confronts us with our responsibility to God as stewards." But we have failed to conserve resources. We have been "selfish and nearsighted." Christians must "assume . . . responsibilit[y]." The SBC then "urge[s] Congress and concerned government agencies to take aggressive action to conserve our diminishing resources" (Southern Baptist Convention 1974).

Evidently, at that time, the SBC still felt positive about federal environmental regulation. It had yet to embrace conservatives' beliefs about the sanctity of private property or their deep-seated aversion to government intervention in the market economy.

Fast-forward to 1990. A new Resolution starts with familiar scriptural citations in support of the idea of stewardship. From the Psalms, "the earth is the Lord's." From Genesis, "God saw that it [the Earth, the Seas] was good." Also from Genesis, God put man "in the Garden of Eden to till it and keep it." The Resolution condemned "the sinfulness of the human race" which "has led to the destruction of the created order . . . as evidenced by the endangerment of the earth by pollution, human extravagance and wastefulness, soil depletion and erosion, and general misuse of creation." It affirmed "our responsibility to God to be better stewards of all the created order."

But then, echoing Beisner (and Paul of Tarsus), the Resolution drew a distinction between "caring for creation" and "worship[ing] the creation." Yes, we care about the environment, but we are not like those environmentalists who, we suspect, are pantheists or neo-Pagans (Southern Baptist Convention 1990).

That concern, that "caring" must not be confused with "worshiping," becomes more pronounced as we move forward in time. Two documents in particular illustrate this trend, the "Policy Statement on Global Warming," the first document we have been able to find that presents the SBC's Ethics and Religious Liberty Commission's views on climate change (Lewis 2005), and the SBC's 2006 Resolution titled "On Environmentalism and Evangelicals" (Southern Baptist Convention 2006). We discuss these documents' views on stewardship, deferring until later a discussion of what they say about climate change.

The ERLC's "Policy Statement on Global Warming" defines stewardship by situating it between two extremes: At one extreme, there are those who "advocate using nature without any reservations, leading to environment destruction." This is labeled "anthropocentric," meaning that it makes "man the center of the universe rather than God." The other end of the spectrum is defined as the view that "puts heavy emphasis on caring for the environment while downplaying or limiting its use for the good of man." The proper Christian perspective rejects heedless environmental destruction—man is "not free to abuse" creation—but it also rejects the other extreme, which it condemns as "pantheistic . . . put[ting] nature at the center of the universe rather than God." So stewardship is not heedless exploitation, but it's certainly not idolatrous veneration, either.

The SBC Resolution passed in 2006 again distinguished between the proper, Christian way to think about stewardship and the views of "neopagan" environmentalists who "reject God the father in favor of deifying mother Earth"; it then moves on to new arguments, ones not seen in the earlier documents. For the first time we see one of climate skeptics' favorite arguments when they make their case against climate change: the science is "questionable." And, in a move that puts quite a bit of distance between this Resolution and the one passed back in 1974 (which called on "Congress and concerned government agencies to take aggressive action"), this Resolution declares its support for "respecting ownership and property rights" and says "we oppose solutions . . . which bar access to natural resources and unnecessarily restrict economic development."

It is worth emphasizing: stewardship, now, is not just distinguished from pantheistic Nature worship; it is further constrained in that whatever is done in its name must not violate the principles of private property and free enterprise. In less than four decades, the SBC's dialogue with environmentalism moved from sympathetic endorsement toward ever greater distancing and disagreement, from fellow-feeling, through labeling secular environmentalism as neo-Pagan, to expressing views (skeptical of science, protective of the rights of private property) that are indistinguishable from those of conservatives who have always opposed environmental regulation and who now form the core of the climate denial movement (Oreskes and Conway 2010).

The 2006 Resolution concludes with the reminder that, although the SBC "reaffirms [the] God-given responsibility of caring for the creation," this doesn't mean that environmental stewardship should be a top priority. Our "Great Commission" is and always will be "to take the Good News of Jesus Christ to people of every tribe, tongue, and nation."

The SBC defines its position on global warming

By the late 1980s, scientists were telling the public with ever greater urgency that climate change would threaten both the planet and the future of humanity. International conferences; national policy debate; increasing media coverage. Faith groups, including some evangelicals, were beginning to take a stand. It seemed inevitable that Southern Baptists would have to go beyond a general discussion of Christian stewardship and state their position on climate change.

The ERLC's "Policy Statement on Global Warming" was their first attempt to do that. It was an oddly constructed document. The first part, which we described above, was a theological discussion of the correct way to think about the environment, situating Christian stewardship between extremes of heedless abuse of Creation, on the one hand, and pantheistic Nature worship, on the other. The statement then shifted abruptly to a completely secular attack on global warming, stating that: "the scientific data is simply not conclusive"; "tens of thousands of scientists agree that there is no conclusive evidence for the manmade global warming theory"; "Records prove that climates have changed in the past without human interaction"; the Kyoto Protocol, if implemented, "would have disastrous effects . . . incredible amounts of lost jobs . . . increased poverty . . . a crippled American economy"; and that, given scientific uncertainty and controversy, and certain to have disastrous economic consequences, implementing policy that aims to decrease the threat of impending climate change would be "rash . . . irrational . . . imprudent" (Lewis 2005).

Such a list simply repeats the then-emerging ensemble of standard denialist talking points. The authorities cited in the statement are telling: Sallie Baliunas and Pat Michaels (easily recognizable names to anyone who studies the climate denier community), the Heritage Foundation and the Cato Institute (conservative think tanks), and the Heartland Institute (a leading climate denier organization).

This was the time when tensions were mounting within the evangelical community, as described earlier: the Evangelical Climate Initiative's "Call to Action"; and the Interfaith Stewardship Alliance's open letter to the National Association of Evangelicals urging its leaders to decline to endorse the ECI's "Call" (Barton et al. 2006). It was in that context that the SBC debated and passed a new Resolution, specifically about climate change, at its annual Convention in 2007.

The new Resolution starts with a restatement of the SBC's position on stewardship, balancing "caring" with "dominion," and it reaffirms "God's [special] concern for the poor," as a kind of preamble. Then on to climate: A series of "whereas" statements deliver what, by then, had already crystallized as a list of standard denialist talking points: Historically, global temperatures have gone up and down. It has been hotter in the past. Many things other than CO_2 contribute to contemporary warming. "The scientific community is divided"; "many scientists reject the idea of catastrophic human-induced global warming"; "many scientists argue that natural causes . . . are more significant . . . than CO_2 emissions." The Kyoto Protocol will cool things only by a small, insignificant amount. At the same time, complying with Kyoto will cause major economic hardship. "Poor people and underdeveloped regions of the world will be impacted the most severely" (Southern Baptist Convention 2007).

The Resolution says the SBC supports climate policy, but only if it will not have "significant negative consequences [for] the U.S. . . . [and] also, and most importantly, on the poor and developing economies." The President and Congress should "reject government-mandated reductions in greenhouse gas emissions."

The Southern Baptist Declaration on the Environment and Climate Change

The 2007 Resolution, SBC's official position on climate change, was passed by the Convention "overwhelmingly" (Baptist Press Staff 2008). If other Southern Baptists reached a different conclusion about climate, closer to that of the EEN's

Evangelical Climate Initiative, they would have to find another venue in which to express their views.[3] Hence Jonathan Merritt's decision to post the "Southern Baptist Declaration on the Environment and Climate Change" online.

The Declaration says the Earth has been "damaged by human activity . . . often reckless, preventable and sinful." One of its four statements declares that "Christian moral conviction and our Southern Baptist doctrines demand . . . environmental stewardship." Although scientists are not unanimous, there is "general agreement," enough "to make informed decisions about the future." "It is prudent," the statement adds, "to engage this issue without any further lingering."

When Merritt went live on the web with the Declaration in 2008, he had already gotten endorsements from over 40 Southern Baptists, including several past presidents of the Southern Baptist Convention, and professors, deans, and presidents of several Baptist theological seminaries (Southern Baptist Environment and Climate Initiative 2008). When last updated in August 2011, the website claimed that the number of signatures had reached 750.

We take this as evidence that, although the official SBC position did represent the majority of Southern Baptists' opinion, the opposite, green, position also had substantial support.

Land and the ERLC oppose the Declaration

Even as the Declaration staked out a position on climate that differed significantly from that of the Convention, it emphatically affirmed its loyalty to Southern Baptists' "deep and lasting commitments to moral issues like the sanctity of human life and biblical definitions of marriage." Signers of the Declaration proclaim that they will not "compromise [their] convictions nor attenuate [their] advocacy" about abortion or about gay marriage, issues which, the Declaration says, are still "the most pressing moral issues of our day" (Southern Baptist Environment and Climate Initiative 2008).

If that was meant to reassure or to mollify those within the SBC who might be upset about the Declaration, it would prove to be of no avail. Publicly, during a teleconference discussing the launch of the Declaration, Jonathan Merritt was diplomatic, saying that "the ERLC had provided helpful inputs to reshape the statement" (Baptist Press Staff 2008). Privately, in our personal interview, Merritt painted a darker picture:

> Some of the leaders in the [ERLC] . . . began to catch wind of this. . . . They had taken a strong stand really against climate change and against, quite frankly, any type of practical creation care at all. And they are something of a powerhouse within the Religious Right in America. . . . They began to engage. They began to contact signatories that they could find and asked them to remove their names.

According to Merritt, there were attempts to coopt him, and then, when that did not work, "they began to make threats." "My character was being questioned.

. . . I had denominational figureheads in Nashville who were writing emails, for example, to Fox News correspondents with slanderous, crazy things. . . . [A]s a sophomore in Seminary I am totally unequipped to combat a political machine like this. . . . I had kicked the hornet's nest."

Land and the ERLC campaign against climate change

Undeterred by the presence of significant minority sentiment for a positive engagement with climate change, Land and the ERLC pushed on. Aside from ritualized expressions of concern for the poor, who, they said, would be the greatest victims if the government did anything to curtail the emission of greenhouse gases, and except for equally ritualized affirmations of stewardship, Land's actions and words were, by now, completely indistinguishable from those of any secular climate denier.

Land closely coordinated his anti-climate activities with Beisner. In 2008, he joined the Beisner/Cornwall "We Get It!" campaign. In an endorsement prominently displayed on the "We Get It!" web page, Land praised the campaign for its "Biblical view of stewardship and a responsible regard for the needs of the poor," and for "Rejecting ill-conceived calls for drastic action in response to poorly understood, hypothetical risks" (Strode 2008; We Get It! 2008). In 2010, Land's name was prominent among those listed as endorsing Cornwall's December 2009 "Evangelical Declaration on Climate Change."[4] Land is featured in one of Cornwall's Green Dragon DVDs and promotional video.

Land and the ERLC mobilized the ERLC's base, repeatedly urging their people to contact their senators and their representatives to tell them to vote against "cap and trade" legislation (Carlson 2009; Land 2009a, 2009b; Stoddard 2009). Cap and trade would be "the largest tax increase in American history . . . a job-killer." Over 2 million jobs would be lost. Prices would skyrocket. "[H]ardest hit would be the poor." And for what? The "expected benefit—little if anything." The science is "not even settled." In fact, "we have been experiencing a cooling trend." Tremendous costs, no benefits: all in all, cap and trade is "a monstrosity" (Land 2009a, 2009b). Discussing cap and trade legislation in one of his radio programs, Land called climate change a "hoax" and a "scam" (Allen 2008). In his 2011 legislative alert, Land reported that "the Senate chose wisely not to act on global warming legislation in the 111th Congress" (Land and Duke 2011).

Summary

Over the last four decades, we see a fairly straightforward trajectory. Early on, there is a positive response to the growth of environmental awareness in the 1970s. But then the social and political context changes. Politics in the United States becomes ever more polarized. Evangelicals and Baptists go from relative disengagement in politics to becoming, especially through "culture war" issues such as abortion, teaching creationism versus evolution, and gay marriage, completely engaged. In fact, they become the most reliable constituency base of an increasingly conservative, ideologically driven Republican Party. This transformation has a profound impact on these faith communities' relationship to environmentalism and, eventually, to their relationship to the issue of climate change, moving them from an initial sympathy to the point where its most prominent official spokespeople are allied with secular climate deniers and the extractive resources industries that fund them.

The situation today

Stepping back, we seem to have written a narrative of complete victory for Beisner and Cornwall, for Land and the ERLC, and for climate denial in the conservative sector of American Christianity. But we wish to note the presence of certain other data points that suggest that the story is not over and that the wheel might yet turn again.

Although Land and the ERLC lobbied hard against energy and cap and trade legislation—Warner–Lieberman in the Senate and Waxman–Markey in the House—the Baptist faith did not speak with one unified voice about those bills. The Baptist Center for Ethics, a recently organized grouping of politically liberal and progressive Baptists, lobbied equally vigorously on the other side (EthicsDaily.com Staff 2008; Kaylor 2009; Parham 2009).

And evangelicals continue to be divided in spite of Beisner and Cornwall's anti-climate campaigns. The EEN is still very active. In 2011 the EEN supported the Environmental Protection Agency's regulation of mercury emissions from cement plants, saying that the unborn need to be protected from exposure to mercury pollution (Sheppard 2011). The EEN has been sponsoring an annual Global Day of Prayer and Creation Care (Gryboski 2012). In 2012, the third annual observance was titled Global Day of Prayer for Creation Care and the Poor. EEN partners with global development NGOs, Plant with Purpose, Food for the Hungry, World Relief, and World Vision, underlining the EEN's contention that Christian concern about the poor requires climate *action*, not climate *denial*. Interestingly, and perhaps surprisingly, given the story we have been telling, the National Association of Evangelicals was one of the co-sponsors of the Global Day of Prayer and Creation Care.

There is an underlying reason for the continuing organization-level conflict: continuing disagreement among rank-and-file evangelicals and Southern Baptists. Recall the Pew survey which showed that about a third of American evangelicals believe climate change is caused by human action and that there is "solid evidence the earth is warming." Those numbers are likely to grow; several of our sources express the opinion that Southern Baptists are in the midst of a generational shift and that their younger generation is less uniformly conservative about the classic "culture war" issues and is more concerned about the environment and climate (Lamb 2010; Merritt 2011; Stein 2008).

There does seem to be, then, a substantial mass base for continuing to green Southern Baptists and the larger evangelical community, especially among younger cohorts. There are even signs that the SBC itself can, under the right circumstances, sound surprisingly green. Consider the 2010 Resolution "On the Gulf of Mexico Catastrophe." As a blown British Petroleum well continued to disgorge crude oil into Gulf waters, the 2010 Convention called "on the governing authorities to act determinatively . . . to end this crisis . . . to ensure full corporate accountability for damages, clean-up, and restoration . . . and to promote future energy policies based on prudence, conservation, accountability, and safety" (Southern Baptist Convention 2010; see Young Evangelicals for Climate Action 2012).

Regrettably, if the climate crisis is *not* addressed assiduously and *soon* (and there is little indication, at present, that it will be so addressed) societies across the globe will experience a series of increasingly severe shocks: extreme weather, drought, flooding, hunger, thirst, illness, social stress, disorganization, conflict, migration. Although right now the politically powerful leadership structures of conservative Christianity seem to have been successful at beating back the green insurgency, worsening social and environmental conditions may well reverse these trends, strengthening again those who truly believe that Christians are called to care for human well-being and to care for nature, and that those are inseparable.

Acknowledgments

Andrew Szasz wishes to thank Samantha Shattuck and Andrea Rojas, University of California, Santa Cruz Environmental Studies majors, for their research on controversy in the Southern Baptist community.

Notes

1 http://www.pbs.org/moyers/moyersonamerica/print/beisner_print.html (accessed 27 October 2012).
2 www.youtube.com/watch?v=rGEOFipx70U (accessed October 27, 2012).
3 How many Southern Baptists think climate change is real and is serious? The numbers are difficult to pin down, because polls rarely identify respondents by denomination, preferring instead to use broader categories like "evangelical" or "born again." That introduces substantial uncertainty, because, although most Southern Baptists consider themselves to be evangelicals, adherents of other faiths also define themselves as evangelical. Still, those polls are the best available data. These polls show that evangelicals are less likely than the rest of the American public to agree that climate change is real, serious, and caused by human action, as opposed to being simply natural variation in Earth's climate. It is also true, though, that a quite substantial minority of evangelicals *do* believe those things about climate change (Grossman 2008; Pew Forum on Religion and Public Life 2009). A different indicator: When we interviewed Jonathan Merritt, he was able to cite from memory quite a number of individuals and congregations, around the U.S., where, locally, Southern Baptists were engaged in some form of climate activism (Merritt 2011).
4 http://www.cornwallalliance.org/blog/item/prominent-signers-of-an-evangelical-declaration-on-global-warming/ (accessed October 27, 2012).

References

Allen, Bob. 2008. "Studies Show Rise in Greenhouse Gases in 2007." *Baptist Standard: The Texas Baptist Newsjournal*. Available at http://www.abpnews.com/archives/item/3693-studies-show-rise-in-greenhouse-gases-in-2007 (accessed September 24 2012).

Baptist Press Staff. 2008. "Seminary Student's Climate Change Project Is Not SBC's." Available at http://www.sbcbaptistpress.org/BPnews.asp?ID=27582 (accessed September 24, 2012).

Barker, David C. and David H. Bearce. 2013. "End-Times Theology, the Shadow of the Future, and Public Resistance to Addressing Global Climate Change." *Political Research Quarterly* 66, 267–279.

Barton, David, E. Calvin Beisner, Charles W. Colson, James C. Dobson, John C. Hagee, D. James Kennedy, Richard Land and Donald E. Wildmon. 2006. "Interfaith Stewardship Alliance's Letter to the National Association of Evangelicals on the Issue of Global Warming." Interfaith Stewardship Alliance. Available at http://www.pbs.org/moyers/moyersonamerica/green/nae_response.pdf (accessed September 24, 2012).

Beisner, E. Calvin. 1993. "Are God's Resources Finite? A Group of Christian Leaders Claim They Are, but Does the Claim Square with the Evidence?" *World* 8.27, 10–13.

——. 1997a. "Shining a Light in a Dark Place." *Religion and Liberty* 7.2, 8–10.

——. 1997b. *Where Garden Meets Wilderness: Evangelical Entry into the Environmental Debate*. Grand Rapids, MI: Acton Institute for the Study of Religion and Liberty and W.B. Eerdmans.

Berry, Wendell. 1993. *Sex, Economy, Freedom and Community: Eight Essays*, 1st edn. New York: Pantheon Books.

Brumley, Albert E. 1938. *Albert E. Brumley's Book of Radio Favorites: A Collection of Sacred, Sentimental, and Folk Songs*. Powell, MO: Stamps-Baxter.

Carlson, Doug. 2009. "The 'Cap-and-Tax' Fiasco." *erlc.com*. Available at http://erlc.com/article/the-cap-and-tax-fiasco/ (accessed September 24, 2012).

Casciato, Tom (dir.). 2006. "Is God Green?" In *Moyers on America*, ed. F. Firestone and J.D. O'Neill. U.S.A.: PBS.

Catton, William R., Jr. and Riley E. Dunlap. 1980. "A New Ecological Paradigm for Post-Exuberant Sociology." *American Behavioral Scientist* 24.1, 15–47.

Cooper, John W. 2006. *Panentheism, the Other God of the Philosophers: From Plato to the Present*. Grand Rapids, MI: Baker Academic.

Cooperman, Alan. 2006. "Evangelicals Will Not Take Stand on Global Warming." *Washington Post*, February 2. Available at http://www.washingtonpost.com/wp-dyn/content/article/2006/02/01/AR2006020102132.html (accessed September 24, 2012).

De Vos, Peter, Loren Wilkinson and Calvin Center for Christian Scholarship. 1980. *Earthkeeping: Christian Stewardship of Natural Resources*. Grand Rapids, MI: Eerdmans.

DeWitt, Calvin B. 1994. *Earth-Wise: A Biblical Response to Environmental Issues*. Grand Rapids, MI: CRC Publications.

EthicsDaily.com Staff. 2008. "Baptist Leaders Urge Senate to Pass 'Strongest Possible' Bill on Climate Change." *EthicsDaily.com*. Available at http://www.ethicsdaily.com/baptist-leaders-urge-senate-to-pass-strongest-possible-bill-on-climate-change-cms-12572 (accessed September 24, 2012).

Evangelical Climate Initiative. 2006. "Climate Change: An Evangelical Call to Action." *Christiansandclimate.org*. Available at http://christiansandclimate.org/statement (accessed September 24, 2012).

Fowler, Robert Booth. 1995. *The Greening of Protestant Thought*. Chapel Hill: University of North Carolina Press.

Friedman, Richard Elliott. 1995. *The Hidden Face of God*. San Francisco, CA: Harper San Francisco.

Glacken, Clarence J. 1967. *Traces on the Rhodian Shore: Nature and Culture in Western Thought from Ancient Times to the End of the Eighteenth Century*. Berkeley: University of California Press.

Goodstein, Laurie. 2006. "Evangelical Leaders Join Global Warming Initiative." *New York Times*, February 8. Available at http://www.nytimes.com/2006/02/08/national/08warm.html (accessed September 24, 2012).

Grossman, Cathy L. 2008. "Evangelicals Less Worried about Global Climate Change." *USA Today*. Available at http://www.usatoday.com/news/religion/2008-09-18-baylor-environment_N.htm(accessed September 24, 2012).

Gryboski, Michael. 2012. "Evangelical Leader Sees Connection between Climate Change and Poverty." *Christian Post*, April 26. Available at http://www.christianpost.com/news/evangelical-leader-sees-connection-between-climate-change-and-poverty-73951/ (accessed September 24, 2012).

Interfaith Council for Environmental Stewardship. 2000. *The Cornwall Declaration on Environmental Stewardship*. Available at http://www.cornwallalliance.org/articles/read/the-cornwall-declaration-on-environmental-stewardship/ (accessed October 15, 2012).

Kaylor, Brian. 2009. "Baptist Leaders Endorse 'American Clean Energy and Security Act,'" *EthicsDaily.com*. Available at http://ethicsdaily.com/baptist-leaders-endorse-american-clean-energy-and-security-act-cms-14405 (accessed September 24, 2012).

Lamb, Gregory M. 2010. "Conservative Evangelicals Embrace God and Green." *Christian Science Monitor*. Available at http://www.csmonitor.com/Environment/2010/0325/Conservative-Evangelicals-embrace-God-and-green (accessed September 24, 2012).

Land, Richard. 2009a. "House Votes for Cap and Trade." Ethics and Religious Liberty Commission. Available at http://erlc.com/article/cap-and-trade-vote-today/ (accessed September 24, 2012).

——. 2009b. "Tell Senators to Oppose Global Warming Tax." Ethics and Religious Liberty Commission. Available at http://erlc.com/article/tell-senators-to-oppose-global-warming-tax/ (accessed September 24, 2012).

Land, Richard and Barrett Duke. 2011. "Legislative Agenda for 2011." Ethics and Religious Liberty Commission. Available at http://erlc.com/article/legislative-agenda-for-2011/ (accessed September 24, 2012).

Larsen, David. 2005a. "Evangelical Environmental Network." In *Encyclopedia of Religion and Nature*, ed. B. Taylor, pp. 624–625. London: Continuum International.

——. 2005b. "Interfaith Council for Environmental Stewardship." In *Encyclopedia of Religion and Nature*, ed. B. Taylor, pp. 849–850. London: Continuum International.

Lewis, Andy. 2005. "Policy Statement on Global Warming." *Erlc.com*. Available at http://erlc.com/article/policy-statement-on-global-warming/ (accessed September 24, 2012).

Little, Amanda. 2005. "Cizik Matters: An Interview with Green Evangelical Leader Richard Cizik." *Grist*. Available at http://grist.org/article/cizik/ (accessed September 24, 2012).

Marrs, Texe W. 1990. *Texe Marrs Book of New Age Cults and Religions*. Austin, TX: Living Truth Publishers.

McCammack, Brian. 2007. "Hot Damned America: Evangelicalism and the Climate Change Policy Debate." *American Quarterly* 59.3, 645–668.

Merritt, Jonathan. 2011. Personal interview by Andrew Szasz, February 4.

National Association of Evangelicals. 2004. "For the Health of the Nation: An Evangelical Call to Civic Responsibility." National Association of Evangelicals, Washington, DC.

Oreskes, Naomi and Erik M. Conway. 2010. *Merchants of Doubt: How a Handful of Scientists Obscured the Truth on Issues from Tobacco Smoke to Global Warming*, 1st U.S. edn. New York: Bloomsbury Press.

Parham, Robert. 2009. "Supporting House Climate Bill Gives Concreteness to Moral Imperatives." *EthicsDaily.com*. Available at http://www.ethicsdaily.com/supporting-house-climate-bill-gives-concreteness-to-moral-imperatives-cms-14280 (accessed September 24, 2012).

Pew Forum on Religion and Public Life. 2006. "Many Americans Uneasy with Mix of Religion and Politics." *www.pewforum.org*. Available at http://www.pewforum.org/Politics-and-Elections/Many-Americans-Uneasy-with-Mix-of-Religion-and-Politics.aspx (accessed October 2, 2012).

——. 2009. "Faith in Global Warming: Religious Groups' Views on Earth Warming Evidence." Available at http://pewresearch.org/pubs/1194/global-warming-belief-by-religion (accessed September 24, 2012).

Schaeffer, Francis A. 1970. *Pollution and the Death of Man: The Christian View of Ecology*. Wheaton, IL: Tyndale House Publishers.

Sheppard, Kate. 2011. "Evangelicals Rally for Mercury Rules." *Mother Jones*. Available at http://www.motherjones.com/blue-marble/2011/09/evangelicals-rally-mercury-rules (accessed September 24, 2012).

Southern Baptist Convention. 1970. "Resolution on the Environment." Available at http://sbc.net/resolutions/amResolution.asp?ID=452 (accessed September 24, 2012).

——. 1974. "Resolution on Stewardship of God's Creation." Available at http://sbc.net/resolutions/amResolution.asp?ID=453 (accessed September 24, 2012).

——. 1990. "Resolution on Environmental Stewardship." Available at http://sbc.net/resolutions/amResolution.asp?ID=485 (accessed September 24, 2012).

——. 2006. "On Environmentalism and Evangelicals." Available at http://sbc.net/resolutions/amResolution.asp?ID=1159 (accessed September 24, 2012).

——. 2007. "On Global Warming." Available at http://sbc.net/resolutions/amResolution.asp?ID=1171 (accessed September 24, 2012).

——. 2010. "On the Gulf of Mexico Catastrophe." Available at http://sbc.net/resolutions/amResolution.asp?ID=1207 (accessed September 24, 2012).

Southern Baptist Environment and Climate Initiative. 2008. "A Southern Baptist Declaration on the Environment and Climate Change." Available at http://baptistcreationcare.org (accessed September 24, 2012).

Stein, Adam. 2008. "Southern Baptists Shift Stance on Climate." *Terrapass.com*. Available at http://www.terrapass.com/politics/southern-baptist-leadership-shifts-stance-on-climate-change/ (accessed September 24, 2012).

Steinfels, Peter. 1996. "Evangelical Group Defends Laws Protecting Endangered Species as a Modern 'Noah's Ark,'" *New York Times*, January 31. Available at http://www.nytimes.com/1996/01/31/us/evangelical-group-defends-laws-protecting-endangered-species-amodern-noah-s-ark.html (accessed September 13, 2012).

Stoddard, Ed. 2009. "U.S. Conservative Christians Sound 'Cap and Trade' Alarms." *FaithWorld on Reuters Blogs*. Available at http://blogs.reuters.com/faithworld/2009/06/29/us-conservative-christians-sound-cap-and-trade-alarms/ (accessed September 24, 2012).

Strode, Tom. 2008. "'We Get It!' Environmental Campaign Launches with Goal to Be Biblical, Factual." *Baptist Press*. Available at http://www.bpnews.net/printerfriendly.asp?ID=28076 (accessed September 24, 2012).

Tillich, Paul. 1951. *Systematic Theology*, vol. 1. Chicago, IL: University of Chicago Press.

Van Biema, David. 2008. "The Greening of the Baptists." *Time*. Available at http://www.time.com/time/nation/article/0,8599,1721132,00.html (accessed October 18, 2012).

We Get It! 2008. *WeGetIt.org* (accessed September 24, 2012).

White, Lynn. 1967. "The Historical Roots of Our Ecologic Crisis." *Science* 155.3767, 1203–1207.

Womersley, Mick. 2005. "National Religious Partnership for the Environment." In *Encyclopedia of Religion and Nature,* ed. B. Taylor, pp. 1158–1159. London: Continuum International.

Young Evangelicals for Climate Action. 2012. *Young Evangelicals for Climate Action.* Available at http://www.yecaction.org/ (accessed October 18, 2012).

15 Christian and Muslim climate activists fasting and praying for the planet

Emotional translation of "dark green" activism and green-faith identities

Maria Nita

Introduction

Scholars have noted that environmental activism often has a spiritual dimension (Kearns and Keller 2007; Oelschlaeger 1996) and, furthermore, that environmental spirituality may be regarded as a religion in its own right (Taylor 2010). Bron Taylor (2010) identified these seemingly diffuse forms of nature spiritualities as "dark green religion," a fast-growing form of religiosity with a coherent ideology and praxis despite its informal organization. As he further explores, many practitioners of dark green religion hope to create a new biocentric myth that will replace the anthropocentric Judeo-Christian worldview (2010: 99–100). A powerful motivator of environmentally salutary behaviors, dark green religion "seeks to end the world now unfolding, in order to prevent the end of the world as we have known it" (2010: 195).

My research investigated Christians and Muslims[1] involved in the Climate Movement and the Transition Towns Movement in Britain. Based on participant observation conducted between 2007 and 2010, I examine the processes and practices that allowed these activists to stay simultaneously involved in two ideologically distinct communities: that of their respective faith tradition and the "dark green" territory found in the Climate and Transition Towns movements.

My research shows that Christian and Muslim activists, whom I call "faith activists" in order to distinguish them from participants in climate events who did not belong to organized religious institutions, are in fact primarily motivated in their activism by their religious faith. That is, most of my religious informants attended to their ecological concerns as a way of practicing their faith, rather than because of civic or secular green values. Their faith had become imbued with ecological beliefs and practices, or perhaps hybridized with dark green beliefs and practices, through their involvement in the Climate Movement.

This process was not straightforward, however. Instead, the Christian and Muslim activists in my research struggled to negotiate their identities as religious people on the one hand and environmental activists on the other, a struggle I will refer to as a "cultural commute"[2] between identities. This struggle was amplified by the opposition toward institutionalized religion found in some of the networks that partake in the Climate Movement. Notwithstanding such pressures, faith activists often succeeded in merging their faith and activist identities in unpredictable and radical ways. I will endeavor to show here that the faith activists in my study were engaged in a process of cognitive and emotional translation of dark green elements into the language, prayers and rituals of their respective tradition.

The research field and methodology

Both the Climate Movement and Transition Towns started to take shape in 2005, in Britain, and within a few years achieved global representation (see Hopkins 2008: 8–15). The Climate Movement started with the Global Day of Climate Action, which has since become an annual event worldwide. In Britain, protest activities and protest camps focused around airport expansion (Heathrow in 2007), the coal mining industry and coal power stations (Drax in 2006, Kingsnorth in 2008 and Merthyr Tydfil in 2009). Historically speaking, the Climate Movement may be considered to be the most recent incarnation of the Green Movement, incorporating previous foci (such as animal rights, anti-nuclear, road protests and GM crops) alongside climate-related concerns, such as the need to reduce carbon emissions and the use of fossil fuels.

If the Climate Movement was predicated on protest, the Transition Towns Movement represented its lifestyle counterpart, focusing on community building as a means of laying the foundations for a zero carbon society. The Transition Towns Movement is a movement of intentional communities created in order to prepare for the dual catastrophes of climate instability and peak oil. It is, therefore, situated within the web[3] of networks associated with climate activism. It has a similarly countercultural ethos, being somewhat related to previous movements from the New Age spectrum, such as the Human Potential Movement.

To investigate how members of institutionalized religions (specifically, Christianity and Islam) participated in these movements, I conducted a multi-site ethnographic research and employed a mixed methodology, using both qualitative (in the form of unstructured and semi-structured interviews and participant observation) and quantitative methods (surveys) of data collection. I chose mixed methods, because research in this field was scarce and I wished to understand the macro-level of this field alongside the personal activist narratives that would unfold during interviews.

I participated in 39 climate- and transition-related events ranging from a day to over a week in duration. During this time I conducted a total of 38 interviews with Christians, Muslims, Buddhists, Druids and Pagans, as well as religiously unaffiliated activists.[4] I collected data according to the preferences of the networks I studied. The Muslim network met for monthly public forums and bi-monthly meetings, where I conducted participant observation in 2008 and 2009. In one of the Christian networks I studied, GreenSpirit, gatekeepers denied me participation as a researcher in its group activities or retreats, and so I gathered all my data by in-depth interviews with individual members.

I also administered two surveys (N=78) at two climate protest camps in London and Wales. This was a non-random, self-selected sample, as the surveys were made available in key areas around the protest camps, such as the welcoming tent, the communal kitchen and the main meeting tent. Geographically, I crisscrossed the country, moving from London to Nottingham, from Plymouth to Lancaster, from Cardiff to Totnes, from Bristol to Kent and from Suffolk to Milton Keynes. This was due to the necessity of following events as they unfolded at different places. In addition to the empirical data, I also made use of the various media (leaflets, magazines, recordings) published by faith groups and climate activists from many other, intersecting, networks.

After having discovered the existence and participation of Christian and Muslim groups in various climate campaigns through websites and internet blogs, I began to contact them during collective climate events. I met some of my Christian informants at the Global Day of Climate Action in December 2007, and I then met the representatives of the Muslim network at the Global Day of Climate Action the following year, in December 2008. My Christian informants came from different denominations, including Anglican, Catholic, Salvation Army and Quaker, and they almost always remained involved in their own home churches. Most of the Christians in my study belonged to the Christian Ecology Link (CEL), which is a national organization, but I also interviewed members of the GreenSpirit[5] network and Isaiah 58.[6] All these organizations draw members from multiple denominations, with the main one being the Church of England. Many Christian activists were involved with other networks and movements, such as the Student Christian Movement, Operation Noah, A Rocha and Christian Aid. Only a small number of the Christian participants came from evangelical churches. CEL had been in operation for the past 25 years and had over a hundred active members, and a much larger readership via its bi-monthly publication *Green Christian*. Not only were CEL members involved in local Transition initiatives, but also as an organization it sought to offer support to the Transition Towns Network, in the form of a flanking campaign called Churches in Transition (launched in 2009) that aimed to introduce other Christian churches to the Transition Towns Movement.

All the Muslim participants in my study were either second-generation immigrants in Britain or converts to Islam. Some activists publicized forthcoming ecological events in home mosques, and they had attempted to introduce some environmental literature to other local Islamic centers, yet these actions were not always successful, as their activism and radical ecological approach were regarded with some suspicion by mainstream practitioners. They were all members of the London Islamic Network for the Environment (LINE), a network that officially started in 2004 and changed its name in 2010 to Wisdom in Nature (WIN). The network had a smaller number of active members (fewer than 20), with a larger online or web community. They organized a monthly forum, which I attended as a participant observer. I was also invited to participate in two of their closed meetings, where the main members planned forthcoming events, publications and other media. They provided discussions, open forums, performances (poetry, music and drama), food sharing, and gardening and produced a monthly leaflet which was mainly distributed online. Both CEL and LINE/WIN, the two main networks in my study, wished to function as links between their respective faiths and the Climate and Transition movements.

Religion and identity

During the course of my research, identity, and particularly religious identity, emerged as a central concept governing faith activists' participation in the Climate and Transition Towns movements. While scholars have alternated between understanding identity as a stable property internal to the individual and understanding it as something that evolves organically through social interaction (see Giles and Coupland 1991: 196), in this chapter I draw on new approaches that tend to understand the self as relational (Andersen and Chen 2002; Fogel 2001; Hermans and Hermans-Jansen 2001). In this approach, "the self can be conceived of as comprised of multiple identities varying in salience" (Stryker, 1997, cited in Klandermans 2000: 246). Thus the relational model views the self as shaped by its relations with others. The identities that make up the relational self vary in salience, just as the relationships that forged these identities have varying degrees of significance for the individual. New relationships can extend the self in new directions, yet the newly emergent identities would tend to follow previous patterns of relating (see Andersen and Chen 2002: 638).

New theoretical approaches to identity also maintain that identity is an emotional and relational (rather than primarily cognitive) experience (Andersen and Chen 2002; Fogel 2001; Hermans and Hermans-Jansen 2001). Relational identity theorists maintain that identity arises through developing a sense of continuity and uniqueness through emotional experience (Kunnen et al. 2001: 227). Hence a sense of identity "comes into existence when the emotional experience that accompanies a currently ongoing relationship stabilizes into a primary orientation toward the relationship" (Kunnen et al. 2001: 207). Emotions have come to represent in recent interdisciplinary studies an area of central importance for understanding both identity and religious experience, as Douglas Davies (2011) advances in his *Emotion, Identity and Religion*, since "different religious traditions prefer certain patterns of emotion, fostering and managing them at individual and community levels of identity" (Davies 2011: 1).

The relational model of identity is particularly useful in making sense of my research, because the activists in my study were simultaneously involved in their faith networks as well as the networks operating in the Climate Movement, where they had to form and negotiate multi-layered and overlapping identities (see Wuthnow 2005: 276–278). Although their Christian or Muslim identity is not taken here to be rigidly defined or monolithic, by taking a view of identity as a relational, emotional and dynamic self-organizing process I investigate how activists who hold a broad traditional religious identity operate inside the Climate Movement or in the dark green territory of this movement. I therefore enquire into the processes that caused these activists to change, adapt or hybridize these identities so that they could become Green Christians or Green Muslims, given the conflicts between anthropocentric and biocentric values, and the subversive discourse that opposes institutionalized religion in these countercultural movements. My research indicates that their faith identity represents their primary or most salient social identity, and hence I propose here that religious environmentalists need to maintain a continuous sense of identity by extending their faith in a new (green) direction while still maintaining an identity centered in their faith. In this way faith activists capitalize on their existing relationships with individuals or communities and also form new relationships with other networks in the Climate and Transition movements and even a more reflective relationship between the self and the planet.

Using this relational model of identity, I will describe how the extension of faith activists' identities takes the form of an affective re-engagement by which green concerns become central to the expression of their faith. By examining green prayers I show that Christian prayer already contains the inner mechanisms to enable this change of direction by stimulating affective processes[7] and encouraging a re-engagement from a personal relatedness with God to a similar relationship towards the planet. I also suggest that activists aim to induce a love for the planet and its inhabitants through ecological rituals that draw from traditional religious practices. Similarly, through ritualized protest actions and performances, activists aim to provoke an emotional or cathartic transformation in their audience: the police, the media and the unengaged public. This is relevant, since new research shows that risk assessment and decision making concerning climate change are first and foremost an affective process rather than a primarily cognitive activity (Leiserowitz 2006).

Identity conflict and the "cultural commute" between activist and faith identities

Almost half of a total of 78 respondents who participated in my surveys (conducted at the London and Wales Climate Camps for Direct Action in 2009) self-identified as "atheists," "non-religious" or even "anti-religious."[8] My data suggests that such self-designations indicated that respondents opposed institutionalized religion. Indeed, I observed opposition to traditional forms of institutionalized religiosity on a variety of fronts. First, many of the Climate and Transition Towns activists with whom I spoke carefully avoided official associations with any particular religious group. In other cases respondents offered a disclaimer when they described their religious affiliation, making a point of respecting the religious neutrality of the camp. For example, one respondent wrote, "My religious beliefs *aren't affiliated to this camp* [my emphasis], but I revere nature and celebrate the seasons through ceremonies with the Order of Bards, Ovates and Druids." This suggests a tension between religious affiliation and participation in the climate events. Second, faith groups were discouraged from expressing their faith identity and forthrightly denied entry as a group in some climate protest camps. For example, at a London climate protest camp, in August 2009, camp rules explicitly stated that no religious groups should be allowed on site, which was explained by organizers as a strategy for maintaining unity. Finally, non-faith activists built on a common narrative that tended to exclude those who identified with traditional religious institutions. Frequently this resulted in identifying with groups that had experienced political or religious oppression, such as the seventeenth-century Diggers Movement.[9] In the climate activists' version of the Digger's Song (London, 2009), originally a folk ballad, the attitude towards organized religion is explicitly conveyed: "They make the laws / To chain us well / The clergy dazzle us with heaven / Or they damn us into hell / We will not worship / The god they serve." Hence a barrier to participating in climate-related activism that members of organized religions faced was the conflict between their own faith identities and the Climate Movement's identity.

The anti-religious ethos of the Climate Movement (which was less prominent yet still present in the Transition Movement) meant that Christians and Muslims often experienced identity conflicts. For example, faith activists were sometimes "closet Christians." One informant told me that his fellow transitioners knew he was Christian and they were "fine about it," a remark that suggested he was grateful to be tolerated. Faith activists often alternated between embracing and downplaying the contribution their faith made to their activism. For example, a Christian activist whom I will call Matt told a Christian congregation that was shortly going to join the Global Day of Climate Action march in December 2008 that "I am sick and tired of these environmentalists saying to me, 'Finally you've decided to join the bandwagon' . . . and I always tell them '*We* were the first environmentalists'" (my emphasis). Yet, only hours later, in an address to a mixed crowd of religious and non-religious activists in a public square in London, Matt explained that, whether people were coming into the Climate Movement from a faith perspective, like himself, or not, "*we* are *all* in this together" (my emphasis). His choice of words implies that, while he privately felt that his Christian background was an asset, he publicly felt the need to downplay its uniqueness in order to assure other participants that it would not get in the way of the Climate Movement's collective goals.

Identity conflict appeared to be a challenge within the Muslim community as well. One Muslim informant felt that the eco-spirituality that was put forward in collective climate events lacked the wisdom of religious traditions. To her this was like "re-inventing the wheel," since in her opinion Islam already had a rich spiritual tradition which eco-activists ignored. In an online environmental blog post entitled "Why Is It So Hard to Be Black and Green?" environmentalist Sylvia Arthur mused:

> Sometimes I think I'm just too black[10] to be green. With my required international travel (to see the relatives) and my hereditary love of meat it seems being black and green are two incompatible states. There are other times when I almost feel too green to be black. Trying to convince friends and family of the urgency of global warming is like trying to persuade the BNP [British National Party] that immigration is a good thing. As Muzammal Hussain, founder of the London Islamic Network for the Environment says, it's a constant "cultural commute."
>
> (Arthur 2009)

Arthur clearly struggled with maintaining the two identities simultaneously. The difficulty of doing so led to a feeling of "commuting" between identities—an experience that seemed to be especially common among Muslim activists.

The centrality of this commute to the Muslim experience is also demonstrated by other scholarly studies on the greening of Islam in Britain. As Gilliat-Ray and Bryant (2011) note, much of this greening has taken place through establishing gardening projects associated with mosques. While they argued that participating in such projects encouraged "a new sense of agency, belonging and ownership of local spaces" (2011: 284), my own research suggests that Green Muslims, just like Green Christians, were not particularly successful at getting their environmental message across to the mainstream of their respective institutions. Hence gardening in their local mosque was actually an indirect (and spatially peripheral) way of drawing attention to environmental issues in a practical way rather than by preaching this message to religious officialdom.

In the same way that faith activists were marginalized during climate-related activities, Christians and Muslims often described feeling marginalized in their respective faith communities and congregations, this time for being green. In one instance Green Christians taking part in a public event were asked to remember their experience of "coming out" as a Christian in the Green Movement as well as that of "coming out" as a Green in their church. The allusion to having a hidden, non-normative identity powerfully expresses the marginalization that these individuals perceived in both faith and climate communities. One Christian activist told me that in her Salvation Army congregation some people "ducked when they saw her," because they did not want her speaking to them about climate change.

Many Christian activists reported feeling rejuvenated by taking part in faith retreats with fellow Green Christians, where they could freely voice and express their ecological concerns, instead of participating in regular church activities where they felt they had to downplay these concerns. Some Church of England Christians told me in interviews that this disconnect had led them to place themselves "on the questioning side of the church," in order to constructively challenge the church's environmentally indifferent structure and organization.

In sum, Christian and Muslim climate activists faced a number of barriers to joining climate activism efforts, both in the form of disapproval and marginalization from the movement they wanted to be part of, and from within their own faith communities. Nevertheless, I found that they responded to the challenge by finding creative ways to merge the environmental and religious aspects of their identities.

Merging and extending faith identities

Christians and Muslims who become involved in the Climate and Transition movements adopted several strategies for incorporating these experiences into their religious identities.

First, Christian faith activists often adopted a biocentric outlook by merging "Nature" and "God" at a linguistic level. For example, in a lecture on Transition Towns given in 2010, a Christian vicar who was also a transitioner explained the core principles of transition living (in the context of a transition community) by saying: "You find out what God is doing and you get out of the way" and "You let God do the heavy lifting." In these statements he seemed to equate God to natural processes: God is Nature or works through Nature. Furthermore, Christian activists often spoke of the planet as "God's creation," or in some cases as "the body of the Lord." By merging "God" and "the planet," Green Christians could merge their faith and ecological concerns. Since for many other participants in the Climate Movement references to "nature" or "the planet" were tied to invocations of the earth as a deity[11]—meanings that would be distinctly uncomfortable to members of religions that reject pantheism as heresy—this linguistic merger seems to have helped smooth over a potential identity conflict, allowing them to use similar language to the non-religious climate activists, while also retaining a clearly theistic terminology. Interestingly, Muslim activists almost never referred to the planet as "God's creation," which may explain why they more frequently referred to experiencing "a cultural commute" and were less able to incorporate dark green experiences into their religious identities. Instead, they appeared to have appropriated the biocentric discourse of the Climate and Transition movements without alteration. This may be because the newly acquired biocentric discourse was in English, which they already used as a "secular" language, whilst they used Arabic for prayer.

Christians and Muslims also merged their green and faith identities by attaching religious significance to their climate activism. For example, Christians sometimes spoke about "transitioning" (that is, participating in a Transition Town Movement) as "witnessing." This was not a random association, as both transitioning and Christian witnessing shared a common emphasis on "preaching" through living an exemplary lifestyle, rather than through active evangelism. In other cases a new green practice would be done "in Jesus' name," a further means of fusing it with tradition. The tree in Figure 15.1 comes from a conference on Transition Towns held in 2009 by CEL in a church building. Note the biblical dictum at the top: "Jesus said, 'a tiny seed growing into a big tree will provide shelter and food.'" This was the only Christian text contained in the poster; the rest of it was entirely concerned with environmental news or facts. The biblical reference could be seen as a way of seeking to legitimize the green content by using language that would already be familiar to the Christian attendants. The lower branch of the tree had a prayer pocket attached to it, in which people were encouraged to place their own prayers, promises and thoughts. Placing a personal prayer on the poster might enable the reader to express his or her acceptance of the poster's message and thus become an active contributor rather than remain a detached observer.

In addition to attaching religious significance to their climate activism, activists also attached environmental significance to their religious practices. Muslim activists organized and took part in such practices as "fasting for the planet," which were similar to existing religious practices. As an annual group event, fasting for the planet during the day and breaking fast in the evening were in my informants' view a "transformative approach to caring for the Earth."

Figure 15.1 Transition tree at a Christian Ecology Link conference of Transition Towns, Devon, 2009

My research shows that my informants were not moved to adopt ecological lifestyles or become climate activists simply by learning environmental facts; their religious identity was of paramount importance in motivating them. Lucy, a Christian climate activist, described her journey to becoming a Green Christian in the following manner:

> I first came into contact with other Green Christians at an Operation Noah meeting [in London]. . . . It is amazing when you inject faith into the subject of climate change; suddenly you're not allowed to be resigned. Before I was an environmental officer at university. [Yet] they were separate issues [environmentalism and religion]. But when it becomes your faith, you have to believe you can change things.

Her occupation did not make Lucy an activist. Instead, she became an activist when climate change became enmeshed with her Christian identity. Another activist, called Helen, told me, "When you truly understand climate change *in your heart* [my emphasis], once it truly hits you, you have no choice but to act." Such references to the importance of emotional engagement with climate change were common occurrences in my field of research.

Prayers often provided activists with the opportunity to restructure their faith by retaining the external form of the prayer and inserting an environmental message in this template, thus literally "injecting faith into the subject of climate change," as my informant Lucy (quoted above) put it. Moreover prayer and ritual enabled activists to engage affectively with these green issues and frame them into the pattern of emotions of their respective tradition. This restructuring of faith could be so powerful as to cause a permanent schism between the old beliefs and practices and the new ones, leading participants into a possible "eco-reformation."[12] The following section will look at this process more closely.

Green prayer

Prayers provided faith activists with a safe way to connect with their own respective traditions, while simultaneously enabling them to reformulate their faith by introducing a new, environmental direction. Christian activists were quite experimental with the prayers they used both during specific worship occasions and during conferences, retreats and gatherings. They used prayers both to deplore the ecological crisis and to envision a new ecologically aware time, thus progressing from grieving to celebration. My data indicates that the lament–celebration axis functions as a paradigm for the ecological rituals and prayers I have examined. Religious fasting for the planet also moved participants from lament (in the form of bodily deprivation) to celebration (in the form of eating). In essence, lament and celebration represent the expression of key emotions, namely grief and joy, which are intensified through ritual means (see Davies 2011: 37–40). Douglas Davies considers grief "a pivotal emotion" that can provoke "identity depletion" (Davies 2011: 95–97), a depletion that may in turn pre-empt the development of a new identity.

In his functionalist study of biblical psalms, Walter Brueggemann (1980) distinguished between psalms that served as *psalms of disorientation*, that is, psalms which lamented, rejected or questioned the present order, and *psalms of re-orientation*, which offered a new order, and indicated "a movement from the disorientation marked by the lament" (Brueggemann 1980: 82). The ecological prayers I encountered during my fieldwork combined these two themes, as they often began with some kind of lament, suggesting the disorientation motif that Brueggemann described, and followed by showing what it was necessary for the participants (and often their audience) to do in the face of the climate crisis, resembling psalms of re-orientation. By such means green prayers provided a matrix for provoking a change of direction and for coping with a changing reality.

Green prayers demonstrated a high degree of innovation through their ecological content while also drawing from traditional resources. In the case of prayers recited at climate protest camps, the content tended to be very political, and thus innovative, yet in form the prayers made use of traditional Christian prosody, imitating the rhythm, stress and intonation of traditional prayers. For instance, the person leading a prayer might say "With corporate greed," and the group would answer "we would not comply," and so on. Participants thus used the traditional form and structure of prayers to preserve their links to a religious identity, which helped anchor the group in their respective tradition. At the same time, the new ecological content of the prayers made them relevant to climate activism. Prayer therefore seemed to function as a familiar and thus safe frame for learning new attitudes and behaviors (see Schechner 1993: 25).

Christian groups often demarcated themselves as Christian by offering prayer services that ran concomitantly with the main event, yet took place in a close proximity.[13] When I asked a group of Christian activists about their ritual at the gates of Kingsnorth power station in Kent in 2008 a few acknowledged that, by running their ritual concomitantly with the speeches, it competed for attention. Yet they felt the ritual was both meaningful and necessary. One Christian activist distinguished the prayers and ritual from more "playful" activities the group had considered doing, such as wearing dramatic costumes or reading the "Last Rites"[14] sacrament at the gates of the power station, as a symbol of its imminent demise. They had decided against this strategy, however, because

> We didn't want what we did to be a play. I mean we did think about costumes and things like that, but we wanted to be there praying, in spirit and in truth. . . . We wanted to do something that would *set us apart* [my emphasis] and bring Jesus in the situation.

Significantly, during one protest camp the Christian group held all its services inside the Christian Café tent, thus further "setting them apart" and leaving them spatially separated from the rest of the camp. On the other hand, non-religious activists participated in rituals or performances that were happening outside, on the main field or in the kitchens. This spatial separation underscores the tensions that existed between faith and non-faith activists, while also demonstrating how committed activists were to retaining their faith identities.

By preserving the liturgical *form* of prayers, groups were freer to innovate in terms of *content*, which allowed them another means of merging faith and climate activist identities. The prayer below, called a "Prayer of Confession," was part of a closing service during a Christian conference on sustainable transport, held in 2009. The format was that of a regular Christian service, with responses from the congregation and formal proceedings indicated by a prayer guide. The content of this prayer, however, was ecological:

> *Leader:* Giver of Life
> In the midst of a plundered earth/ poisoned water/ polluted air/ mountains of waste . . .
> We groan with creation
> *All:* Have mercy on us.

By lamenting the plundered earth and associating the congregation (and, by extrapolation, the Christian community or humanity) with the rest of creation, this prayer may represent an attempt to reform the anthropocentric division between humans and the rest of earth's inhabitants.

A Christian hymn sung by a Christian Ecology Link congregation on the eve of the 2009 Copenhagen Summit also contained this ecological message of interdependence:

> Oh mother earth
> We take to Copenhagen
> Our last, our deepest hope of all,
> To live within our means, no more delay, gone are the heady days
> When men could steal and stall
> Now look ahead, you delegates of nations,
> And see our children, what their lives are worth—
> To live at last in knowing we're relations,
> Christ's family and offspring of our mother earth.

We may note here that most written references to "the earth" or "mother earth" were not capitalized, which was likely intended to preserve the distinction between "God" and "creation." However, since prayers are typically addressed to God, by addressing "mother earth" instead, Green Christians are relating personally with the earth. The fusion of the traditional religious practice of prayer with such poignant ecological concern provides evidence that ritual and prayer serve a crucial role in the ecological reform some Abrahamic religionists have become engaged in. Furthermore, considering Davies's position whereby religious traditions promote different patterns of emotion (Davies 2011), green prayer provided religious activists with the opportunity to situate their ecological concerns onto a familiar emotional template.

Conclusions

Even though the Christian and Muslim activists in my study experienced identity conflicts as a result of being involved in both their faith communities and the Climate and Transition movements, it seems they were able to calibrate these seemingly incompatible traditions through a process of extending participants' faith identities. This resulted in conflating what some would argue were doomed to remain incongruous concepts, such as the transcendent Judeo-Christian fatherly divine and Mother Nature or "mother earth."

I suggested here that Christian and Muslim activists deployed a variety of tools, including religious language, prayer and ritual, to demarcate themselves from the other activist networks and maintain their primary faith identities. Moreover faith activists used prayers and rituals to lament the old and celebrate the new, and were thus able to change direction and make ecological concerns relevant to their faith. This, I argued, extended their primary identities and was in essence a process of emotional translation or "a transformative approach to *caring* for the Earth [my emphasis]," as the Muslim activists actively practiced and promoted. Lastly, religious practices provided faith activists with an exemplary model of how to relate to the planet as one entity.

Despite their heterodoxy and fringe status, the heterogeneous religious groups in my research gathered activists from such a diverse Christian spectrum as to indicate that a broader ecological re-orientation may be at hand. I speculated that the lament–celebration axis in prayer and ritual could provide traditional religions with a powerful mechanism for framing green concerns in familiar patterns of emotions whilst allowing religionists to articulate a new green religion.

Notes

1 I originally intended to provide a comparative study of Christian, Muslim and Jewish grassroots environmental organizations. In the end, I limited my research to Christians and Muslims, because Judaism was not represented by grassroots groups active in the Climate Movement at the time I undertook my fieldwork (2007–10).

2 The phrase "cultural commute" was first used by Muzammal Hussain, chair of Wisdom in Nature.

3 Dominic Corrywright's (2004) web model for alternative spirituality networks allows researchers to identify central and intersecting networks in a web of networks and is well suited to looking at the highly heterogeneous field of climate activism.

4 Interviewees identified themselves as belonging to the following networks: 5 belonged to the London Islamic Network for the Environment/Wisdom in Nature; 11 belonged to Christian Ecology Link; 5 belonged to Isaiah 58; 6 belonged to other Christian networks (such as various evangelical communities); 5 belonged to the GreenSpirit network; and finally 6 were activists I interviewed at Climate Camp, among which 2 identified as Pagan and 1 as a Buddhist, with the remaining 3 not religiously affiliated.

5 GreenSpirit was a national network, with members across the UK, and represented the British counterpart of the worldwide Creation Spirituality network.

6 At the time my research started Isaiah 58 was a small network of Christian anarchists. The group did not have any official literature, and their activities were not made public.

7 See Douglas Davies's (2011: 37–40) discussion on ritual as an intensifier of emotion.

8 This figure is based on my interpretation and on emic descriptors that respondents gave in response to open-ended questions about their religious background.

9 The Diggers or True Levellers were agrarian communists led by Gerrard Winstanley in England in 1649 (see Campbell 2010: 129). Although the Diggers experienced opposition from some church officials there are no historical accounts to confirm their own directed dissent against the clergy.

10 Here the author equates being black to being Muslim.

11 "Nature doesn't do bailouts!" was a frequent slogan at protest camps, implying that nature does not have a special relationship with humanity.

12 I suggest an analogy between the ecological re-orientation of certain Christian organizations and the Protestant Reformation in sixteenth-century Europe, which led to a schism in Western Christianity and the creation of Protestantism.

13 This may constitute a spatial claim or indicate a "disputed territory." See for example Marion Bowman's (2004) discussion of concomitant mirror processions held by Christians and Goddess worshipers in Glastonbury as means of asserting identity.

14 The Last Rites is one of seven Christian sacraments, which is administered to the dying or gravely ill.

References

Andersen, Susan M. and Serena Chen. 2002. "The Relational Self: An Interpersonal Social-Cognitive Theory." *Psychological Review* 109, 619–645.

Arthur, Sylvia. 2009. "Why Is It So Hard to Be Black and Green?" *Environment Blog: The World's Leading Journalists on Climate Energy and Wildlife*. Available at http://www.earth-stream.com/Earth/Community-and-Politics/Why-is-it-so-hard-to-be-black-and-green-%7C-Sylvia-Arthur_18_193__221515.html.

Bowman, Marion. 2004. "Procession and Possession in Glastonbury: Continuity, Change and the Manipulation of Tradition." *Folklore* 115.3: 273–285.

Brueggemann, Walter. 1980. "Psalms and the Life of Faith: A Suggested Typology of Function." *Journal for the Study of the Old Testament* 17.1, 3–33.

Campbell, Heather M. 2010. *The Britannica Guide to Political and Social Movements That Changed the Modern World*. New York: Britannica Education Publishing.

Corrywright, Dominic. 2004. "Network Spirituality: The Schumacher–Resurgence–Kumar Nexus." *Journal of Contemporary Religion* 19.3, 311–327.

Davies, Douglas J. 2011. *Emotion, Identity and Religion: Hope, Reciprocity and Otherness*. Oxford, U.K.: Oxford University Press.

Fogel, Alan. 2001. "A Relational Perspective on the Development of Self and Emotion." In *Identity and Emotion: Development through Self-Organisation*, ed. Harke A. Bosna and Saskia E. Kunnen, pp. 93–115. Cambridge, U.K.: Cambridge University Press.

Giles, Howard and Nikolas Coupland. 1991. *Language: Contexts and Consequences*. Buckingham, U.K.: Open University Press.

Gilliat-Ray, Sophie and Mark Bryant. 2011. "Are British Muslims 'Green'? An Overview of Environmental Activism among Muslims in Britain." *Journal for the Study of Religion, Nature and Culture* 5.3, 284–306.

Hermans, Hubert J.M. and Els Hermans-Jansen. 2001. "Affective Processes in a Multivoiced Self." In *Identity and Emotion: Development through Self-Organisation*, ed. Harke A. Bosna and Saskia E. Kunnen, pp. 120–141. Cambridge, U.K.: Cambridge University Press.

Hopkins, Rob. 2008. *The Transition Handbook: From Oil Dependency to Local Resilience*. Totnes, U.K.: Green Books.

Kearns, Laurel and Catherine Keller, eds. 2007. *Ecospirit: Religions and Philosophies for the Earth*. New York: Fordham University Press.

Klandermans, Bert. 2000. "Social Movements: Trends and Turns." In *The International Handbook of Sociology*, ed. Arnaud Sales and Stella Quah, pp. 236–254. London: Sage.

Kunnen, Saskia E., Harke A. Bosna, Cor P.M. Van Halen and Matty Van der Meulen. 2001. "A Self-Organisational Approach to Identity and Emotions: An Overview and Implications." In *Identity and Emotion: Development through Self-Organisation*, ed. Harke A. Bosna and Saskia E. Kunnen, pp. 202–231. Cambridge, U.K.: Cambridge University Press.

Leiserowitz, Anthony. 2006. "Climate Change, Risk Perception and Policy Preferences: The Role of Affect, Imagery and Values." *Climatic Change* 77.1, 45–72.

Oelschlaeger, Max. 1996. *Caring for Creation: An Ecumenical Approach to the Environmental Crisis*. New Haven, CT: Yale University Press.

Schechner, Richard. 1993. *The Future of Ritual: Writings on Culture and Performance*. London: Routledge.

Taylor, Bron. 2010. *Dark Green Religion: Nature, Spirituality, and the Planetary Future*. Berkeley: University of California Press.

Wuthnow, Robert. 2005. *America and the Challenges of Religious Diversity*. Princeton, NJ: Princeton University Press.

16 "Healing the Land" in the Canadian Arctic

Evangelism, knowledge, and climate change

Noor Johnson

> When I shut up the heavens so that there is no rain, or command locusts to devour the land or send a plague among my people, if my people, who are called by my name, will humble themselves and pray and seek my face and turn from their wicked ways, then will I hear from heaven and will forgive their sin and will heal their land.
>
> (2 Chr. 7:13–14[1])

On a Wednesday evening in July 2009, a special church service was held in Clyde River, an Inuit community of around 1,000 people on Baffin Island, north of the Arctic Circle, in the Canadian territory of Nunavut. Usually, services are held on Sundays, but this week the church community was welcoming a group of visitors from Fiji. I was in Clyde River to conduct anthropological research, and I had heard through the general excited gossip of the community about the men and women from Fiji who were returning to Clyde River for the third time in as many years. I knew that they practiced an evangelical form of Christian worship, with song and dance, which contrasted with the relatively formal Anglican services that were normally held in the church. And so, when a friend invited me to accompany her to the service, I tagged along, curious to see what the fuss was about.

The pews of the small, plain church hall were filled that evening. As we waited for the service to begin, children ran up and down the aisles shouting and laughing, and women sat talking, their babies hidden inside their *amautiit*.[2] After welcome speeches from the *Inuk*[3] church minister and the mayor, a large *qallunaat* (non-Inuk or white) man in neatly ironed slacks and a button-down shirt went to the podium. He introduced himself as Roger Armbruster from Canada Awakening Ministries in Niverville, Manitoba.

"I'm so pleased to be back in Clyde River," he said. "When I got off the plane, I was greeted by the sight of tall, green grass and flowers. There was a great light reflected from the ground. I can see that the Lord's work is alive and well in Clyde River and that all the hard work you have been doing to heal the land and to heal yourselves has been working." Armbruster was accompanying two men from Fiji, Pastor Vuniani Nakauyaca and his son, Savenanca Nakauyaca, or Savi. When the younger Nakauyaca, a heavy-set man in his 30s, came to the podium to preach and testify, he shared more Good News with the congregation. They had just come from the Inuit hamlet of Pangnirtung, he said, where community members had been experiencing many changes through the power of prayer and healing. "In Pangnirtung, they told me that there are two new species of fish they have never fished before that they are now able to catch. And, in addition to new fish, Pangnirtung is now mining diamonds and gold," Nakauyaca Jr. said.

This was my introduction to Healing the Land (HTL), a Gospel-based movement founded by Nakauyaca Sr. that has spread through evangelical networks from Fiji to Papua New Guinea to the Canadian Arctic and beyond. Many of the observations of change that the group attributed to prayer, such as more grass and flowers and new species of fish, were also changes that had been observed in the Arctic by scientists researching the impacts of climate change. In contrast to sentiments of skepticism about the value of climate change research I had encountered in the community, there was clearly something very compelling about HTL's vision of change, which had attracted more than 100 individuals to attend the service that evening. Many in attendance came forward and testified to the power of the movement and its impacts, sharing personal observations of environmental changes that they related to the prayer and healing work.

HTL offered an alternate narrative of environmental change to more dominant narratives that were also present in Clyde River. Over the previous decade, the community had hosted a number of researchers and other visitors interested in climate change issues. These projects included: research on glaciation patterns (Briner, Davis and Miller 2009), sea level rise and coastal erosion, freshwater availability, and Inuit observations of climate change (Fox 2004; Gearheard et al. 2006; GN 2005); a community-based sea ice monitoring initiative (Huntington et al. 2009); the collection of testimony for a legal petition on climate change and human rights (Watt-Cloutier 2005); filming for a BBC production entitled *Frozen Planet*; and a "Global Warming 101" dogsledding and education expedition led by Arctic explorer Will Steger. The way that Clyde River residents understood environmental change was therefore shaped by engagements with visitors who arrived with very different environmental change narratives, as well as by observations of residents themselves.

In this chapter, I compare the narratives of environmental change offered by HTL with other perspectives on change, including scientific and Inuit knowledge. For scientific researchers, particular manifestations of change, such as melting glaciers or change in sea ice dynamics, were indications of global and regional trends linked to anthropogenic climate change. For Inuit, understandings of environmental dynamics were based on careful observation and extensive engagement over time. They were also rooted in a social cosmology that traditionally viewed human and animal agency as interconnected. Their understandings have often contrasted with perspectives from ecology and wildlife biology that have contributed to the establishment of hunting quotas on particular species, leading to a mistrust of scientific research. Furthermore, scientific and media discourses about climate change are often alienating to Inuit, who feel left out of the production of scientific knowledge, and who may prioritize other, more immediate areas of concern for research and policy. Among these are challenges of health and healing of trauma associated with the fast pace of cultural change and with the impact of colonial governance practices. In this context, the spiritual approach of HTL presented an alternate way of engaging with change that resonated with local needs and priorities.

This chapter draws on ethnographic research focusing on climate change and institutions conducted in Clyde River in 2009 (July–September) and 2010 (March–May). This research included in-depth semi-structured interviews with community members and leadership from the HTL movement, and observation of church services. Other sources included the *Manual for Healing the Land* (Nakauyaca and Ani 2007), a DVD of a healing ritual from Clyde River shared with me by Armbruster, and blog posts and updates about Healing the Land from the Canada Awakening Ministries website.

Christianity and evangelism in the Canadian Arctic

Anglican missionaries first introduced Christianity to the eastern Canadian Arctic at the end of the nineteenth century (Laugrand and Oosten 2010). Prior to the introduction of Christianity, shamans or *angakkuit* served as intermediaries between the human world and the world of non-human beings, assisted by *tuurngait* or helping spirits. The shamans played an important role as healers of physical illness, and they used their powers to help address transgressions that caused fissures in human–environment relations (Tungilik and Uyarasuk 1999). As Christianity spread, shamans sent away their helping spirits, since they were no longer needed (Laugrand and Oosten 2010; Tungilik and Uyarasuk 1999).

The introduction of the Healing the Land (HTL) ministry to Inuit communities in the mid-2000s followed several decades of growth for evangelical churches in the Canadian Arctic. Armbruster had been a central figure in promoting evangelism in the region since the 1970s, working with a loosely affiliated network of church groups in the United States and Canada (Laugrand and Oosten 2010). In 2004, Armbruster travelled with Nunavut political leader Tagak Curley to Fiji, where they met Nakauyaca Sr. In 2006, they brought Nakauyaca Sr. to Nunavut for the first time, organizing a regional Bible workshop in the town of Pangnirtung, which several church leaders from Clyde River attended. In 2007, Armbruster and a team from Fiji arrived in Clyde River to share the Good News, returning the following two summers. As of 2011, HTL had held healing services in at least nine Nunavut settlements, as well as a number of Inuit communities in the Nunavik region of northern Quebec.

While evangelical Christian churches enjoyed a strong following in some communities of Nunavut, Clyde River did not have a formal evangelical congregation. Although the HTL service was well attended and featured testimony from residents of different ages, this could partly be explained by HTL's entertaining worship style with its focus on song and dance. On a Wednesday night with little else happening in Clyde River, many people may have attended out of curiosity more than religious conviction. On the whole, Clyde River elders tended to be more religious than younger community members; some younger adults shared with me their discomfort with the way that some forms of Christianity denigrated traditional Inuit beliefs and told me they were against formal religion. Nevertheless, the high level of participation and testimony offered from residents during the service indicated that HTL's narrative of change resonated with participants.

Healing the Land theology and rituals of stewardship through repentance

After attending HTL's public worship service, I decided to take Armbruster and Nakauyaca Jr. up on their open invitation to visit them at the house where they were staying. When I arrived, they were eating lunch, but they courteously invited me to sit with them in the living room, where they answered my questions about HTL's theology.

"You should read Leviticus 26," Nakauyaca Jr. advised me. "It explains how God cursed the land because of the sins of the people." Armbruster offered me a copy of *A Manual for Healing the Land* (Nakauyaca and Ani 2007) (hereafter referred to as the Manual), a guide that outlined a series of teachings drawn from the scriptures.

The ethic of stewardship in the Manual centered on several key points. The earth and all its creatures were viewed as "belonging to God" who is the Creator (Ps. 24:1). Stemming from this principle, the Manual described the responsibility of stewardship: "Those to whom land has been entrusted are responsible to be good stewards, for one day they will be held accountable" (Matt. 25: 14–30). Healing the Land's vision of stewardship was based on avoidance of sin, since "the sinner defiles the ground" (Lev. 18: 20–28). Sin also caused animals to suffer (e.g., Jer. 14: 5–7). Since environmental degradation was understood to result from sin, the basic responsibility of stewardship was therefore to receive the Gospel and repent.

According to Nakauyaca Jr., there were four major sins that defiled the land: bloodshed (murder); adultery or immorality; breaking the covenant; and idolatry. According to the Manual, immorality included fornication, prostitution, pornography, homosexuality, abuse, and bestiality (Nakauyaca and Ani 2007: 31). "Breaking the covenant" referred to divorce (2007: 33), although Armbruster also used the term to refer to broken treaties between the government of Canada and Inuit and First Nations communities. The Manual explained that idolatry, which is related to witchcraft and "demonic oppression," included "worshipping . . . anyone or anything else other than God" (2007: 29). According to this theology, Inuit were guilty of idolatry before they converted to Christianity, although this was never explicitly stated in the interviews I conducted with HTL leaders or in church services.

"There were some good shamans and some that worked with the devil," Armbruster explained. "The good shamans," he continued, "were the ones who paved the way for the arrival of Christianity. But the bad ones, those who refused to convert, used their powers for evil" and were implicated in the sin of idolatry.[4] By recognizing both "good" and "bad" shamans, Armbruster tried to strike a balance of condemning idolatry while also respecting Inuit traditions.

One of the central tenets of HTL was that defilement from sin was transmitted from one generation to the next until it was "healed." As Nakauyaca Jr. declared from the pulpit during the worship service: "The sins of the forefathers can cause a curse upon the land that will last until the third or fourth generation." In order to "heal" intergenerational sins and restore proper communal relations, these sins had to be identified and named publicly.

The ritual process of this work was clearly delineated in the Manual. First, sins were identified through conducting outreach and household visits, and particularly by talking with elders who had a longer memory of past events. Specific locations on the land were then chosen for healing rituals based on their association with past events. In Clyde River, for example, a healing ritual was conducted across the inlet where the community was located before the government moved it to its current site in the 1960s.

The next step involved a public confession of sin. During the ritual in Clyde River, community members recounted events and behavior that they associated with the former community location and considered to be defiling or sinful; participants then asked for forgiveness from the individuals and families who had been affected. Once the parties involved had repented, the facilitator made a statement rededicating the land to God. An anointing salve made up of oil, salt, and water was poured on the ground. As Armbruster explained, at this point, the sin that defiled the land was "transformed and released."

These activities required the intermediating practice of intercession in which the group as a whole prayed to God on behalf of those who had sinned. Intercession was central to Christian missionary work and practices of conversion, but it also had its parallels in pre-Christian practices in the eastern Arctic. Transgression of social norms and taboos resulted in acts of retaliation by the non-human world, which could be healed only through confession. It was the role of the shaman to seek out or even force confessions that were not forthcoming and to then conduct the proper rituals to restore balance (Laugrand and Oosten 2010). As such, the rituals of HTL, while particular to the evangelical Christian context through which they were introduced, also had parallels in pre-Christian practice.

God's miracles: HTL and scientific interpretations of change

HTL theology placed a significant emphasis on miraculous manifestations of communal prayer and healing. Miracles took different forms in different places, but HTL narratives focused on miracles associated with the environment. Nakauyaca Jr. told me, for example, that: "In my own village, after the healing work we did, a river that was heavily polluted, that the scientists said we shouldn't drink from, became restored overnight."

Armbruster shared an example from Fiji about coral reef destruction from bleaching, which is the process of warm water temperatures causing coral to expel the symbiotic algae that gave them their color (Kaufman 2004).

Substantial bleaching had occurred in 2000 and 2002, yet some reefs had apparently grown new algae and come to support life again. "Scientists can't explain it," Armbruster said. From his perspective, these reefs were restored through the HTL process.

While Armbruster's theory of causation was faith based, he adopted the scientific community's interpretation of coral bleaching as ecosystem decline and regrowth as ecosystem renewal. His engagement with scientific narratives points toward the complex ways that individual actors engage with multiple frames to make sense of environmental change. While evangelical leaders may reject science in certain contexts, they also reference scientific knowledge at other times, reinterpreting it to fit into faith-based understandings.

In the Arctic, HTL's restoration narrative directly contradicted scientific interpretations of climate change indicators, particularly those relating to plant abundance and the northward shift of the tree line. The Arctic Climate Impact Assessment, a well-respected, comprehensive study assessing the impacts of climate change on the Arctic, stated that "arctic [plant] diversity is very likely to respond strongly and rapidly to high-latitude temperature change" (ACIA 2005: 257). Scientists have described how rapid warming and the early onset of the spring ice melt had resulted in a longer growing season (Hudson and Henry 2009), with plants flowering up to 20 days earlier (Høye et al. 2007; Post et al. 2009). Ranges of low Arctic trees and shrubs have extended northward (Post et al. 2009), and a climate modeling study predicted that up to 40 percent of Canada's northern tundra might be replaced by shrubs and boreal forest by the end of the century (Feng et al. 2011). Scientists therefore viewed plant growth as an indication of larger, systemic change that they predicted would be detrimental to ecosystem stability and would add stressors to certain species, including marine and terrestrial mammals that Inuit hunt.

In contrast, HTL leaders interpreted plant growth as an indicator of restoration and renewal attributed to their healing work in Arctic communities. When the HTL team arrived in Clyde River in late July, Arctic cotton (*Eriophorum*), a type of sedge grass with a white, cottony head, was blooming so profusely that some areas around town looked as though they were carpeted in a soft, white blanket. In his opening remarks for the worship service, Armbruster referred to the plentiful Arctic cotton as a "great light reflected from the ground," evidence of God's grace and forgiveness. He expanded further on his vision of environmental restoration on the Canada Awakening Ministries website:

> I am told that now in some places, even in the High Arctic, edible berries are beginning to grow. . . . There has never been a berry season in the High Arctic—only desolate wilderness . . . trees are now starting to appear above the ground. . . . Could it be that the treeless Arctic tundra was once covered with trees, and was once like a garden, and that it will once again be covered with trees, and become like a garden once again?
>
> (In Laugrand and Oosten 2007: 248)

In this example, Armbruster evoked an Edenic metaphor to reinforce his claim that abundance of plants on the tundra was evidence of restoration to a past state of glory. His interpretation of plant growth and the northern expansion of the tree line directly contradicted scientific explanations that equated these phenomena with anthropogenic climate change; instead, they were interpreted through the HTL narrative as "miraculous" manifestations of the healing power of prayer.

While HTL leaders attributed a variety of observed environmental changes to their healing work, the miracles that featured most prominently in Clyde River testimonies were those relating to "abundance of animals." HTL participants testified that they had caught more ptarmigan (small game birds), seals, narwhals, and polar bears. One woman described the transformation of fish in a particular lake from skinny and healthy to fat and healthy. In another testimonial a male hunter explained:

> When I was a child, there used to be a lot of animals around like ptarmigan. But for a long time we never had them around very much anymore. And then after the workshop, you noticed more people bringing you ptarmigan or saying they had got enough ptarmigan so, if anybody wants some, just come around.

During the Clyde River prayer service, Armbruster shared another dramatic example of animal abundance that he related to healing work in the nearby community of Pond Inlet. In late November of 2008, a pod of narwhals became trapped in an area of open sea near the community when the winter sea ice closed in over them. The Department of Fisheries and Oceans Canada and the Nunavut Wildlife Management Board agreed with the Pond Inlet Hunters' and Trappers' Organization (HTO) that the community should be allowed to harvest the animals, since they would likely have died from being trapped under the ice. Initially, hunters believed the pod to number around 200, but more than 600 were harvested by the end of the cull.

"After the harvest, the head of the HTO came to church and gave thanks to God for the blessings," Armbruster told the worshippers. "He said he could not only feel the presence of Christ but could also smell the aroma of Christ."

This episode and Armbruster's interpretation of it reveals a wider site of tension within the Canadian Arctic between Inuit and scientific approaches to wildlife management. Inuit traditional knowledge and Western scientific knowledge have conflicted over stewardship practices, conservation, and the foundations of belief about the relationship between humans and animals. One source of the conflict between Inuit knowledge and Western science lay in different beliefs about stewardship practices. One way to "care" for animals in a scientific management context was to place quotas on hunting and human use of animals. In contrast, Inuit hunters traditionally believed that animal souls were reincarnated, and that the best way to ensure abundant animals was to hunt them (Campbell 2004; Gombay 2010). During an interview I conducted with an active hunter, a man in his 60s who had spent his childhood and early adulthood living on the land, he shared this understanding of population dynamics with me:

> With the natural way of the world, if people catch something, if more people try to catch fish, the fish will become more abundant. . . . [At a local fishing lake], when not many people used to fish there, the fish were small. And now lots of people fish there and there are big fish now. Whenever you try to catch more of some type of animal, the population of them will grow.

Inuit and other First Nations people have traditionally viewed animals as part of an extended network of sociality. While these views vary across different communities and households and may be changing somewhat in younger generations, conversations I had with Clyde River residents reflected a widespread belief in and respect for animal intelligence, which has also been documented in social science literature (Gombay 2010; Stairs and Wenzel 1993). From their perspective, wildlife management initiatives were often a mechanism to control and manage people rather than animals, since animals could manage themselves perfectly well without human interference (Gombay 2010; Nadasdy 2011).

The different beliefs about animal behavior and stewardship held by Inuit and government scientists have led to disagreements about the resilience and health of animal populations, including large marine mammals, such as belugas and narwhals, as well as caribou and polar bears. Biologists who study wildlife dispersed across the vast Arctic tundra often rely on aerial and boat surveys to make their population estimates. Inuit have maintained that this is an unreliable way of tracking populations, suggesting that their longstanding knowledge of animals offers a more accurate basis for estimating population numbers.[5] Inuit have also been concerned about some of the practices that wildlife managers use to study animals, suggesting that helicopter surveying and the use of tranquilizers and radio collars can be harmful (Aupilaarjuk et al. 2002; Kunuk and Mauro 2010).

While Inuit knowledge and scientific knowledge about animals have sometimes conflicted, Inuit and scientists agreed that caribou numbers had been declining in recent years on Baffin Island and across Nunavut as a whole (N.A. 2010). Studies of wild reindeer and caribou have pointed toward a worldwide decrease in population by 33 percent since their most recent peak in the early 2000s, suggesting that climate change may be making it harder for caribou populations to recover after a decline (Gunn et al. 2009).

Caribou has historically been an important and popular seasonal food for Inuit. Clyde River residents were therefore concerned about caribou and hoped to see greater numbers in the near future. Thus it was probably not a coincidence that the Arctic HTL miracle that Armbruster cited most often was a story of caribou herds returning to Resolute, a community located on Cornwallis Island, one of the two northernmost communities in Canada. Promotional videos for HTL showed herds of caribou running across the tundra, while a narrator described how, after prayer and healing work, "tens of thousands of caribou returned to Resolute." Armbruster explained that caribou "circled the spot where they had prayed and where the healing took place. It was as if they knew that something had happened there. And it was enough to convince even the non-believers."

Through these stories, HTL emphasized a spiritual connection between human emotional wellness and the well-being of animals, which resonated with traditional Inuit conceptions of human–animal relations (Stairs and Wenzel 1993). As one HTL participant described her perception, since HTL "the animals even seem happier. If we are not at peace, then we could contaminate the land and the animals. But if we become at peace with ourselves and the land, the animals are more at peace, too." These narratives contrasted with Western scientific and medical conceptions of human and animal health, which limit their focus to the biological organism and its physical environment (Wenzel 1981). Social ties between humans and animals are not part of the biomedical or scientific models of health. By addressing Inuit hunters' desire for animal abundance, HTL leaders engaged concerns that Inuit had about the hegemony of science-based policymaking and wildlife management. Though ultimately denigrating some aspects of traditional beliefs, this appeared to be an important part of HTL's success in the region.

Continuity and disruption of traditional relations with the land and animals

HTL's conception of defilement of the land suggested that sin was the dominant form of human pollution, a narrative that also contradicted Western scientific perspectives. This narrative may have resonated to a certain extent with traditional Inuit beliefs about the power of particular places in the landscape and the proper way of relating to these places. As discussed previously, HTL defines stewardship as avoidance of sin, based on the belief that sin defiles the land. A Clyde River resident who had been involved with HTL rituals reframed this theological tenet through the language of traditional belief, stating: "Inuit have a belief that, if you do bad things on the land, bad things will happen to you." She told me that there was a lake near Clyde River where a young woman had been crushed by an iceberg and killed. "This was due to the negativity that was left on that land."

"What happened in that place?" I asked.

"Have you seen the film *Atanarjuat: The Fast Runner*?[6] That's the same lake where it happened, on the ice there, where the people were killed."

"How is *Atanarjuat* related to that woman being killed by an iceberg?" I asked her, confused.

"Because of the negativity people had towards it, because people had not forgiven that part of the land, so negativity is put on the land." She explained that HTL involved forgiving and releasing grudges held against the land that form because of negative associations, such as the death of a loved one, or the experience of hunger in a certain place.[7]

"How do you heal those things?" I asked.

"We have to apologize and pray at the same time, like forgiving the land. You don't even actually have to go there; you can just talk to another human being about it and then pray together."

While this individual's description reflected her involvement in HTL, traditionally Inuit did identify particular places on the land that required special respect or care. These places were understood to be inhabited by spirits or invisible beings with the capacity to harm humans and were to be avoided or approached very carefully. Burial places were respected as the home of spirits, and these places were governed by particular rules about how humans should approach them and how they should act. Awareness of space and place included knowledge of how to respect and respond to these beings (Laugrand and Oosten 2010). Thus, while HTL and traditional Inuit beliefs both identify particular places in the landscape as holding power to harm Inuit, HTL describes these as places of human sin mediated by God, rather than the more diffuse conception of non-human spiritual agency reflected in traditional beliefs.

In spite of the synergies between HTL and traditional Inuit ways of relating to land and animals, the movement also disrupted these relationships. As Joel Robbins has noted, scholars tend to explain the global appeal of Pentecostal and charismatic Christianity by emphasizing the similarities between the "enchanted and ecstatic" rituals of these Christian traditions and indigenous beliefs and rituals (Robbins 2004: 126). He notes that this position often fails to take into account the ways in which these Christian forms "accept local enchanted cosmologies only to attack them, thus profoundly altering the way they are understood" (2004: 127). This was certainly the case with HTL in Clyde River, as illustrated by the testimony offered by a hunter during one of the prayer services:

> One time, I came upon a polar bear, and it ran away when it saw me. When it was gone, I saw it had killed a baby seal and left it there—it hadn't eaten it. When I took it back to camp, my wife asked me if I had thanked the polar bear, and I said no, and we agreed that all thanks and praise should go to God for everything.

This hunter's statement that "all thanks should go to God" can be understood as a response to HTL's preaching against idolatry. In the HTL worship services, Nakauyaca Sr. emphasized the sin of idolatry, stating emphatically: "It would be a sin to worship the whales!" While traditional Inuit belief in the intelligence of animals might suggest that the bear had left the seal under its own agency, the Christian version emphasizes God's complete and total authority. The direct relationship between the hunter and the animals he hunted was thus redirected in this evangelical Christian framework, with God now positioned as the provider and mediator of all relations.[8]

Conclusion: What's to be learned from the local appeal of Healing the Land?

My research suggests that knowledge categories that are often viewed as separate from one another in a scientific or even religious context may be more fluid and dynamic than they appear on the surface. HTL leaders reference scientific understandings of change in some contexts even as they reject them in others, and Inuit in Clyde River use hybrid knowledge traditions, including local knowledge, science, and religion, to make sense of social and environmental change. They draw on the variety of worldviews and perspectives available to them, emphasizing those that support their ability to pursue agendas that they deem important.

Yet, because of the history of conflict with biologists and environmentalists over animal populations, Inuit are at times skeptical when scientists claim expertise about the Arctic environment, particularly when they do so in ways that diminish Inuit understandings and ways of knowing. Over the past decades, through negotiating land claims and establishing the Nunavut government, Inuit have become increasingly attuned to the ways that non-local perceptions and understandings can impact their lives at the local level. Inuit are concerned, for example, that research on polar bear populations and climate change, disseminated through academic journals and environmentalist networks, may result in reduced hunting quotas (Dowsley and Wenzel 2008). Many Inuit feel that their knowledge is still treated as less valuable than scientific knowledge, in spite of the establishment of co-management boards and protocols for local input into wildlife management.[9]

Scientific researchers, attuned to processes of change that are global in nature, often arrive in northern communities with research agendas that do not correspond to local priorities. Situated far from the centers of decision making and industrial production where emissions targets are set and the bulk of emissions are created, many Inuit feel that there is little they can do to address global climate change directly. From such a perspective, pursuing research on climate change is wasteful and rarely leads to concrete action.

HTL sets itself apart from these dynamics by offering a theory of agency that connects group engagement and action with local environmental impact. With its ritualized group healing ceremonies, HTL may appeal to Clyde River residents because it offers an immediate way to take action, encouraging a sense of collective agency in dealing with change. Over the past century, Inuit have navigated many changes, including the transition from a semi-nomadic life on the land to living in permanent settlements, which largely took place in the 1950s and 1960s. They associate many social challenges of the present time, such as the struggle for healthy food, a lack of good housing, high rates of drug and alcohol use, and a high suicide rate, to the legacy of these changes, including the colonial practices that have guided government engagement with Inuit.

In this context, HTL's emphasis on local practices of prayer and healing offers a framework through which community members can directly engage with processes and outcomes of change. HTL's integrated vision links the social world with environmental resilience, offering a way for community members to engage with social issues while also, according to the movement's leaders, supporting the well-being of animals and healing the land. By doing so, participants also reframe environmental change as an issue that can be addressed through local agency.

Of course, many would argue that this sense of control is false. From a secular perspective, the idea that social and environmental problems can be healed through prayer is misguided or even delusional. HTL's conception of stewardship as avoidance and repentance of sin redirects action toward spiritual ideals and practices and away from addressing the political, economic, and social causes of environmental issues. To suggest that prayer will miraculously restore caribou populations, for example, sidelines the potential for Inuit and scientists to use their significant bodies of knowledge to cooperatively address this area of mutual concern. Finally, by labeling certain practices of the past—such as shamanism—as "sinful," HTL reinforces a discourse that denigrates pre-Christian Inuit beliefs and implicitly suggests that Inuit identity is inferior to *qallunaat* identity.

Social and natural scientists and environmentalists concerned about global climate change have much to learn from being attentive to the conditions and processes by which alternative narratives of change, such as HTL, gain persuasive power. In Clyde River, the appeal of HTL points simultaneously toward gaps in social policy as well as a sense of alienation from the norms and processes of Western science. Most noticeably, the dominant position of scientific knowledge in national and international contexts means that local people rarely set the terms of engagement. Yet, from the perspective of community members, what is most important is not so much the acquisition of scientific knowledge as the ability to address the issues that they find most pressing in their own lives, such as the availability of game or the maintenance of social harmony.

Notes

1 The New International Version (NIV, 1973) of the Holy Bible is used throughout this chapter.

2 An *amauti* (plural: *amautiit*) is a traditional parka worn by women that features a large hood in the back used to carry infants and toddlers.

3 *Inuk* is the singular form of Inuit.

4 In interviews, community members also made a distinction between "good" and "bad" shamans, although they did not always associate good shamans with Christianity. Good shamans used their powers to heal and assist people in times of need, while bad shamans used their powers to inflict harm.

5 For examples and discussion of some of these conflicts about wildlife, see Armitage (2005), Collings (1997), Dowsley (2007), Dowsley and Wenzel (2008), and Suluk and Blakney (2009). Space does not allow a full discussion of the relevant conflicts in Clyde River in this chapter.

6 *Inuk* director Zacharias Kunuk (2001) directed this film based on an Inuit legend. Atanarjuat is a young man who must overcome a curse that caused an evil spirit to wreak havoc in his community. The film won the Camera d'Or for best first feature at the Cannes film festival. Although *Atanarjuat* was filmed in Igloolik, elders in Clyde River claim the nearby landscape as the place where the events recorded in legend actually happened.

7 This story has some synergy with Julie Cruikshank's (2005) description of the Tlingit of British Columbia's belief in the agency of glaciers.

8 This discourse about God as mediator between humans and animals was also present in Anglican Christianity (Laugrand and Oosten 2007).

9 Co-management boards were established as part of the Nunavut Land Claims Act of 1993. For more on co-management, see the references in note 5. For a discussion of how these institutions are likely to constrain Inuit adaptation to environmental change, see Wenzel (2009).

References

ACIA (Arctic Climate Impact Assessment). 2005. *Arctic Climate Impact Assessment*. Cambridge, U.K.: Cambridge University Press.

Armitage, D. 2005. "Community-Based Narwhal Management in Nunavut, Canada: Change, Uncertainty, and Adaptation." *Society and Natural Resources* 18, 715–731.

Aupilaarjuk, M., P. Suvaksiuq, F. Pisuk, P. Tapaqti, L. Iluittuq, L. Nuliajuk, O. Itinnuaq and J. Angutinngurniq. 2002. *Inuit Perspectives on the 20th Century*, vol. 4: *Inuit Qaujimajatuqangit: Shamanism and Reintegrating Wrongdoers into the Community*. Iqaluit, NU: Nortext/Nunavut Arctic College.

Briner, J.P., P.T. Davis and G.H. Miller. 2009. "Latest Pleistocene and Holocene Glaciation of Baffin Island, Arctic Canada: Key Patterns and Chronologies." *Quaternary Science Reviews* 28, 2075–2087.

Campbell, C. 2004. "A Genealogy of the Concept of 'Wanton Slaughter' in Canadian Wildlife Biology." In *Cultivating Arctic Landscapes: Knowing and Managing Animals in the Circumpolar North*, ed. D.G. Anderson and M. Nuttall, pp. 154–171. New York: Berghahn Books.

Collings, P. 1997. "Subsistence Hunting and Wildlife Management in the Central Canadian Arctic." *Arctic Anthropology* 34, 41–56.

Cruikshank, J. 2005. *Do Glaciers Listen? Local Knowledge, Colonial Encounters, and Social Imagination*. Vancouver: University of British Columbia Press.

Dowsley, M. 2007. "Inuit Perspectives on Polar Bears (Ursus Maritimus) and Climate Change in Baffin Bay, Nunavut, Canada." *Research and Practice in Social Sciences* 2, 53–74.

Dowsley, M. and G.W. Wenzel. 2008. "'The Time of the Most Polar Bears': A Co-Management Conflict in Nunavut." *Arctic* 61, 177–189.

Feng, S., C.H. Ho, Q. Hu, R.J. Oglesby, S.J. Jeong and B.M. Kim. 2011. "Evaluating Observed and Projected Future Climate Changes for the Arctic Using the Köppen–Trewartha Climate Classification." *Climate Dynamics* 38, 1359–1373.

Fox, S. 2004. "When the Weather Is Uggianaqtuq: Linking Inuit and Scientific Observations of Recent Environmental Change in Nunavut, Canada." Ph.D. dissertation, University of Colorado, Boulder.

Gearheard, S., W. Matumeak, I. Angutikjuaq, J. Maslanik, H.P. Huntington, J. Leavitt, M.D. Kagak, G. Tigullaraq and R.G. Barry. 2006. "'It's Not That Simple': A Collaborative Comparison of Sea Ice Environments, Their Uses, Observed Changes, and Adaptations in Barrow, Alaska, and Clyde River, Nunavut, Canada." *AMBIO: A Journal of the Human Environment* 35, 203–211.

GN (Government of Nunavut). 2005. *Inuit Qaujimajatuqangit on Climate Change in Nunavut: A Sample of Inuit Experiences of Recent Climate and Environmental Changes in Clyde River, Pond Inlet, Resolute Bay, Grise Fiord, Nunavut*. Iqaluit, NU: Environmental Protection Division, Department of Environment.

Gombay, N. 2010. *Making a Living: Place, Food and Economy in an Inuit Community*. Saskatoon, SK: Purich Publishing.

Gunn, A., D. Russell, R.G. White and G. Kofinas. 2009. "Facing a Future of Change: Wild Migratory Caribou and Reindeer." *Arctic* 62, iii–vi.

Holy Bible, New International Version. 1973. Grand Rapids, MI: Zondervan.

Høye, T.T., E. Post, H. Meltofte, N.M. Schmidt and M.C. Forchhammer. 2007. "Rapid Advancement of Spring in the High Arctic." *Current Biology* 17, R449–451.

Hudson, J.M.G. and G.H.R. Henry. 2009. "Increased Plant Biomass in a High Arctic Heath Community from 1981 to 2008." *Ecology* 90, 2657–2663.

Huntington, H.P., S. Gearheard, M. Druckenmiller and A. Mahoney. 2009. "Community-Based Sea Ice Programs and Indigenous and Local Sea Ice Knowledge." In *Handbook on Field Techniques in Sea-Ice Research: A Sea-Ice System Services Approach*, ed. H. Eicken, R. Gradinger, M. Salganek, K. Shirasawa, D. Perovich and M. Leppäranta, pp. 345–364. Fairbanks: University of Alaska Press.

Kaufman, L. 2004. "Fiji's Rainbow Reefs." *National Geographic* 206, 54.

Kunuk, Z. (dir.). 2001. *Atanarjuat: The Fast Runner*. Igloolik, NU: Isuma Productions.

Kunuk, Z. and I. Mauro (dirs.). 2010. *Qapirangajuq: Inuit Knowledge and Climate Change*. Igloolik, NU: Isuma Productions.

Laugrand, F. and J. Oosten. 2007. "Reconnecting People and Healing the Land: Inuit Pentecostal and Evangelical Movements in the Canadian Eastern Arctic." *Numen* 54: 229–269.

——. 2010. *Inuit Shamanism and Christianity: Transitions and Transformations in the Twentieth Century*. Montreal, QC and Kingston, ON: McGill-Queen's University Press.

N.A. 2010. "Baffin Island Caribou Sightings Few: Inuit." *Canadian Broadcasting Corporation (CBC)*, January 11. Available at http://www.cbc.ca/news/canada/north/story/2010/01/11/baffin-caribou.html (accessed July 15, 2011).

Nadasdy, P. 2011. "'We Don't *Harvest* Animals; We *Kill* Them': Agricultural Metaphors and the Politics of Wildlife Management in the Yukon." In *Knowing Nature: Conversations at the Intersection of Political Ecology and Science Studies*, ed. M.J. Goldman, P. Nadasdy and M.D. Turner, pp. 135–151. Chicago, IL: University of Chicago Press.

Nakauyaca, V. and W. Ani. 2007. *A Manual for Healing the Land*. Toowoomba, Australia: Toowoomba City Church.

Post, E., M.C. Forchhammer, M.S. Bret-Harte, T.V. Callaghan, T.R. Christensen, B. Elberling, A.D. Fox, O. Gilg, D.S. Hik, T.T. Høye, R.A. Ims, E. Jeppesen, D.R. Klein, J. Madsen, A.D. McGuire, S. Rysgaard, D.E. Schindler, I. Stirling, M.P. Tamstorf, N.J.C. Tyler, R. van der Wal, J. Welker, P.A. Wookey, N.M. Schmidt and P. Aastrup. 2009. "Ecological Dynamics across the Arctic Associated with Recent Climate Change." *Science* 325, 1355–1358.

Robbins, J. 2004. "The Globalization of Pentecostal and Charismatic Christianity." *Annual Review of Anthropology* 33, 117–143.

Stairs, A. and G.W. Wenzel. 1993. "'I Am I and the Environment': Inuit, Community and Identity." *Journal of Indigenous Studies* 3, 1–12.

Suluk, T.K. and S.L. Blakney. 2009. "Land Claims and Resistance to the Management of Harvester Activities in Nunavut." *Arctic* 61, 62–70.

Tungilik, V. and R. Uyarasuk. 1999. *The Transition to Christianity: Inuit Perspectives on the 20th Century*, vol. 1, ed. J. Oosten and F. Laugrand. Iqaluit, NU: Nortext/Nunavut Arctic College.

Watt-Cloutier, S. 2005. "Petition to the Inter-American Commission on Human Rights Seeking Relief from Violations Resulting from Global Warming Caused by Acts and Omissions of the United States." December 5. Available at www.ciel.org/Publications/ICC_Petition_7Dec05.pdf.

Wenzel, G.W. 1981. "Inuit Health and the Health Care System: Change and Status Quo." *Etudes/Inuit/Studies* 5, 7–16.

——. 2009. "Canadian Inuit Subsistence and Ecological Instability: If the Climate Changes, Must the Inuit?" *Polar Research* 28, 89–99.

Part III

The transnational context

17 An investigation of perception of climate change risk, environmental values and development programming in a faith-based international development organization

Carolyn Peach Brown, Douglas R. Brown and Christopher Shore

Introduction

There is widespread agreement that the mobilization and coordination of faith-inspired actors is essential to maximize the impact of aid and development agencies worldwide (Bergmann 2009; Spencer, White and Vroblesky 2009; World Bank 2011b). Empirical evidence suggests that, in some countries, faith-based organizations (FBOs) provide a large share of the education and health services used by the local population (World Bank 2011c). Over the past decade partnerships have been developed between major development agencies and FBOs. Such assistance is all the more essential in light of climate change, which threatens to undo years of development work undertaken by humanitarian non-governmental organizations (NGOs) (McGray, Hammill and Bradley 2007; World Bank 2009). Climate change makes the vulnerable even more vulnerable, threatening not only livelihoods but more specifically the security and stability of the food systems on which they depend (FAO 2008). Changing precipitation patterns, floods, droughts, increased intensity of storms, sea level rise, and changing disease patterns are already reducing the availability of drinking water, the health and well-being of children, and physical safety among those vulnerable to disaster. Such changes may become even more challenging in the future if dire predictions about "tipping points" hold true (Stern 2007; World Bank 2009).[1] Such problems will be particularly severe for those groups that development agencies have long targeted: those with less capacity to adapt, such as those who live in poverty in the Global South (Füssel 2009; IPCC 2007; Lobell et al. 2008; Toulmin 2009). This has led to FBOs being active in both climate change advocacy and project implementation (Mitchell and Tanner 2006; Parris et al. 2009; World Bank 2011a), alongside other development aid agencies and NGOs (McGray, Hammill and Bradley 2007; Mitchell and Tanner 2006; Mitchell and van Aalst 2008; Parris, Lansley and Finnigan 2009).

Some have suggested that faith communities are uniquely placed to bring about the transformation that is needed to address the roots of the climate change problem (Bergmann 2009). Yet, despite over a decade of cooperation between major development agencies and FBOs, little work has been done to examine what role FBOs are playing in this most practical of arenas when it comes to addressing climate change. In particular, given the controversy surrounding climate change, it seems increasingly worthwhile to investigate what barriers, if any, have arisen within FBOs to implementing climate change programming. Within organizations, the setting of priorities and decision making related to climate change may be affected by the understandings and perceptions of risk of those in decision-making roles. People's perceptions of the risks climate change poses are shaped by psychological, social, cultural and institutional processes (Lowe and Lorenzoni 2007), as well as how they perceive the risks posed by other environmental or social problems (Pidgeon, Lorenzoni and Poortinga 2008; Shisanya and Khayesi 2007; Whitmarsh 2008). Internal differences in risk perception could potentially prevent an organization with economic, cultural and political differences from dealing with difficult issues, such as allocating resources to climate change.

Similarly to other international non-governmental organizations (INGOs), World Vision (WV), an FBO, exhibits such complex challenges owing to its structure as an international federation of national organizations, whose international partnership office serves to facilitate communication and cooperation among the member offices in diverse countries and cultures around the world (McCormick 2005). WV is a Christian relief, development and advocacy organization dedicated to working with the world's most vulnerable people, especially children, to help them improve their well-being. The employees of WV, 40,000 staff in nearly 100 countries, reflect the broad spectrum of global Christianity, including evangelical and mainline denominations. Established in 1950 to care for orphans in Asia, WV has grown to embrace the larger issues of community development, humanitarian relief, and advocacy for the poor in its mission to help children and their families build sustainable futures. In light of the potential of climate change to negatively impact child well-being (CWB),[2] its key indicator of success, WV's international partnership office began addressing climate change issues in 2006 through what is now called the Climate Change Response Initiative (CCRI) (Shore 2009; World Vision International 2010b). Using a case study approach, we investigated the differences in risk perceptions and attitudes to climate change and its relationship to implementation of programming in WV.

Given that FBOs are poised to play a pivotal role in addressing climate change, we argue that internal activities and organizational learning about climate change should be sure to include those key decision makers and staff who have less direct contact with those impacted by climate change.

Survey methodology

Our preliminary case study used a purposive sampling approach, targeting specific employee groups within WV whose opinions were likely to influence decision making and/or who might have diverse perceptions of the subject matter. Given the diffuse nature of the organization, and that no centralized list of staff exists, we chose not to attempt a random sample. While our approach therefore allowed us to explore the relationship between risk perceptions and attitudes to climate change and its relationship to implementation of programming in WV, subsequent research using a random sampling approach would be necessary to provide more conclusive results.

To obtain our sample, we sent a link to a web-based survey, including both closed and open-ended questions, to directors of country-level, regional and global offices of the organization, asking them to complete the survey themselves and to forward it to at least ten staff members within their jurisdiction based on the following criteria. Directors of the NGO's offices in developed countries (Support Offices—SOs), which raise and administer funds to support the work in developing countries, were asked to include staff who support programming in the field, those involved in marketing and communications with the donor constituency, and senior-level management. Directors of the NGO's offices located in developing countries (National Offices—NOs) were asked to include program staff from the country's head office as well as field-level staff and senior management. Regional Office (RO) and international partnership office (Global Centre—GC) staff were to include both senior management and staff who provide technical support to programs at the country level. Survey recipients were also invited to forward it to other interested staff.

NO staff (both head office and field) have primary responsibility for program design and implementation. They prepare proposals and submit them to SOs for approval and funding. They receive guidance from SO staff who support programming in the field and technical staff at the Regional Office level. Senior NO management set strategic goals and priorities for programming. Similarly, senior SO management set strategic goals for both fund-raising staff and those who support the work of NOs. Marketing staff have no role in program design. However, they have an indirect impact, as they raise funds for projects that they feel will be of interest to donors. Finally, the GC and ROs function as a coordinating body in strategy development and budget allocation. Therefore, similar to some other INGOs, WV is not a top-down hierarchical organization with a single locus of decision making but is a partnership of quasi-independent offices around the globe with decentralized decision-making processes.

The survey was open to respondents for a period of three months, beginning in mid-May 2009. The majority of respondents completed the survey by the end of June. A total of 555 individuals completed some or all of the survey questions,[3] including about half from developing country NOs (54.8 percent) and nearly one-third (30.8 percent) from SOs, with the majority of SO respondents being located in Canada or the U.S. Of these SO respondents, 18.8 percent had a close connection to the field-level work of the organization, and 36.1 percent held a management role. The survey reached a wide variety of staff in the different areas of responsibility within the organization, including Transformational Development programming (TD), Humanitarian and Emergency Affairs (HEA), sponsorship, advocacy, marketing, communications, senior management and technical sector specialists. For roles related to TD and HEA this included field-level staff in NOs as well as program officers in NOs and SOs. Another 24.5 percent of respondents were grouped as "Other," including people in administration, information technology (IT), finance, human resources and other management roles. For the most part, they are in roles not closely connected with the field or programming.

To create a nuanced measure of people's opinions, we created four blocks of questions. In the first and second blocks, we measured attitudes and perceived risk related to climate change using a bank of 14 attitude and 12 risk statements developed and tested by Whitmarsh (2008). In these blocks we used a Likert scale (McIver and Carmines 1981) to measure the degree of agreement/disagreement with each of the statements. Open-ended questions gave respondents an opportunity to clarify their responses. In the third block, respondents were also asked the degree to which they agreed or disagreed with each of seven statements about the possible impact of climate change on CWB and WV's efforts to improve it, since children are the primary target group of its programming around the world (World Vision International 2008b, 2009b). In the fourth block, the respondents replied to a series of 15 statements commonly used to measure environmental beliefs or an ecological worldview which was developed and tested by Dunlap et al. (2000), used by Whitmarsh (2008) and collectively referred to as the New Ecological Paradigm (NEP) scale. In this scale, respondents stated their level of agreement with 15 items on five different facets of an environmental worldview (reality of limits to growth, anti-anthropocentrism, fragility of nature's balance, rejection of exemptionalism,[4] possibility of an ecocrisis) (Dunlap and Van Liere 2008; Dunlap et al. 2000). Finally, respondents were asked about their role within the organization as well as basic demographic questions.[5]

As reported elsewhere (Brown, Brown and Shore 2012), to analyze the data we first used a one-way analysis of variance to determine whether there were statistically significant differences in attitudes, risk perceptions, ecological worldview and opinions about the impact of climate change on CWB among the groups of respondents with respect to office type and role within the organization. Then each of the four blocks of questions was combined into an index using factor analysis. Using the same three blocks of questions as Whitmarsh (2008), we created the Climate Change Attitude Index (CC-AI) (Cronbach's alpha = 0.8683), the Climate Change Risk Perception Index (CC-RPI) (Cronbach's alpha = 0.9008) and the NEP Index (NEPI).[6] In a similar manner, responses to statements of the relationship between climate change and CWB were used to develop the Climate Change Child Well-Being Index (CC-CWBI) (Cronbach's alpha = 0.8822). Finally, we used linear regression to assess the degree to which variations in the indices were explained by respondents' experience with climate change, their office type and role within the organization, and their demographic characteristics (gender, age, length of service, education). Given the highly diverse nature of the organization, we anticipated that opinions might differ regionally by office type (e.g., developed versus developing country, or global coordination office versus country-specific office). However, we also felt that such a classification might be overly simplistic, knowing that there might be other factors which were associated with differences in attitudes, perceptions and values related to climate change, environment and CWB. Complete details of survey results for the attitude, risk perception and CWB questions are reported elsewhere (Brown, Brown and Shore 2012), but are summarized here.

General awareness of climate change

Among respondents the general awareness of climate change and its causes was high. Fully two-thirds of respondents (67.4 percent) indicated that they had noticed changes in the climate in the area that was the focus of their work during the past five years. In explaining their understanding of climate change, almost all respondents (94 percent) included elements from the IPCC definition and indicated that, while climate change had natural causes, most of the changes now occurring were anthropogenic in origin. While two-thirds had observed changes in the climate, they were nearly evenly split (49.9 percent versus 50.1 percent) in terms of awareness of climate change predictions for the areas that were the focus of their work. On the other hand, 87.3 percent of respondents were optimistic that there were concrete things that could be done to tackle the challenges presented by climate change.

When asked for examples, many cited activities that WV was already doing to respond to climate change beyond the extensive response to natural disasters. These included efforts in long-term sustainable development through reforestation projects, introduction of drought-resistant crops, alternative sources of energy for cooking, and building capacity in disaster risk preparedness.

A respondent in Bangladesh described the work in an area development program 200 kilometers from the capital city, Dhaka: "It is low and marshy—land are 6–7 month under water in every year [*sic*]. There are frequently floods affected and cyclone is another disaster for the area. We already introduce floating vegetable gardening and cage fish culture as adaptation mechanism. It has already replicated in the area."

In Ethiopia, WV's Humbo area development project has been hailed as a highly successful example of a reforestation project that benefits the environment through improved natural resource management and increased biodiversity. It also contributes to poverty alleviation by creating a new community-based income stream through the generation of carbon offset credits under the Clean Development Mechanism (Brown et al. 2011; World Vision International 2008a).

Attitudes about climate change and perception of climate change risk

Most respondents' choices on the CC-AI questions conveyed an attitude that climate change is an important issue (see Table 4, p. 8 in Brown, Brown and Shore 2012 for a complete list of questions and results). With respect to the CC-AI itself, most respondents had a low index value, indicating an attitude of concern about climate change. However, approximately 5 percent of individuals had much higher scores, indicating an attitude that climate change was largely irrelevant to them and the work of the organization. One respondent stated:

> I am very disappointed that WV spends any time, energy, resources to this issue. Whether you "believe" in it or not, it is a distraction from the core focus of this organization. It is also polarizing. Let us focus on the things that unite us, faith in Christ, helping the poor.

One-way analysis of variance of the CC-AI scores showed no significant difference in the overall attitude by office type, even though there were some significant differences for individual attitude statements. On the other hand, there was a significant difference in overall attitude depending on whether or not the respondent had observed changes in the climate over the past five years. Those who had observed changes were of the opinion that climate change was a very important issue, whereas those who had not noticed any changes felt that it was not very important.

Turning to the CC-RPI, responses indicated that most respondents perceived climate change to be an important issue that presents considerable risks (see Table 5, p. 9 in Brown, Brown and Shore 2012). Most cited concerns about extreme weather patterns, drought, increasing temperature and sea level rise, water shortages and environmental degradation, which they felt would increase the poor's vulnerability. Analysis of variance on the value of the CC-RPI index itself (Brown, Brown and Shore 2012) indicated that, in some cases, there were significant differences in opinion between respondents from different office types. Respondents from SOs felt that climate change presented less of a risk than either those who had a closer connection to the field (NOs) or those who had a more global picture (GC) of the areas where the organization worked (Table 17.1). One respondent stated: "It is not possible to say that particular events are caused by climate change, as there are always a number of factors involved—vulnerability, poverty, etc." As with the CC-AI, there was a significant difference in the CC-RPI depending on whether or not the respondent had observed changes in the climate over the past five years. Those who had observed changes were of the opinion that climate change presented very serious risks, whereas those who had not noticed any changes felt that it presented few risks at all.

Table 17.1 Mean Climate Change Risk Perception Index (CC-RPI) value for the four office types

Office type	*Mean CC-RPI value*	*Standard deviation*
National Office	0.1432	0.8073a**
Support Office	–0.1933	1.0975a**, b*
Regional Office	–0.0698	1.3989
Global Centre	0.3350	0.6154b*
Total	0.0334	0.9636

Note: Rows with the same letter have mean values that are significantly different; ** and * indicate that the difference is significant at the 5% and 10% levels of confidence respectively.

Impacts on child well-being

CWB is central to World Vision's mission. Therefore, whether respondents linked CWB to climate change is an important indicator of whether they will take steps to account for its impacts and/or minimize its risks in their programmatic activities. Interestingly, we found that, as in the analysis above, NO and GC staff members were more likely to perceive a direct connection than were SO staff members. For example, one-way analysis of variance showed that SO staff were more likely than NO staff to disagree with the statements that climate change would make it more difficult to improve CWB, and that it would have a serious impact on development programming (see Table 7, p. 11 in Brown, Brown and Shore 2012). Furthermore, SO staff were also more likely than NO and GC staff to disagree with the statement that climate change would have a serious impact on community health and nutrition. Finally, GC staff were more likely than either NO or SO staff to disagree with the statement "Climate change is unlikely to affect CWB." One respondent from India stated: "We do see the changing patterns of rainfall, floods, droughts, increased intensity of storms, sea level rise, and changing disease patterns are affecting the availability of drinking water, crop yields, and physical safety . . . climate change is taking place and is going to affect child well-being outcomes."

Factor analysis of these questions produced a third index, the CC-CWBI, which we developed as an indicator of the perceived connection between climate change and CWB (Brown, Brown and Shore 2012). The majority of respondents from all office types felt that climate change poses significant risks to both CWB and organizational efforts to bring about improvements. However, one-way analysis of variance showed a significant difference in the mean value of the CC-CWBI by office type (Brown, Brown and Shore 2012). As with the indices presented above, SO staff stood out, in this case by being much less likely than either GC or NO staff to make a strong connection between climate change and the work of the NGO. Similarly, there was a significant difference in perceived impact on CWB between those who had and those who had not observed changes in the climate over the preceding five years. Those who had observed changes felt that climate change was likely to have a serious impact on efforts to improve CWB, whereas those who had not noticed any changes felt that it presented few risks to efforts to improve CWB at all.

Environmental values as measured by the NEP Index

Environmental values might also be expected to influence one's views about climate change, the risks it presents and the level of potential impact on organizational goals related to CWB. In ten out of the 15 statements, respondents' choices indicated that they agreed with a pro-ecological worldview (Table 17.2). Responses to open-ended questions indicated a link between concern for climate change, concern for the environment and respondents' Christian faith. Conversely, for three of the items (that humans have the right to modify the natural environment; that the earth has plenty of natural resources; and that humans were meant to rule over

Table 17.2 Percentage of respondents agreeing or disagreeing with statements about ecological worldview

NEP question/statement	*A pro-ecological view is implied by*	*Strongly agree*	*Agree*	*Neither agree nor disagree*	*Disagree*	*Strongly disagree*	
1. We are approaching the limit of the number of people the earth can support.	Agreement	12.05%	23.19%	**29.52%**	26.20%	9.04%	–
2. Humans have the right to modify the natural environment to suit their needs.	Disagreement	7.23%	**31.33%**	16.57%	30.42%	14.46%	–
3. When humans interfere with nature it often produces disastrous consequences.	Agreement	22.29%	**43.67%**	21.39%	11.14%	1.51%	**
4. Human ingenuity will insure that we do *not* make the earth unlivable.	Disagreement	8.13%	26.20%	**30.12%**	26.51%	9.04%	**
5. Humans are severely abusing the environment.	Agreement	37.65%	**50.00%**	7.23%	3.01%	2.11%	–
6. The earth has plenty of natural resources if we just learn how to develop them.	Disagreement	28.31%	**41.57%**	14.16%	12.65%	3.31%	***
7. Plants and animals have as much right as humans to exist.	Agreement	34.94%	**38.86%**	15.06%	9.04%	2.11%	**
8. Nature is strong enough to cope with the impacts of modern industrial nations.	Disagreement	4.22%	9.94%	15.96%	**44.58%**	25.30%	**
9. Despite our special abilities humans are still subject to the laws of nature.	Agreement	39.46%	**43.67%**	11.75%	5.12%	0.00%	**
10. The so-called ecological crisis facing humankind has been greatly exaggerated.	Disagreement	6.02%	6.02%	20.78%	**42.77%**	24.40%	–
11. The earth is like a spaceship with very limited room and resources.	Agreement	13.55%	**34.94%**	24.40%	23.49%	3.61%	–
12. Humans were meant to rule over the rest of nature.	Disagreement	12.65%	**35.24%**	18.37%	21.69%	12.05%	–
13. The balance of nature is very delicate and easily upset.	Agreement	21.69%	**44.88%**	20.48%	12.35%	0.60%	–
14. Humans will eventually learn enough to be able to control nature.	Disagreement	6.63%	25.00%	26.81%	**30.12%**	11.45%	***
15. On the present course, we will soon experience a major ecological catastrophe.	Agreement	34.34%	**41.57%**	16.27%	5.12%	2.71%	**

Notes: N = 332; *** indicates that differences in the level of agreement between office types are significantly different at 1%, and ** at 5%; percentages in bold indicate the category chosen by the largest percentage of respondents. Agreement with the eight odd-numbered items and disagreement with the seven even-numbered items indicate pro-NEP responses.

the rest of nature) the majority of respondents disagreed with a pro-ecological worldview. Finally, for two of the items respondents on average were divided on the issue.

To determine if office type was affecting ecological worldview, we performed an analysis of variance. We found that in some cases there were significant differences of opinion on individual questions between respondents from different office types (as indicated by the starred items in Table 17.2). One-way analysis of variance of the NEPI values themselves showed no significant difference in the overall NEPI score by office type. However, there was a significant difference (p = 0.0000) in ecological worldview depending on whether or not the respondent had observed changes in the climate over the past five years. Those who had observed such changes (n = 222) were more pro-ecological in their worldview, whereas those who had not noticed any changes (n = 48) leaned more towards the opposite worldview—what Dunlap (2008) has called the "dominant social paradigm".

In summary, one-way analyses of variance revealed that there were significant differences in the mean values of the CC-RPI and the CC-CWBI by office type. Even more striking is that, in all four of the indices, we found significant differences in attitudes based on whether or not the respondent had observed changes in the climate over the past five years.

Sources of variation in perceptions of climate change

While it is clear from the preceding analyses that there is significant variation in results among the different office types, they do not reveal the source of this variation. Is it best explained by differences in perspectives between offices that fund the NGO's work in the field (SOs) and those that carry it out (NOs), or are there other factors at play? Does one's ecological worldview help to explain the variations in the perceived relationship between CWB and climate change? To answer these questions we regressed the first three indices (CC-AI, CC-RPI and CC-CWBI) against a series of explanatory variables, including experience with climate change, office type, the type of work carried out within the organization, and several demographic variables (see Table 8, p. 13 in Brown, Brown and Shore 2012 for these regression results). The NEPI was regressed against the same set of explanatory variables as well as being included in regressions for the first three indices (see Table 17.3).

As the regression results show, those who had not observed changes in the climate in the area where they worked over the past five years scored higher on the CC-AI, lower on the CC-RPI, higher on the CC-CWBI and higher on the NEPI. This indicates that they were less concerned about climate change, perceived it to pose a lower level of risk, perceived it to have less of an impact on CWB, and had a worldview more aligned with the dominant social paradigm. This appears to indicate that experience with climate change may have an impact on the relative strength of one's attitudes and risk perceptions and also on one's ecological worldview. However, this should not be assumed to indicate causality, because, for example, those with an ecological worldview may already be more concerned

Table 17.3 Regressions results for the NEPI

Variable name	*NEPI*	*CC-AI*	*CC-RPI*	*CC-CWBI*
Attitude to climate change				**0.26*****
Perceived risk from climate change				**–0.20** ***
Ecological worldview (NEP)		**0.66*****	**–0.63*****	**0.27*****
Noticed changes in climate dummy (Yes is omitted)				
No	**0.72*****	**0.58*****	**–0.51*****	**0.35*****
Don't know	0.21	–0.03	0.05	**0.23****
Office type dummy (National Office is omitted)				
Support Office	0.03	–0.16	–0.05	–0.01
Regional Office	–0.05	0.22	–0.16	–0.20
Global Centre	**–0.43****	–0.06	0.04	–0.23
Role level dummy (ADP or field-level staff is omitted)				
Office staff	0.22	**–0.23***	–0.01	0.05
Office management	0.14	**–0.25****	–0.11	0.01
Area of work dummy (Development programming (TD) is omitted)				
Emergency response programming (HEA)	–0.14	0.02	0.14	–0.01
Sponsorship	0.09	0.18	–0.01	0.16
Policy and advocacy	–0.36	0.28	0.01	0.11
Marketing	0.28	**0.34***	–0.29	**0.30***
Communications	–0.33	0.26	–0.26	**0.53*****
Senior management	0.23	0.12	0.05	**0.31****
Technical/sector specialist	–0.14	**0.27***	–0.07	0.05
Other	0.07	0.21	–0.12	**0.45*****
Gender dummy (Female is omitted)				
Male	0.13	**0.16***	0.07	–0.03
Age dummy (34 or under is omitted)				
35–49	0.04	–0.03	–0.04	–0.04
50–64	0.02	–0.03	–0.06	0.08
65 or over	0.71	0.44	**–1.69****	0.40
Education dummy (Secondary school is omitted)				
Bachelor's degree	0.16	**–0.35****	**0.30***	0.23
Master's degree	–0.07	**–0.30***	**0.41****	0.25
Doctorate	–0.42	**–0.80*****	**0.49***	**0.61****
Other	**0.43***	–0.29	0.17	0.14
Years worked for WV	**0.02****	**0.01***	0.00	0.01
Intercept	**–0.54****	0.15	–0.15	**–0.54*****
Number of obs.	319	319	319	319
F	3.75	15.98	13.97	18.76
Prob. > F	0	0	0	0
R-squared	0.2344	0.5768	0.5438	0.6351
Adj. R-squared	0.1719	0.5407	0.5049	0.6012

Note: *** indicates that coefficients are statistically significant at the 1%, ** at the 5% and * at the 10% level of confidence; coefficients highlighted in bold are statistically significant at the 1%, 5% or 10% levels of confidence.

about climate change and therefore perceive extreme weather events as being more salient.

What is also interesting is the insight to be gained from the significance and sign of some of the regression coefficients. Respondents connected with the GC, the NGO's international partnership office, showed CC-AI and CC-CWBI scores (Table 8 in Brown, Brown and Shore 2012) that were significantly higher on average than those in NOs, indicating that respondents from the GC were much more concerned about climate change and its impact on CWB than those from any of the other office types. While these were no longer significant in the regressions which included the NEPI (Table 17.3), this is most likely because respondents connected with the GC were also more pro-ecological in their worldview.

The significance and sign of the coefficients on the dummy variables for those in marketing, communications, senior management and other[7] roles help to explain the underlying source of the variation between office types that showed up in the one-way analysis of variance (see Table 8 in Brown, Brown and Shore 2012). Marketing staff saw climate change as less relevant, with lower perceived risk, and less likely to impact CWB than those who worked in the area of development programming and, for that matter, in most other areas (the first three regressions in Table 8 in Brown, Brown and Shore 2012). Respondents with responsibilities in the areas of communications, senior management and the "other" category see less of a connection between climate change and CWB (the third and fourth regressions in Table 8 in Brown, Brown and Shore 2012). The same applies to the NEPI (the first regression in Table 17.3). Respondents with responsibilities in the areas of marketing, communications, senior management and the "other" category see less of a connection between climate change and CWB.

While all of the NGO's office types have employees working in marketing, communications and senior management roles, the majority of marketing and communications staff are found in SOs. This suggests that some of the variation by office type that our analyses uncovered was due to the increased presence of this category of employee in the SOs. Perhaps most important for the purposes of this analysis, however, is the impact of respondents in the senior management role, given that they can be assumed to exert the greatest influence on the organization's strategic vision and programming. Given the fact that there are many more NOs than SOs, it is not surprising that only 13 percent of respondents in the senior management role were from SOs, while 73 percent were from NOs, and the balance were from ROs and the GC. More important, however, is that our results indicated that senior management respondents from SOs saw less of a connection between CWB and climate change than staff in other roles in the organization in general and, more specifically, than those at the field level with responsibilities for programming. While few in number, their CC-CWBI scores were high enough above the mean for the regression coefficient on this dummy variable to be significant. Similar conclusions can be drawn for respondents in the "other" category, 37 percent of whom were associated with NOs and 40 percent with SOs. Again, the mean CC-CWBI score for respondents in the "other" category from SOs was far enough above the mean for the CC-CWBI score for all respondents in the "other" category for the regression coefficient to be significant. As for senior management, those in the "other" category in SOs (a group of people which includes some in fairly key management, finance and administrative roles) do not see a strong connection between climate change and CWB in comparison with other respondents. This suggests that those in a variety of senior management roles in the SOs tended to have views that were quite dissimilar to those found in the rest of the organization—views that overall indicated less concern about climate change.

Discussion and conclusions

The results show a widespread acknowledgement among WV employees of the impact of climate change on CWB and recognition of the need to mainstream climate change into WV's programming. Respondents indicated a serious concern about climate change, perceived serious risks from it and saw an important connection between it and the key outcome goals of the organization—CWB. Overall, respondents' views of climate change were close to the scientific consensus—much closer than is generally found in surveys of the U.S. population at large, in fact (Kohut 2009; Maibach, Roser-Renouf and Leiserowitz 2009). Therefore the disagreement about climate change, prevalent particularly in the U.S., is less evident in this exploratory case study. However, although the Global Centre of WV has a presence in the U.S., and WV U.S. is a large Support Office in the organization, WV cannot be considered a U.S.-based or U.S.-run organization. The results are in line with the attitudes of those in the general public who generally feel that climate change is a threat (GlobeScan 2011; World Public Opinion.org 2009). Our results are also consistent with the attitudes of the global Christian community, the majority of whom feel climate change is an important issue (Barna Group 2008; Kerber 2010; Spencer, White and Vroblesky 2009).

This research supports Whitmarsh's (2008) finding that environmental values are strongly linked to attitudes and actions on climate change. Yet, in contrast to her findings for flood victims, respondents from WV who had observed the effects of climate change were more likely to be concerned about climate change, to perceive the risks it presents and to take personal action. They also were more pro-ecological in their worldview. In this case study, this appears to be connected to the fact that many respondents linked their concern for climate change and their concern for the environment with their Christian faith. This linkage is consistent with statements made by Christian groups about climate change which cite biblical imperatives of care for creation, neighbor care, and a commitment to justice for those who are most vulnerable as a basis for action (Kerber 2010; Spencer, White and Vroblesky 2009; Wilkinson 2010).

The variation in perspectives shown among office and employee categories is an important finding because of their role in decision making within the organization. Those working in communications are responsible for the organization's message to its donor constituency. Those in senior management make decisions around resource allocation priorities. Some respondents in the "other" category hold fairly senior positions in administration and finance. In each case, respondents in these categories are in positions of importance within the organization when it comes to setting priorities and communicating with the donor constituency.

Given these important differences in perceptions within WV, one might expect that this would have an impact on the adoption of climate policies (World Bank 2009). There is evidence that this might be the case. While the CCRI has been working internally on climate change issues since 2006, climate change as an issue did not figure prominently on the organization's international website (World Vision International 2010c). Although there is evidence of a developing policy position and advocacy on the issue (World Vision International 2010b), the organization's major Child Health Now campaign (World Vision International 2010a) mentions the issue only tangentially in its background document (World Vision International 2009a). There is also a certain amount of variation within the organization. Some SOs feature the subject prominently on their internet portals (World Vision Australia 2009a) and have clearly articulated policy and programming options (World Vision Australia 2009b), while for others this is less so (World Vision U.K. 2009; World Vision U.S. 2010). However, it is important to note that, while the issue may not feature prominently in all WV areas, many who work in the field are already seeking to respond to the effects of climate change at the community level, either through disaster response and risk reduction or through sustainable development projects.

There is also considerable variation in the prominence given to the issue among the larger INGOs. There are some for which this is a major issue (Care International 2010a; Oxfam International 2010b), while others, like WV, give it less importance on their websites (International Save the Children Alliance 2010; Medecins Sans Frontieres 2010). Some organizations have clearly articulated policy positions and give evidence that it is being factored into programming initiatives (Care International 2010b; Oxfam International 2010a) even though it may not appear in a position of prominence on their website (International Save the Children Alliance 2008, 2009). Others are involved in international advocacy efforts (Childreninachangingclimate.org 2010), or appear to have neither programming nor policy clearly targeted at climate change (Medecins Sans Frontieres 2010). It may very well be that some of them are facing similar internal challenges to developing a firm position on the subject.

Our research shows that those with little or no overseas experience beyond a brief visit or two to a developing country and/or a less direct connection with the well-being of children in the field may be less likely to perceive the actual connections between climate change and the work of the organization. Therefore, one possible area for action and organizational learning is to find ways to help those in decision-making roles in Support Offices, less connected with the on-the-ground reality, to gain more direct experience with the challenges climate change presents to children in the developing world. If climate change issues are to receive the necessary attention in the face of competing and urgent priorities, such targeted awareness raising within FBOs and other international humanitarian and development agencies is essential (Mitchell and Tanner 2006).

Acknowledgments

We would like to thank all employees of World Vision who took time from their very busy schedules to respond to this survey.

Notes

1 Tipping points are the points beyond which ecosystems begin to change and rapid, irreversible changes are unavoidable.
2 Children's well-being includes positive relationships, healthy individual development (involving physical and psycho-social health, cognitive, social and spiritual dimensions), and contexts where all children experience safety, social justice, and participation in civil society (World Vision International 2009b).
3 We do not know why some individuals chose not to complete the survey, which has potential to bias results, but there is a similar risk in any survey design, random or otherwise. However, we expect that interest in the subject matter was not a significant determining factor in limiting response, as individuals were asked by senior staff to complete the survey and one would normally expect compliance with such a request.
4 Exemptionalism holds that the intelligence and technological prowess of humankind make humanity exempt from the natural processes that govern other species.
5 For further details on the specific procedures followed in data collection and analysis please see Brown, Brown and Shore (2012).
6 Factor analysis was used to determine whether or not this series of statements was able to serve as a measure of ecological worldview—NEPI (Dunlap 2008; Dunlap et al. 2000). Reliability analysis showed that it was internally consistent (Cronbach's alpha = 0.7271), and correlation analysis showed that none of the responses were highly correlated. The first factor extracted had an eigenvalue of 2.84240 and explained 62.19 percent of the variation in the data. While the eigenvalue on the second factor was greater than one, and might therefore indicate the possibility of a second factor, following Dunlap (2008) we concluded that the first eigenvalue was sufficiently larger than that for the second (and other) factor that it was reasonable to treat it as a single scale. This conclusion was supported by the observation that the signs on the factor scoring coefficients alternated between negative and positive as expected, indicating that the individual items contributed to the overall pro-ecological measure of the scale.
7 The "other" category included people in administration, IT, finance, human resources and other management roles.

References

Barna Group. 2008. "Evangelicals Go 'Green' with Caution." Available at http://www.barna.org/barna-update/article/13-culture/23-evangelicals-go-qgreenq-with-caution?q=global+warming_ (accessed July 29, 2011).

Bergmann, S. 2009. "Climate Change Changes Religion: Space, Spirit, Ritual, Technology—Through a Theological Lens." *Studia Theologica* 63, 98–118.

Brown, D.R., H.C.P. Brown and C.A. Shore. 2012. "A Case Study of Perceptions of Climate Change Impacts and Child Well-Being: Implications for Development Programming." *Environmental Hazards: Human and Policy Dimensions.* Doi: 10.1080/17477891.2011.635187.

Brown, D.R., P. Dettman, T. Rinaudo, H. Tefera and A. Tofu. 2011. "Poverty Alleviation and Environmental Restoration Using the Clean Development Mechanism: A Case Study from Humbo, Ethiopia." *Environmental Management* 48.2, 322–333.

Care International. 2010a. "Care International: Home Page." Available at http://www.care-international.org/ (accessed April 7, 2011).

——. 2010b. "Climate Change Information Centre." Available at http://www.careclimatechange.org/ (accessed April 7, 2011).

Childreninachangingclimate.org. 2010. "Children in a Changing Climate: Home Page." Available at http://www.childreninachangingclimate.org/ (accessed April 7, 2011).

Dunlap, R.E. 2008. "The New Environmental Paradigm Scale: From Marginality to Worldwide Use." *Journal of Environmental Education* 40.1, 3–18.

Dunlap, R.E. and K.D. Van Liere. 2008. "The 'New Environmental Paradigm': A Proposed Measuring Instrument and Preliminary Results." *Journal of Environmental Education* 40.1, 19–28.

Dunlap, R.E., K.D. Van Liere, A.G. Mertig and R.E. Jones. 2000. "Measuring Endorsement of the New Ecological Paradigm: A Revised NEP Scale." *Journal of Social Issues*, 56.3, 425–442.

FAO. 2008. *Climate Change and Food Security: A Framework Document*. Rome: Food and Agriculture Organization of the United Nations.

Füssel, H.-M. 2009. "An Updated Assessment of the Risks from Climate Change Based on Research Published since the IPCC Fourth Assessment Report." *Climatic Change* 97, 469–482.

GlobeScan. 2011. "Climate Concern Falls Sharply, but Public Remains Conflicted." Available at http://www.globescan.com/findings/?id=16 (accessed July 28, 2011).

International Save the Children Alliance. 2008. *In the Face of Disaster: Children and Climate Change*. London: International Save the Children Alliance. Available at http://www.savethechildren.net/media/publications/2008/IntheFaceofDisaster.pdf (accessed April 7, 2011).

——. 2009. *Feeling the Heat: Child Survival in a Changing Climate*. London: International Save the Children Alliance. Available at http://www.savethechildren.net/alliance/what_we_do/drr/publications/200910_feelingtheheat.pdf (accessed April 7, 2011).

——. 2010. "International Save the Children Alliance: Home Page." Available at http://www.savethechildren.net/alliance/index.html (accessed April 7, 2011).

IPCC (Intergovernmental Panel on Climate Change). 2007. *Climate Change 2007: Impacts, Adaptation and Vulnerability. Contribution of Working Group II to the Fourth Assessment Report of the Intergovernmental Panel on Climate Change*. Cambridge, U.K.: Cambridge University Press.

Kerber, G. 2010. "Caring for Creation and Striving for Climate Justice: Implications for Mission and Spirituality." *International Review of Mission*, 99.2, 219–229.

Kohut, A. 2009. *Fewer Americans See Solid Evidence of Global Warming*. Washington, DC: Pew Research Center for People and the Press.

Lobell, D.B., M.B. Burke, C. Tebaldi, M.D. Mastrandrea, W.P. Falcon and R.L. Naylor. 2008. "Prioritizing Climate Change Adaptation Needs for Food Security in 2030." *Science* 319, 607–610.

Lowe, T.D. and I. Lorenzoni. 2007. "Danger Is All Around: Eliciting Expert Perceptions for Managing Climate Change through a Mental Models Approach." *Global Environmental Change* 17, 131–146.

Maibach, E., C. Roser-Renouf and A. Leiserowitz. 2009. *Global Warming's Six Americas 2009: An Audience Segmentation Analysis*. New Haven, CT and Fairfax, VA: Yale Project on Climate Change and the George Mason University Center for Climate Communication.

McCormick, J. 2005. "The Role of Environmental NGOs in International Regimes." In *The*

Global Environment: Institutions, Law, and Policy, ed. R.S. Axelrod, D.L. Downie and N.J. Vig. Washington, DC: CQ Press.

McGray, H., A. Hammill and R. Bradley. 2007. *Weathering the Storm: Options for Framing Adaptation and Development*. Washington, DC: World Resources Institute.

McIver, J.P. and E.G. Carmines. 1981. *Unidimensional Scaling*. Newbury Park, CA: Sage.

Medecins Sans Frontieres. 2010. "MSF International Website: Home Page." Available at http://www.msf.org/ (accessed April 7, 2011).

Mitchell, T. and T. Tanner. 2006. *Adapting to Climate Change: Challenges and Opportunities for the Development Community*. Teddington, U.K.: Institute of Development Studies and Tearfund.

Mitchell, T. and M. van Aalst. 2008. *Convergence of Disaster Risk Reduction and Climate Change Adaptation: A Review for DFID*. London: Institute of Development Studies.

Oxfam International. 2010a. "Climate Change Campaign." Available at http://www.oxfam.org/en/climate (accessed April 7, 2011).

——. 2010b. "Oxfam International: Home Page." Available at http://www.oxfam.org/ (accessed April 7, 2011).

Parris, B., D. Lansley and G. Finnigan. 2009. *Climate Change and Poverty: Implications for Children and for World Vision*. Melbourne: World Vision Australia.

Pidgeon, N.F., I. Lorenzoni and W. Poortinga. 2008. "Climate Change or Nuclear Power—No Thanks! A Quantitative Study of Public Perceptions and Risk Framing in Britain." *Global Environmental Change* 18, 69–85.

Shisanya, C.A. and M. Khayesi. 2007. "How Is Climate Change Perceived in Relation to Other Socioeconomic and Environmental Threats in Nairobi, Kenya?" *Climatic Change* 85, 271–284.

Shore, C.A. 2009. "Why Climate Change Matters for Children." *Global Future* 9.3, 19.

Spencer, N., R. White and V. Vroblesky. 2009. *Christianity, Climate Change, and Sustainable Living*. Peabody, MA: Hendrickson Publishers.

Stern, N. 2007. *The Economics of Climate Change: The Stern Review*. Cambridge, U.K.: Cambridge University Press.

Toulmin, C. 2009. *Climate Change in Africa*. London: Zed Books.

Whitmarsh, L. 2008. "Are Flood Victims More Concerned about Climate Change than Other People? The Role of Direct Experience in Risk Perception and Behavioural Response." *Journal of Risk Research* 11.3, 351–374.

Wilkinson, K. 2010. "Climate's Salvation? Why and How American Evangelicals Are Engaging with Climate Change." *Environment* 52.2, 47–57.

World Bank. 2009. *World Development Report 2010: Development and Climate Change*. Washington, DC: World Bank.

——. 2011a. "Development Dialogue on Values and Ethics: Climate Change." Available at http://go.worldbank.org/JQ6ACADP90 (accessed October 29, 2011).

——. 2011b. "Development Dialogue on Values and Ethics: Dialogue with Faith Leaders." Available at http://go.worldbank.org/LLXAK8JMJ0 (accessed October 29, 2011).

——. 2011c. "Development Dialogue on Values and Ethics: July 2009 Faith and Development Leaders Meeting." Available at http://go.worldbank.org/2IIS2PJN80 (accessed October 29, 2011).

World Public Opinion.org. 2009. *Public Attitudes toward Climate Change: Findings from a Multi-Country Poll*. Washington, DC: World Bank. Available at http://www.worldpublicopinion.org/pipa/pdf/dec09/ClimateChange_Dec09_rpt.pdf (accessed April 7, 2011).

World Vision Australia. 2009a. "World Vision Australia: Home Page." Available at http://www.worldvision.com.au/Home.aspx (accessed April 7, 2011).

——. 2009b. "World Vision Australia: Issues: Climate Change." Available at http://www.worldvision.com.au/Issues/Climate_Change.aspx (accessed April 7, 2011).

World Vision International. 2008a. "Case Study: World Vision's First Climate Change Mitigation Project." Available at http://www.wvi.org/wvi/wviweb.nsf/maindocs/588A47FC5C58ACD7882574CC0061E44B?opendocument (accessed July 28, 2011).

——. 2008b. *World Vision Overview: Hope for the Most Vulnerable*. Monrovia, CA: World Vision International.

——. 2009a. *Child Health Now: Together We Can End Preventable Deaths*. Monrovia, CA: World Vision International. Available at https://childhealthnow.com/docs/pdf/Child_Health_Now-Report.pdf (accessed February 17, 2011).

——. 2009b. *Partnership Policy on Children's Well-Being*. Monrovia, CA: World Vision International.

——. 2010a. "Child Health Now." Available at https://childhealthnow.com/ (accessed February 17, 2010).

——. 2010b. "World Vision International: Climate Change Response Initiative." Available at http://www.wvi.org/wvi/wviweb.nsf/maindocs/C799F54BADDD6FA1882574BF007F82CB?opendocument (accessed April 7, 2011).

——. 2010c. "World Vision International: Home Page." Available at http://www.wvi.org/wvi/wviweb.nsf (accessed April 7, 2011).

World Vision U.K. 2009. "World Vision U.K.: Home Page." Available at https://www.worldvision.org.uk/ (accessed April 7, 2011).

World Vision U.S. 2010. "World Vision U.S.: Home Page." Available at http://www.worldvision.org (accessed April 7, 2011).

18 International advocacy for climate justice

Guillermo Kerber

Introduction

Do faith traditions speak about climate justice? How have these two concepts—climate and justice—come together? What kind of actions have religious organizations taken in pursuit of climate justice?

The purpose of this chapter is to try to respond to these questions, exploring how the concern for the environment and climate in particular has been seen in relationship to justice issues and what kind of concrete activities have been implemented to advocate for climate justice. Taking the World Council of Churches (WCC) as its focus, this chapter will show how climate and justice are considered in the WCC's policy and action. Interfaith initiatives in which the WCC has been a participant will also be discussed. The chapter will conclude with some lessons learned from the years of international advocacy by religious organizations (Kerber 2010).

The World Council of Churches is a worldwide, ecumenical fellowship of 349 Christian churches, denominations and church fellowships in more than 110 countries and territories throughout the world. Representing over 560 million Christians, it includes most of the world's Orthodox churches, and scores of Anglican, Baptist, Lutheran, Methodist and Reformed churches, as well as many United and Independent churches and some Evangelical and Pentecostal churches. The Roman Catholic church is not a member church of the WCC, though it participates officially in various programs of the council. While the bulk of the WCC's founding churches were European and North American, today most member churches are in Africa, Asia, the Caribbean, Latin America, the Middle East and the Pacific.[1]

The WCC was an outgrowth of the ecumenical movement, a movement developed especially since the beginning of the twentieth century among various Christian denominations. The organization has served as a venue for greater unity, common witness and service among Christian churches. Created in 1948, the WCC was one of the first recognized nongovernmental organizations with general consultative status at the United Nations (UN), through its Commission of the Churches on International Affairs. Through this Commission the WCC has had a long history of international advocacy work at the UN (Nurser 2005). The General Secretariat of the WCC is based in Geneva, Switzerland, where about 150 staff coordinate worldwide programs on the WCC and ecumenism in the twenty-first century. These programs today focus on: Christian unity, mission, evangelism and spirituality; public witness; justice, diakonia (service) and responsibility for creation; education and ecumenical formation; and inter-religious dialogue and cooperation. Other WCC offices are based in Jerusalem and New York, while a major HIV AIDS program operates all over Africa. The WCC is governed by the Assembly, which meets every seven years. The purpose of the Assembly is to review programs and determine the overall policies of the WCC, as well as to elect presidents and appoint a central and executive committees that serve as the governing bodies of the WCC between Assemblies. Assemblies and central and executive committees adopt statements which constitute the policy of the WCC, though they do not have a binding character in relationship to member churches.

In terms of environmental concern, the WCC held a series of meetings in relation to sustainability in the 1970s that led in the 1980s to the process labeled as *Justice, Peace and the Integrity of Creation* (JPIC). The relevance of this process is that, on the one hand, it has clearly linked environment degradation with justice issues and, on the other, it has gone beyond the WCC constituency such that many Christian groups and organizations still identify themselves with the JPIC premises. The WCC began referring to climate change (then called "global warming"[2]) as early as 1988. A working group on climate change was created and started working for the WCC and in preparation for the Rio Summit in 1992. The WCC has been present at the United Nations Framework Convention on Climate Change (UNFCCC) annual conferences (called Conferences of the Parties—COPs) and has increasingly worked with faith groups outside of the Christian tradition on advocacy and action in response to the common threat of climate change.

Toward a working definition of climate justice

The WCC has increasingly linked the issues of climate change and global justice. An example of the rationale for this linkage is expressed in a joint declaration from the Lutheran World Federation and WCC delegates to a UN Climate Conference (COP 16, 2010):[3]

> In the churches' perspective, justice must be the basic criterion of applied ethics in all decisions concerning the measures to cope with climate change. Although climate change is a global issue affecting all peoples and nations, those who are and will increasingly be affected by negative climate change consequences are the vulnerable communities who have contributed the least to global emissions. These include women and children, indigenous peoples, poorest communities, people with disabilities and inhabitants of coastal low lying areas. Vulnerable communities and states are also much more dependent on natural resources for their subsistence and do not have the means to mitigate emissions and to adapt to climate change. Their survival is at risk, and justice requires that the nations most responsible historically for the adverse ecological conditions should take the greatest responsibility towards the adaptation of these vulnerable communities and nations.

Such advocacy for climate justice is rooted in the affirmation of churches and other religious organizations that climate change has to be addressed holistically. This has meant recognizing the interconnectedness of the various dimensions of the climate crisis—including environmental, economic, social and cultural. The climate justice advocacy agenda has also implied linking what is experienced at the grassroots level within congregations, parishes and communities around the world with theological reflection and international advocacy.

A case study: the conciliar process on *Justice, Peace and the Integrity of Creation*

The early association of environment and justice by faith groups is a story that is not told often enough in the environmental justice literature (Gibson 2004). The JPIC conciliar process, which culminated in 1990, is a clear example of early efforts by WCC to link justice and ecological sustainability.[4] The JPIC process was developed after the 6th WCC Assembly in Vancouver, Canada in 1983. The Assembly stated that "to engage member churches in a conciliar process of mutual commitment (covenant) to justice, peace and the integrity of all creation should be a priority for World Council programmes" (Niles 1992: 2). While the two first concepts of the motto—justice and peace—have strong biblical roots, the third concept, "integrity of creation," became a new ecumenical concept that needed further exploration.

During the years following the Assembly in Vancouver, the WCC made various attempts to explore the challenges posed by the ecological crisis. Then WCC director of church and society Wesley Granberg-Michaelson (1991, 1994) described a number of these efforts. For instance, a meeting of theologians and scientists (including James Lovelock, who formulated the Gaia hypothesis) was held in 1987 and issued a report on *Reintegrating the Creation*. In the same year, another consultation linked the integrity of creation with the theme of liberation applied not just to humanity but to all of life; these papers were collected in Birch, Eakin and MacDaniel (1990). Finally, in 1988 a major consultation in Granvollen, Norway enabled new and comprehensive attention to the threats posed to the global environment, and to the challenges presented in reframing a theology of creation. While the JPIC process emphasized the interrelationship between the integrity of creation and the issues of justice and peace, some participants at the consultation in Granvollen—including indigenous peoples, women and those from the Pacific—argued that in fact the sequence of JPIC should be reversed, and begin with the integrity of creation, for this was foundational to justice and peace. The report from this consultation began with theological and biblical affirmations and was followed by a section with stories and experiences of various groups of people. As Granberg-Michaelson (1991) stressed, Orthodox participants made an important contribution, and Granvollen demonstrated that Orthodox theology offered vital and influential perspectives in this area. The report concluded:

> The drive to have "mastery" over creation has resulted in the senseless exploitation of natural resources, the alienation of the land from people and the destruction of indigenous cultures. . . . Creation came into being by the will and love of the Triune God, and as such it possesses an inner cohesion and goodness.
>
> (Hall 2002: 274)

The Granvollen consultation ensured the space for a growing ecumenical discussion around the integrity of creation in the years leading to a big gathering called the JPIC World Convocation, which took place in Seoul, South Korea in 1990. Various regional JPIC meetings provided further points of emphasis. Interestingly, the ecumenical tension between justice and creation was by no means resolved. Paralleling a similar debate in society at large, the ecumenical movement, the WCC, and churches struggled to understand the relationship between social justice and the environment. For many years activists struggling for justice looked suspiciously at those concerned about the environment. In the Global South, the environment and ecology were seen as Northern concerns, which prevented a more critical interpretation of the social and political realities and concerted action. In the JPIC process there were those who said that *justice* (understood as economic and social equity) should be the more important issue. As a matter of fact, two different programs of the WCC independently organized forums immediately prior to the JPIC World Convocation for the purpose of declaring that justice must be the priority and the starting point in the JPIC process. Others—and here the influence of pacifists and the historic peace churches (Mennonites, Brethren and others) should be recognized—would stress the primacy of peace, while a third group would emphasize the crucial importance of environment and ecological concerns. But, despite these internal tensions within the churches, the process helped to bring to the same table those who could be labeled as social justice activists, pacifists and environmentalists. These three groups had been on separate if not conflicting tracks for many years.

At the World Convocation in Seoul in 1990, ten affirmations on justice, peace and the integrity of creation were adopted, together with a commitment and concrete actions which, as the final document of the consultation expressed, "represent the basic direction which the Christian commitment to JPIC is to follow" (Niles 1992: 177). The preamble of the Act of Covenanting states that participants came:

> To consider their common response to the threats the present generation faces confronted by new and complexly interwoven threats, among them, the entrenched and deadly forms of injustice, universal violence and the rapid degradation of the environment. The *real danger lies in the interaction of these threats*. Together they represent a global crisis.
>
> (Niles 1992: 164–165)

Threats to justice, peace and the integrity of creation are, therefore, intertwined and constitute a global crisis.

The theological rationale is lengthy. One important point is the sixth of the ten affirmations, which summarizes what "integrity of creation" means and describes its links with peace and justice. The statement of belief that "We affirm the creation as beloved of God" is explained:

> As Creator, God is the source and sustainer of the whole cosmos. God loves the creation. . . . We affirm that the world, as God's handiwork, has its own inherent integrity; that land, waters, air, forests, mountains and all creatures, including humanity, are "good" in God's sight. The integrity of creation has a social aspect which we recognize as peace with justice and an ecological aspect which we recognize in the self-renewing, sustainable character of natural ecosystems.
>
> (Niles 1992: 173)

The peace and justice components of the JPIC triad are therefore seen as inherent aspects of the integrity of creation.

The text continued with a critique of anthropocentrism:

> We will resist the claim that anything in creation is merely a resource for human exploitation. We will resist species extinction for human benefit; consumerism and harmful mass production; pollution of land, air and waters; all human activities which are now leading to probable rapid climate change; and policies and plans which contribute to the disintegration of creation.
>
> (Niles 1992: 173)

Anthropocentrism has been one of the main critiques to the Judeo-Christian traditions since the publication of Lynn White Jr.'s famous article blaming Christianity for the domination of human beings over nature. White stated that, "especially in its Western form, Christianity is the most anthropocentric religion the world has seen" (1967: 1205). Against this background the following commitment of the JPIC Convocation acquired more import:

> we commit ourselves to be members of both the living community of creation in which we are but one species, and members of the covenant community of Christ; to be full co-workers with God, with moral responsibility to respect the rights of future generations; and to conserve and work for the integrity of creation both for its inherent value to God and in order that justice may be achieved and sustained.
>
> (Niles 1992: 174)

Not only did the Convocation reject an anthropocentric orientation and the instrumental valuation of nature, but it extended the concern for justice for future generations, a concern that would continue to be of utmost importance more than 20 years after the Convocation.

To understand the WCC efforts at this time, it is relevant not only to see how the concepts are articulated but to highlight the concrete commitments proposed. Participants agreed to call on churches to lead efforts to reverse the thought patterns that have underpinned unlimited energy consumption and economic growth. Among other means, this was to be done "through monitoring international negotiations regarding the Montreal Protocol [1987], the Helsinki meeting [1989] and the upcoming UN Conference on Environment and Development [1992] like the successful monitoring of the Helsinki Final Act [1975] on human rights" (Niles 1992: 175). For the WCC, international environmental negotiations were already part of the agenda in 1990; participating in them was understood as an explicit part of the faith and witness of the churches.

Together with this international focus, participants also committed themselves "to promote and facilitate the achievement of these goals through our witness and lifestyle . . . and through solidarity with those affected by climate change" (Niles 1992: 189). Interestingly enough, the final document already called for the "establishment of an international fund of atmosphere solidarity supported by a tax on carbon dioxide emissions above the world average, to be used for the transfer to developing countries of technologies which are less harmful to the environment and to projects aimed at combating deforestation and promoting reforestation" (Niles 1992: 188). Climate change and some of its specifics in the international negotiations—such as the discussion on carbon tax and credits, technology transfer, and what are today known as the Green and Adaptation Funds and Reducing Emissions from Deforestation and Forest Degradation (REDD)—had already been raised in WCC discussions in 1990.

During recent years WCC has participated in several other meetings exploring climate-related themes. In May 2011 another convocation was held, in Kingston, Jamaica, to close the Decade to Overcome Violence. It included four themes which updated the challenges to peace: peace in the community, peace with the earth, peace in the marketplace and peace among the peoples (Kerber 2011). Furthermore, the current WCC program on eco-justice, bringing together ecological and economic concerns, has been an effort to address in a concerted manner the challenges coming from these overlapping domains. Consultations on each continent have focused on "wealth, poverty and ecology."[5] Eco-justice, with its dual components of economic justice and ecological justice (and within the latter climate and water justice), has been a priority for many churches and ecumenical organizations like the World Student Christian Federation during recent years.

Advocacy at UNCED 1992

Affirmations and commitments at the JPIC Convocation set the stage for a large ecumenical and interfaith participation at the 1992 UN Conference on Environment and Development (UNCED; also called the Earth Summit) in Rio de Janeiro. UNCED was one of the major conferences organized by the UN in the 1990s. Delegations from 178 countries, heads of state and heads of government of more than 100 countries, and representatives of more than 1,000 NGOs attended the meeting. UNCED adopted various documents, among them: the UN Convention on Biological Diversity (UNCBD), a binding international agreement aimed at strengthening national control and preservation of biological resources; the Forest Principles, a non-legally binding authoritative statement of principles for a global consensus on the management, conservation and sustainable development of all types of forests; and, last but not least, the UN Framework Convention on Climate Change (UNFCCC), an international environmental treaty with the objective of stabilizing greenhouse gas (mainly carbon dioxide and methane) concentrations in the atmosphere at a level that would prevent dangerous anthropogenic interference with the climate. The UNFCCC later included the Kyoto Protocol, a binding instrument aimed at fighting global warming. Under the Protocol, 37 industrialized countries committed themselves to reduce their greenhouse gas emissions by 5.2 percent on average for the period 2008–12, a target that was not achieved. These three agreements are known as the Rio Conventions. Two other important texts were adopted in Rio 1992: Agenda 21, the comprehensive blueprint of actions to be taken globally, nationally and locally by organizations of the UN, governments and major groups; and the Rio Declaration on Environment and Development, which consists of 27 guiding principles for sustainable development.

UNCED had as precedent the United Nations Conference on the Human Environment (UNCHE) held in Stockholm, Sweden in 1972. UNCHE was the first major modern international gathering on human activities in relationship to the environment and led to the founding of the United Nations Environment Programme (UNEP). The International Day of the Environment (annually on 5 June) was established to commemorate the UNCHE inauguration ceremony on that date in 1972.

At the Earth Summit the WCC organized an ecumenical gathering called *Searching for the New Heavens and the New Earth*, as well as daily activities at a WCC tent at the Expo Rio-92 (ECO-92). The ecumenical work at UNCED marked perhaps the first significant effort to advocate for climate justice at the UN level. Churches had been involved in issues of environment and development for many years before UNCED, but it was hoped that UNCED would be a turning point in a long process to bring about ecological sustainability and economic justice.

Churches had been active in the early negotiation stages of UNCED in providing input to what was proposed as an Earth Charter (Rasmussen 1991). Originally conceived as a statement of values and principles to guide sustainable development with explicit reference to justice and peace in the pursuit of ecological stability, the Earth Charter (http://www.earthcharterinaction.org) was not formally adopted. It remains, however, a significant vehicle for guiding faith groups and others to local action (Haluza-DeLay 2012). The Rio Declaration, which was the outcome of those negotiations, was a compromise among the different positions advocated by various countries.

The ecumenical delegation present in Rio 1992 was not happy with the outcomes of UNCED. In their evaluation of UNCED the delegation commented: "For the first time, the world community was gathered to deal with the interrelated crises of ecological destruction and global poverty. However, the results of UNCED are an inadequate response to the seriousness of the crisis" (World Council of Churches'

UNCED Group 1992: 12). Nevertheless the WCC participants considered that many of the Rio principles (e.g., common but differentiated responsibilities, the preventive principle, the polluter pays principle) provided the basis for significant and perhaps even radical change if taken seriously and implemented. Churches were encouraged to use the Rio Declaration to advocate for changes in their own societies and internationally.

On climate change particularly, the WCC evaluation of UNCED recognized that the WCC and many churches had already been deeply concerned. The UNFCCC signed at UNCED was seen as a first step to address the problem. Yet it was not considered completely sufficient. As the delegation explained, "largely because of opposition from the United States, the Convention contains no specific targets or schedules for limiting greenhouse gas emissions" (World Council of Churches' UNCED Group 1992: 13). At this time, climate change (or global warming) was seen as a serious threat to creation caused largely by the polluting emissions of industrialized nations but with the adverse consequences experienced most severely by developing countries.

To effectively address climate change, participants expressed the need for a drastic change in lifestyles within industrialized countries, a major reduction in energy consumption, and a significant reorientation in their economies. On the controversial issue of nuclear energy, the document clearly stated that it was not an appropriate alternative to fossil fuels from either an ecological or an economic perspective and that reducing energy demand was the better approach.[6]

Industrialized countries cannot resolve the threat of climate change by themselves. Though in 1992 the so-called "emerging economies" like Brazil, China and India were not seen as main actors in the climate change negotiations, the document warned Southern countries not to repeat the mistakes of the North. Aware that limiting the increase in emissions is complicated by many economic factors, the document already called for international cooperation and technology transfer so that the South could pursue what the document called "sustainable and just development models." The issue of justice, therefore, is included in the way development is perceived.

The situation was perceived as alarming. The delegates of the meeting sent a "Letter to the Churches" in which they clearly affirmed: "The earth is in peril. Our only home is in plain jeopardy. We are at the precipice of self-destruction. For the very first time in the history of creation, certain life support systems of the planet are being destroyed by human actions" (World Council of Churches' UNCED Group 1992: 10).

Despite the critical situation, participants attempted to convey a message of hope in their letter. Citing biblical passages, the "Letter" asserted: "Our God is a God of life, and the power of the Holy Spirit permeates all creation." It called for the development of a "spirituality of creation," which was spelled out in the following way: "All that fosters life, such as justice, solidarity, and love, and all that defends life, such as the evangelical commitment to stand with the poor, the struggle against racism and casteism, and the pledge to reduce armaments and violence, concretely signifies living according to the Spirit" (World Council of

Churches' UNCED Group 1992: 10). Such hope was seen as an antidote to the fear or despair that sometimes characterizes environmental, justice and peace movements (Gourevitch 2010; Norgaard 2011).

WCC's increasingly interfaith advocacy for climate justice at COPs

After the ecumenical gathering in Rio, the WCC continued to advocate at subsequent UN climate negotiations until 1995, when the UNFCCC was ratified by a sufficient number of countries and entered into force. A WCC delegation has been present at all COPs and at most of them the WCC organized various activities, including side events and statements delivered during the high-level ministerial segment of the COPs (Abramides and Taffi 2012). The statements and activities show the centrality of justice as well as the increasingly interfaith collaboration of the advocacy work.

Despite its long-time commitment, the WCC has not been received as openly as some secular groups at UN climate negotiations. In these negotiations there are nine so-called "major groups," as defined by Agenda 21. According to the Agenda, major groups are those having substantive contributions to make to the achievements of a sustainable future. These nine groups include: business and industry; children and youth; farmers; indigenous people; local authorities; NGOs; the scientific and technological community; women; and workers and trade unions. Religious groups or faith-based organizations (FBOs) are not recognized as a major group. The UNFCCC Secretariat has, however, agreed to allow a statement from the "community of faith," which in most COPs has been delivered by the WCC.

The first COP (COP 1) in 1995 set the pattern for advocacy with delegates during the negotiations and for interfaith gatherings with members of the local faith communities. The following year marked the first time a WCC statement was delivered to the high-level segment of the COP. Its title, *A Profoundly Ethical and Spiritual Challenge*, contended that climate change was not just a matter of technical mediation. Table 18.1 shows the focus of a selection of WCC statements as well as some religious activity at the COPs.

Interfaith coalitions quickly became part of the WCC activity at the COP meetings. An early example was the Interfaith Climate Change Celebration at COP 3 in Kyoto, Japan. The celebration was organized at the Kawaramachi Roman Catholic Cathedral of Kyoto, with the presence of Buddhist, Shinto and Christian authorities and UNFCCC high officers. A process ended with a ceremony at the Shinto shrine. As Table 18.1 notes, interfaith celebrations have been held at many COPs. Another important milestone at the COP meetings was the *Uppsala Interfaith Climate Manifesto* (2008) signed by 26 religious leaders and delivered to the 2008 COP in Poznań, Poland. The Manifesto started with expressing the inter-religious commonalities:

> From religious traditions, with different approaches to religious life, we come together at this time in human history to assure the world of what we have in

Table 18.1 Sample of WCC-involved activities at UNFCCC Conferences of the Parties to show the consistent relationship expressed between climate change and global justice

COP	*Year and location*	*Title of WCC statement*	*Other notes*
2	1996 Geneva, Switzerland	*A Profoundly Ethical and Spiritual Challenge*	First WCC statement at COP meeting.
3	1997 Kyoto, Japan	*A Matter of Justice: An Interfaith Statement*	Interfaith Climate Change Celebration was a major event.
6	2000 The Hague, Netherlands	*The Earth's Atmosphere: Responsible Caring and Equitable Sharing for a Global Commons. A Justice Statement Regarding Climate Change*	
7	2001 Marrakech, Morocco	*The Role of Religions: An Interfaith Statement*	One-day colloquium focused on Islamic and Christian perspectives on environment and climate change.
8	2002 New Delhi, India	*A Call to Action in Solidarity with Victims of Climate Change*	
9	2003 Milan, Italy	*The Earth Does Not Belong to Us: We Belong to the Earth—An Ethics Statement*	Appendix included *Message on Occasion of COP9 of the World Conference on Religion and Peace: Religions for Peace*.
10	2004 Buenos Aires, Argentina	*Moving beyond Kyoto with Equity, Justice and Solidarity*	
13	2007 Bali, Indonesia	*This Far and No Further: Act Fast and Act Now*	Interfaith Celebration on Climate Change held at Nusa Dua International Church.
14	2008 Poznań, Poland	*Faith and Feasibility: Responsibly Searching for a New Heaven and a New Earth*	*Uppsala Interfaith Climate Manifesto 2008*.
15	2009 Copenhagen, Denmark	*A Sign of Hope for the Future of the People of Goodwill*	*Renew the Face of the Earth: Faith-Based Approach to Climate Justice* side event co-organized by the WCC and Caritas Internationalis and co-sponsored by several national churches and the Asian Muslim Action Network. *Countdown to CO_2penhagen: Time for Climate Justice* international campaign organized before COP meetings.
17	2011 Durban, South Africa	*Once More a Plea for Immediate Action: Climate Justice for All*	African interfaith statement *Climate and Sustainable Peace in Africa*. Interfaith side event *Climate Justice and Food Security: Moral, Ethical and Spiritual Imperatives*.

> common. We all share the responsibility of being conscious caretakers of our home, planet Earth. We have reflected on the concerns of scientists and political leaders regarding the alarming climate crisis. We share their concerns.
> (*Uppsala Interfaith Climate Manifesto* 2008: 9)

Signatories not only called for effective leadership but committed themselves:

> to inform and inspire people in our own religious and cultural contexts to take responsibility for and to implement effective measures; to challenge political and business leaders where we live and work to develop comprehensive strategies and action; to focus on the struggle against global warming and draw upon our innermost religious convictions about the meaning of life.
> (2008: 15)

Justice was presented as one of the common principles of religious traditions: "We want to face the climate challenge with defiant optimism to highlight the core principles of all major sacred traditions of the world: justice, solidarity and compassion." Although quite ambitious in its formulation, the Manifesto has not had a significant follow-up by most of the signatories, and has not had a strong impact in their religious communities.

High expectations were generated for COP 15 in Copenhagen, Denmark in 2009. Many observers thought it would be possible to achieve a fair, ambitious and binding treaty to follow the first phase of the Kyoto Protocol, which was to end in 2012. Copenhagen negotiations would have given time to implement a second phase for 2013.

In the run-up to Copenhagen, one of the WCC's major activities was the *Time for Climate Justice* campaign, which achieved over 500,000 signatures globally.[7] This campaign stressed perhaps the central justice dilemma of climate change: that the poor in developing countries suffered the first and worst consequences of climate change yet had done least to cause the problem. It then summarized what the global interfaith movement for climate justice advocated:

> Our leaders must deliver fair and effective international climate action at a level adequate to stop catastrophic climate change and ensure the world's poorest have the right to develop. . . . We want: a strong commitment that builds on the Kyoto protocol and results in binding global cuts in carbon emission at a level adequate to stop catastrophic climate change; long-term sustainable international finance for developing countries, in addition to aid, for them to cut emissions, develop cleanly and adapt to the impacts of climate change.

South Africa's Archbishop Desmond Tutu presented the "Time for Climate Justice" petition to UNFCC executive secretary Yvo de Boer.

Also at Copenhagen, an *Interfaith Declaration on Climate Change*[8] was endorsed by a hundred organizations and hundreds of individuals, including well-known religious figures from around the world. The declaration reemphasized a common theme that "climate change is not merely an economic or technical problem, but rather at its core is a moral, spiritual and cultural one" and affirmed that "assisting vulnerable communities and species to survive and adapt to climate change fulfills our calling to wisdom, mercy, and the highest of human moral and ethical values." The religious statement called on global leaders to develop strong policies and targets based on science. An interfaith rally at the beginning of the Conference and an interfaith celebration took place at COP 18 in Durban, South Africa in 2011. A notable event here was the presentation of *Climate and Sustainable Peace in Africa*, a statement adopted by African religious leaders of several faiths (including Christian, Muslim and indigenous) in Nairobi earlier in the year. Such statements represent continuing and increasing interfaith cooperation on the issue.

Advocacy at other international forums

Despite major international expectations, the Copenhagen negotiations stalled. However, the interfaith work for climate justice has targeted not only the UNFCCC's COPs. Especially after COP 15 in Copenhagen, some religious organizations with advocacy experience at other forums started looking at new avenues to advance the climate justice agenda.

One new venue was the Geneva Interfaith Forum on Climate Change, Environment and Human Rights, which was created in 2010.[9] Founding members of the Forum included the Brahma Kumaris World Spiritual University, the Centre Catholique International de Genève, the Indigenous Peoples Ancestral Spiritual Council and the World Council of Churches. This group, together with representatives of Franciscans International, serves as steering committee of the Forum.

The mission statement of the Forum again reiterated that injustice is inherent in climate change. The mission statement highlighted "vulnerable communities" that will face the most severe consequences of climatic variation. Later it declared that "climate justice should be at the core of the Human Rights debate." The Forum asserted a human right to safe and secure environments and declared that reversing the causes of climate change involves national and international policies together with a change in those lifestyle patterns which have greed as an intrinsic component (Brubaker and Mshana 2010).

Interfaith advocacy and mobilization also took place at the 2012 UN Conference on Sustainable Development (UNCSD 2012), commonly known as Rio+20, as well as at the People's Summit, where a cluster was named "Religions for Rights," and conferences, workshops, celebrations and other activities were organized by different faith traditions, showing the commonalities of faith traditions in addressing issues related to environment and development.

Besides UN-related events, the WCC has also participated in solidarity with other groups and events, such as the various editions of the World Social Forum and the 2010 People's World Conference on Climate Change and the Rights of Mother Earth, held in Bolivia. Regional gatherings around the globe have also been held, especially in Africa, Asia and the Pacific. A particularly important gathering on resettlement was sponsored by the Pacific Council of Churches, since low-lying islands are among the most vulnerable to climate change. Through the diffusion of WCC theological reflection, spiritual materials, and the visibility of WCC advocacy at various events, national churches and local congregations have found inspiration and encouragement to partner in various global mobilizations, which are often led by other civil society organizations. Nevertheless, the impact of faith-based advocacy could stand further social scientific scrutiny.

Lessons learned and the way forward

Four lessons can be gleaned from this history of religious organizations' international advocacy for climate justice:

1 *Advocacy is part of the mission of religious actors.* Religions have strong links with grassroots communities. In almost every single remote location a religious community can be found. This has allowed religious leaders and other religious actors to have first-hand information on what is the situation of the communities, in this case, how they are being affected by climate change. For some religious communities, the international arena is far removed from their regular activities. Over the years, however, it has become clear to many how international decisions and actions (or the lack thereof) can impact local communities. Social action and concern on issues related to the global poor showed that poverty has clear national and international political implications. The same happened with climate change. Religious actors realized that, to be effective in supporting affected communities, national- and international-level advocacy was essential. This process has not been easy owing both to the tensions between justice and sustainability and to the frustrating processes of climate negotiations. For many religious groups, politics or advocacy has negative connotations, but a coalition with more and more religious actors has been developing for some years now.
2 *Religious advocacy is value-added.* The relevance of faith-based organization involvement in interfaith advocacy for climate justice has been raised on various occasions. Government officials at climate negotiations have frequently told religious actors that governments did not expect religious traditions to present new studies on science or politics or technical solutions to the climate crisis, as many such contributions were already available. But religious and secular actors alike (especially country representatives) have perceived ethical and spiritual reflections on the topic as something necessary to go beyond narrow and short-term perspectives, providing a broader picture, and even perhaps unlocking negotiations at UN conferences. At many COPs, for instance, during meetings between diplomats and church leaders, the former requested the latter to stress the ethical dimensions of climate change in order to move beyond self-interested negotiations and toward more lofty goals on the climate agreements.
3 *Interfaith action promotes common ground.* The various activities organized over the years have made it clear that there are many commonalities among

religions when it comes to addressing climate change. As the previous pages have shown, the centrality and sacredness of life, the care for the earth and for the most vulnerable, and the critiques of exploitation and overconsumption are some of the common values many faith traditions share. When coming together to express their convictions and decide common actions, faith leaders and activists have been reinvigorated and a synergy created. Joint initiatives, for instance between Christians and Muslims in Africa, have been developed.[10] The threat of climate change and the pursuit of climate justice created a common ground on which to walk together.

4 *Religious actors can partner effectively with civil society.* Given the gravity of the climate change crisis and the difficulty of the international negotiations, it became clear quite early to religious organizations that they could not effectively address the climate challenge alone. Therefore, in addition to their interfaith work, campaigns and events with other civil society actors were implemented. Examples of this are the Global Call for Climate Action and 350.org, which have over the years organized various global mobilizations in which religious organizations and communities have been involved. More simply, faith-based organizations are an important part of civil society (see also Chapter 6 by Amri, this volume).

As a consequence of this advocacy experience, networking has become a crucial dimension of the interfaith climate change work, magnifying the initiatives of religious actors being done in different regions on a shared agenda. The coordinated efforts at global, regional, national and local levels have, through the years, led to strengthening the climate change advocacy work in which religious groups have engaged.

As has been verified by scientists repeatedly, the threat of climate change is much worse today than when religious actors began their work on the issue in the 1980s. One cannot help but be sobered further by the minimal progress that has been made by industrialized nations in limiting greenhouse gas emissions. In fact, many industrialized countries have significantly increased their emissions since the 1990 base line established by the Kyoto Protocol. Advocacy and lobbying carried out by civil society at large have not been able to alter international policy on climate change, which remains on its "business as usual" trajectory. Civil society—and faith-based organizations within it—has questioned its role in the international arena. While it is true that UN processes normally take decades to be consolidated because of the difficulty of reaching consensus and the persistent reluctance of some states to accept multilateralism, the gravity of a quickly changing global climate requires a much more effective strategy, which for the time being remains elusive. The poor results of the UN Conference on Sustainable Development, held in Rio in June 2012, are an example of the international community's lack of practical commitment to addressing these issues. Civil society organizations have managed to mobilize hundreds of thousands if not millions of people in most parts of the world. But mobilization has not delivered policy change at the international level. The response to such lack of success has been mixed. While some within civil society and the churches have requested a withdrawal from international political negotiations, others affirm the moral voices that public participation (via civil society or faith-based organizations) brings to the international arena. In this context, despite the insufficient results—or perhaps even because of them—religious voices continue to be an important component of global civil society's international advocacy for justice-oriented action on climate change.

During the last decades, advocacy on climate change has been gaining relevance in the work of churches and faith-based organizations like the World Council of Churches. Using the biblical concept of conversion (*metanoia*) in terms of repentance and change, the WCC has called for a change in the international legal framework and the current development pattern: a change of paradigm. While the focus of this chapter has been on the international advocacy of the WCC, analysis of ecological conversion of religious institutions (in terms of lifestyle or greening sanctuaries and collective religious life) would help in understanding and giving credibility and coherence to the advocacy work carried out at the international level.

Notes

1 For an introduction on what the WCC is and what its programs are see http://www.oikoumene.org/en/about-us (accessed June 14, 2012).
2 See for example "An Evaluation of the UNCED Conventions" in *Searching for the New Heavens and the New Earth: An Ecumenical Response to UNCED*, where under the section titled "Climate Change" the term "global warming" is frequently used as a synonym of climate change (World Council of Churches' UNCED Group 1992: 13–14).
3 See the declaration, "Why Are the Churches at the UN Conference on Climate Change in Cancun?" (Lutheran World Federation and World Council of Churches 2010), at http://www.oikoumene.org/en/resources/documents/wcc-programmes/justice-diakonia-and-responsibility-for-creation/climate-change-water/why-are-the-churches-at-the-un-conference-on-climate-change-in-cancun (accessed June 14, 2012).
4 "Conciliar process" refers to preparatory meetings held before and leading up to a major gathering. In the JPIC conciliar process, meetings were held in various regions, and a global convocation occurred in Seoul, South Korea in 1990.
5 Numerous documents and consultation reports can be found at: http://www.oikoumene.org/en/resources/documents/wcc-programs/public-witness-addressing-power-affirming-peace/poverty-wealth-and-ecology/neoliberal-paradigm.html?print=1%20Presbyterian%20Church%20in%20Nigeria%20to%20%22preach%20water%22.
6 There is no official position of WCC regarding nuclear energy, though strong affirmations were made in relation to the crisis at Japan's Fukushima reactor in 2011. See for example http://www.oikoumene.org/en/resources/documents/general-secretary/statements/anniversary-of-the-fukushima-nuclear-disaster (accessed October 18, 2012).
7 See http://climatejusticeonline.org (accessed June 14, 2012).
8 See http://www.interfaithdeclaration.org (accessed June 14, 2012).
9 See for instance the Forum's 2011 Call for Action at http://www.ekklesia.co.uk/node/15452 or its petition for the establishment of a special rapporteur on climate change and human rights at http://www.petitions24.com/sr_human_rights_and_climate_change (accessed June 14, 2012).
10 See for example the work developed by the Program for Christian–Muslim Relations in Africa (PROCMURA) with UNEP on climate justice and sustainable peace in

Africa at http://en.procmura-prica.org/Climate%20Change%20UNEP%20Sending.pdf (accessed September 7, 2012).

References

Abramides, Elias C. and Juan A. Taffi. 2012. "WCC Participation in UN Conferences on Climate Change." In *Spiritual Values for Earth Community*, David G. Hallman, pp. 153–160. Geneva, Switzerland: World Council of Churches.

Birch, Charles, William Eakin and Jay MacDaniel. 1990. *Liberating Life: Contemporary Approaches to Ecological Theology.* New York: Orbis.

Brubaker, Pamela and Rogate Mshana, eds. 2010. *Justice Not Greed.* Geneva, Switzerland: World Council of Churches.

Gibson, William E., ed. 2004. *Eco-Justice: The Unfinished Journey.* New York: State University of New York Press.

Gourevitch, Alex. 2010. "Environmentalism—Long Live the Politics of Fear." *Public Culture* 22.3, 411–424.

Granberg-Michaelson, Wesley. 1991. "An Ethics for Sustainability." *Ecumenical Review* 42.1, 120–130.

——. 1994. "Towards a Theology of Life." *Reformed World* 44.3, 99–110.

Hall, Douglas John. 2002. "Creation." In *Dictionary of the Ecumenical Movement*, 2nd edn., ed. N. Lossky, J. Miguez Bonino, J. Pobee, T.F. Stransky, G. Wainwright and P. Webb, pp. 271–274. Geneva, Switzerland: World Council of Churches.

Haluza-DeLay, Randolph. 2012. "Making Peace with All Creation." *Peace Review* 24.2, 171–178.

Kerber, Guillermo. 2010. "Caring for Creation and Striving for Climate Justice: Implications for Mission and Spirituality." *International Review of Mission* 99.2, 219–229.

——. 2011. "Peace with the Earth in the Context of the Decade to Overcome Violence." *Ecumenical Review* 63.1, 35–43.

Lutheran World Federation and World Council of Churches. 2010. "Why Are the Churches at the UN Conference on Climate Change in Cancun?" Available at www.oikoumene.org/?id=8299 (accessed June 14, 2012).

Niles, D. Preman, ed. 1992. *Between the Flood and the Rainbow.* Geneva, Switzerland: WCC Publications.

Norgaard, Kari. 2011. *Living in Denial: Climate Change, Emotions, and Everyday Life.* Cambridge, MA: MIT Press.

Nurser, John. 2005. *For All Peoples and All Nations: Christian Churches and Human Rights.* Geneva, Switzerland: World Council of Churches.

Rasmussen, Larry. 1991. "Toward an Earth Charter." *Christian Century*, 108.30, 964–967.

Uppsala Interfaith Climate Manifesto. 2008. Uppsala: Church of Sweden. Available at http://onlinelibrary.wiley.com/doi/10.1111/j.1758-6623.2010.00061_4.x/abstract (accessed June 14, 2012).

White, Lynn, Jr. 1967. "The Historical Roots of Our Ecologic Crisis." *Science* 155.3767, 1203–1207.

World Council of Churches' UNCED Group. 1992. *Searching for the New Heavens and the New Earth: An Ecumenical Response to UNCED.* Geneva, Switzerland: World Council of Churches.

Conclusion

19 Climate change and religion as global phenomena

Summing up and directions for further research

Robin Globus Veldman, Andrew Szasz and Randolph Haluza-DeLay

How are the world's religions responding to climate change? The obvious answer is that they are responding in such a wide range of ways that it would be impossible to arrive at a single, simple answer. Nevertheless, in this chapter we attempt to discern some general patterns and draw some tentative conclusions about the various roles religions are playing in the global confrontation with climate change.

Before we do so, it is important to recall that there are serious gaps in this collection (gaps that mirror similar gaps in the rest of the current social scientific literature on this topic), in terms both of the religious traditions and of the geographic regions that are represented. These gaps force us to acknowledge any conclusions as preliminary and tentative. Still, we believe that some larger patterns have started to become visible.

Individual religions' engagement with climate change

We start our analysis by considering each religion (more properly, each "family" of faiths whose members are sufficiently similar to reasonably be considered as members of a single category) individually.

Four questions guide us in our meta-analytical assessment of each religion:

1. *What does the religion say about climate change? What are the reasons given for engagement (or for declining to engage)?* In other words, why is climate change now considered a *religious* matter by that tradition? Since almost no religion made climate change its concern just three decades ago, it has taken some sort of *discursive work* (of which theology is one form) within each tradition to interpret key texts and core beliefs so that interest in or caring about climate is seen as a legitimate subject. How did that discursive work get done and by whom? What factors facilitate or hinder that ongoing discursive work?
2. *What does the religion say its community should be doing or accomplishing? What do they claim they are doing?* For the most part, these are normative claims. As we described in Chapter 1, there is a growing literature making assertions about what particular religions should do in alignment with their theologies and other discursive representations.

3 *What they are actually doing or accomplishing? If there is a mismatch between 2 and 3, what are the causes?* In social behaviors it is not uncommon for there to be a gap, sometimes a substantial gap, between ideals and practice. Is there a difference between what a faith says it should do, what it wants to do, and what it is actually doing? If so, understanding the reasons for that gap can help us understand the forces that reduce effective action on climate change.

4 *What are the prospects for a more successful engagement?* While answering that question necessarily requires a degree of prediction or speculation, the existing trajectories within each particular tradition and the variability of engagement on climate change, coupled with general knowledge drawn from comparative analysis, can give some ideas of what to expect in the near future.

Before we begin the discussion of specific faiths, we must make some general points. What a religion says or does—or does not say or do—about climate change has to be understood in contexts—historical, economic, demographic, cultural, and political. The very way that we have chosen to organize the chapters in this book acknowledges this. Distinguishing between places in the so-called Global North versus the Global South is one way to highlight the long arm of history—especially the legacies of colonialism, industrialization, and global capitalism—that still determines so much of what happens in many of the earth's nations today. There are other ways to talk about roughly the same distinction: First World versus Third; (over)developed versus underdeveloped; core versus periphery. These are all shorthand for stark differences in affluence, political and administrative capacity, location vis-à-vis global economic systems, and other characteristics that influence the dynamics of the respective countries and their relations with other countries. The religions we discuss here all operate from within this larger context; their responses are therefore all profoundly affected by it. For example, we cannot understand the Church of Bangladesh's presence or poverty-alleviating mission in Bangladesh absent the history of British colonialism (see Chapter 5).

It is also important to note that "religion" (which is itself to some degree a Western academic construct [J. Smith 2004]) plays different roles in different societies. In some societies, religion has been, and in many cases still is, embedded in the institutions, patterns, and fabric of everyday life; no significant part of social life exists outside of it or independent of it. In contrast, in many advanced industrialized societies religion is one of several semi-autonomous social institutions, more or less formally separated from economics and politics. While still powerful culturally, religions in these societies are more compartmentalized, no longer as deeply embedded in every aspect of social life. The level of institutionalization also varies between societies and between the religions. Consequently, "religion" has different meanings and different social roles in different cultural contexts. In many situations, such as in research among indigenous peoples, it is therefore not meaningful to separate "culture" and "religion." Similarly, many expressions of Islam do not conceptualize themselves as "religious" institutions that are separate from "secular" ones, as is the case in Western modernity (especially with Western forms of Christianity) (Asad 2003). This means that some research may not have noted the specifically Muslim or Buddhist or other religious aspects of responses to climate change but rather subsumed them into their analyses of cultural, ethnic, or national responses (e.g., Salick and Byg 2007; Salick, Byg and Bauer 2012). It also underlines the importance of being self-critical about culturally specific frameworks that may underpin our analyses, as well as the need to be responsive to members' understandings (Spalek and Imtoual 2008).

Since societal context matters, we cannot arrive at one general, universally true, and accurate schema that in some linear fashion connects religion to the issue of climate change. This is the case not just for "religion" as a whole,[1] but even for any single religion—even one that appears, at first glance, to be monolithic. For example, even though there is only one Pope and his words presumably define the church's position, what the Roman Catholic Church does in the United States may bear little resemblance to what it does, or what meaning its actions have, in an African nation where some of the citizens have converted to Catholicism. Similarly, there is no reason to believe, without empirical investigation, that the role of Sunni Islam in Indonesia is similar to its role in the Ivory Coast. Instead, we should be attentive to the myriad of ways of being "Catholic" or "Sunni," and the resulting possibility that in different settings Catholics, or Sunnis, will engage with climate change in significantly different ways.

With that final caveat in mind, we begin our attempt at synthesis.

Christianity

We divide Christianity into three groups: mainline Protestantism, conservative Protestantism, and Roman Catholicism. This division reflects the typical set of "family resemblances" of the Global North. It does not account for Eastern Orthodox Christianity (approximately one-ninth of the 2.2 billion Christians worldwide),[2] nor does it recognize groups such as the African Independent Churches that have different historical, institutional, and theological origins. Christianity's rapid growth in the Global South makes it vital to attend to these versions of the historically Eurocentric faith.[3]

Mainline Protestant Christianity

The mainline Protestant category includes the major historical branches within post-Reformation Europe, such as Lutheran, Anglican, Presbyterian, Congregationalist, Wesleyan (i.e., Methodists in the U.S.), and some forms of Baptist. These churches tend to be more liberal in both theological orientation and the sociopolitical issues with which they engage, particularly when compared to evangelical Protestants. Mainline churches often focus on interfaith dialogues more than conversion, and participate in ecumenical organizations such as the World Council of Churches. The chapters that discuss mainline Protestants range widely in terms of geographic scale, from the level of transnational action (Chapter 17 by Brown, Brown and Shore; Chapter 18 by Kerber), to national (Chapter 5 by

Pender; Chapter 11 by Lysack; Chapter 13 by Townsend), to local or small (Chapter 7 by Reale; Chapter 15 by Nita). They vary, too, in terms of their location in the world system, from underdeveloped (Chapter 5 by Pender; Chapter 7 by Reale) to advanced industrial (Chapter 11 by Lysack; Chapter 13 by Townsend; Chapter 15 by Nita).

Generally, these faiths have experienced no significant difficulty acknowledging that climate change is real and that it is serious. They accept Western science's *explanation* of climate change, and they all agree that climate change is clearly a *religious* matter. In their case, then, the discursive work that had to be done to turn climate into an issue that is legitimately of concern to the faithful has not been particularly difficult. Ecologically oriented theologians and laypeople have readily found passages in the Bible that are amenable to green interpretation, textually supporting the claim that God cares for creation and calls on Christians to do so, as well. But that strain of argument, so central to the project of greening more conservative strains of Christianity has, here, a somewhat secondary importance. Mainline, liberal Christian denominations base the argument for the legitimacy of concern for the climate much more often on their churches' well-established and relatively noncontroversial commitment to social justice. The *process* of the discursive work in these faith communities is especially well described by Townsend (Chapter 13), Kerber (Chapter 18), and, on a micro level, Nita (Chapter 15).

Why has it been so easy? In the first place, these religions have little problem accepting the general orientation and specific findings of the Western natural sciences. Evolution, for example, is not a problem for mainline Protestants. Second, these faiths are generally liberal in political orientation, with a concomitant history of engagement on social and environmental issues that corresponds to the issues raised by climate change. More specifically, they are deeply committed to social justice activism, especially activism that addresses the problems of the world's poor. Since climate change is predicted to impact the world's poor first and worst, these faiths' commitment to the poor is a large part of the explanation for why they have been able to so readily define climate change as a legitimately religious matter. They have then made pronouncements about what ought to be done, by the faithful in their churches, and by the institutions and political authorities of their societies, as Pender (Chapter 5) and Lysack (Chapter 11) thoroughly describe.

Our contributors find, though, that things get considerably more problematic when the gaze turns from "what they say" to "what they do."

Mainline Protestants are trying to do a lot. Discerning the gravity of climate change, they develop internal policy at the denominational level. They try to influence policy at the level of the nation state. Some have used the power of investing to support development projects that increase resilience in underdeveloped nations. They travel to international treaty conferences, and participate in them, trying to influence the negotiations, as Kerber (Chapter 18) documents. As we learn from Pender (Chapter 5) and Reale (Chapter 7), they also work to enhance local capacity to adapt to climate change. Internally, these churches educate their own membership about climate change and, in addition to trying to generate support for the churches' "external" activities, they encourage their adherents to green their local church's building or grounds and they challenge their adherents to begin to individually change their consuming habits, to become more ethical consumers, and to lower their carbon footprints.

Based on the research collected here, though, their actual accomplishments have been, to date, far more modest. They seem to have been most successful at the local level, helping communities in underdeveloped nations improve their capacity to adapt, greening local churches, and greening behavior in more affluent societies. Still, in those affluent societies, such as the U.S. and Canada, several authors report that, even when the higher administrative bodies of the national church make it very clear that climate change is a vital issue, local ministers, pastors or priests have often been reluctant or unwilling to make climate a real priority (Agliardo tells the same story in the case of Catholicism in the U.S.). And, obviously, at the national and international levels, the political activities of the mainline Protestant churches, though committed and persistent, have done little to move the world toward effective international agreements. Still, religious traditions have certainly been among the more active sectors, apart from environmental organizations (which have not so far spawned a mass climate movement either). It is difficult to tell what will happen in the future, but it seems likely that these religious traditions will need to join forces with more powerful secular actors in order for a meaningful response to emerge (as both Lysack in Chapter 11 and Kerber in Chapter 18 suggest).

Conservative Protestant Christianity

Conservative Christians ("conservative" refers to theological perspective; in some countries this may coincide with political and social conservatism) have been the most studied religious tradition, especially in the United States, and Chapter 14 by Zaleha and Szasz adds to that body of research. Four other chapters all discussed evangelical Christian activities in the Global South (Chapter 7 by Reale; Chapter 8 by Hermesse; Chapter 9 by Chérif and Greenberg; Chapter 10 by Sarfo-Mensah and Awuah-Nyamekye). Chapter 16 by Johnson also discusses the missionary activity of a conservative Christian movement, though it is technically situated in the Global North.

Climate change has arrived at a time that many countries of the Global South are already reeling from the combined post-colonial experiences of poverty, population growth, economic underdevelopment, water and resource pressures, civil conflicts and dislocation, and slum urbanization. Into such circumstances, climate change brings changes in temperature and rainfall patterns that disrupt traditional agricultural practices even further, markedly worsening living conditions. Conservative Protestants have proselytized aggressively in former colonial nations and are rapidly growing in these regions of the world, as is Islam in regions of Africa and Asia.

Our contributors find that these new Christian religions are typically hostile toward the indigenous belief systems that bear encoded within them knowledge and practices that have for millennia guided local peoples' relationship with nature (e.g., Chapter 8 by Hermesse). At the same time, evangelicals and Pentecostals often interpret symptoms of climate change as signs of "the end times." In such cases we see no efforts to do the discursive work necessary to define climate as a legitimate concern for the faithful. Quite the contrary. Scientific discourses are not referenced, and the visible symptoms of climate change are subsumed into—and interpreted as confirmation of—the discourse of end-of-times prophecy. Jacka (2009) mentions the same dynamic in Papua New Guinea.

What do these faiths tell their adherents to do? Nothing that encourages them to take action on climate change, since personal salvation and evangelism are typically the highest priority in a world that may soon end. But doing nothing is not necessarily the same as having no impact. A combination of the interpretation of climate change, prophecy, and active denigration of traditional beliefs (which are already in a difficult position as a result of climate-induced changes) could hinder the development of effective preventative or adaptive efforts. On the other hand, in the Africa chapters at least, we have little sense of whether there have been efforts to educate the public about climate change or how such scientific narratives intersect with religious ones. We can say that, at present, missionaries are more involved in local communities than climate educators, and so they are in a better position than the latter to promote their interpretation of what is to come. It may be—but remains to be seen—that these conservative religious discourses and associated practices will thus thwart the sociopolitical engagement necessary to better prepare adherents for the challenges they are likely to face.

As Carr et al. (2012) describe, eschatology-based fatalism also has some currency among conservative Christians in the U.S. Yet the situation there is more complicated. Conservative Christians wield considerable political power, and the alliance of some parts of the conservative Christian establishment with militant climate deniers has contributed to policy stalemate on the issue, but there are also minorities within evangelicalism who argue that Christians should embrace "stewardship" and "creation care," which for some includes climate activism. These arguments are made out of a Christian concern for the poor and because the earth is God's creation, with a corresponding human duty of care. Still, although the greens were gaining ground in the first decade of the twenty-first century, more recently their "insurgency" seems to have been successfully contained (Kearns 2011).

It is not surprising, in this light, that several of our chapters (Chapter 11 by Lysack; Chapter 17 by Brown, Brown and Shore; Chapter 18 by Kerber) mention that conservative Christians are largely absent from ecumenical climate coalitions. Many conservative Christians—especially those in Pentecostal and fundamentalist denominations—are avoiding the issue altogether. Others are attempting to work on creation care issues without focusing explicitly on climate change. As Zaleha and Szasz (Chapter 14) point out, "creation care" and "stewardship" have become acceptable terminology, even among climate change-denying conservative Christians. Although at present most members of these faiths accede to politically conservative opposition to climate change, the presence of a passionate green minority means that we cannot rule out the possibility of significant change as climate conditions worsen and a new generation of younger, more politically moderate evangelicals comes of age.

Perhaps more than for most religious groups, it is not possible simply to pinpoint conservative Protestant responses to climate change. The anti-climate change evangelical orientation is, on the one hand, in countries like the United States influenced greatly by political and ideological alliances. On the other hand, in underdeveloped regions of the world we see little interest in the topic except insofar as weather changes seem to confirm prophecy. Yet this is not the entire story. For example, Sir John Houghton, lead scientist of the first three Intergovernmental Panel on Climate Change (IPCC) reports, is a prominent evangelical scientist, and major evangelical organizations like the World Evangelical Alliance and World Vision (according to both Chapter 11 by Lysack and Chapter 17 by Brown, Brown and Shore) are paying increased attention to the issue, as it affects their programs.

Roman Catholicism

With about 1.1 billion adherents scattered around the globe, the Roman Catholic Church is the largest Christian church on earth. Thus the research by Hermesse (Chapter 8) and Agliardo (Chapter 12) cannot speak to the many nations and regions where it is a substantial force, and the many places where it is present but not so significant. Agliardo's study covers the American Catholic church, and although the chapter is wide-ranging it is unclear to what extent American Catholicism is representative of other forms of Catholicism.

From popes to American bishops, the Roman Catholic Church describes climate change as a major moral issue. Some other national bishops' conferences also have climate change-related statements. Also, there has been a great deal of theological writing on climate change (Braun, Hellwig and Byrnes 2007). But implementation is very uneven and, as frequently pointed out, how statements translate into on-the-ground action (or why they do not) is an empirical question that needs social scientific investigation.

In American Catholic groups and in some dioceses, climate matters a great deal; in other dioceses and in many local parishes, it matters less. All in all, the situation sounds very similar to what we find among liberal, mainline Protestants: an excellent level of understanding, caring, and commitment at the level of formal rhetoric; and a modest and uneven amount at the level of concrete action. In addition, although not examined by Agliardo, there are a number of active Catholic denialist websites and organizations, mirroring the situation within American evangelicalism. This is not surprising, for the Roman Catholic Church is a wide tent with many institutional and ideological divisions within it. In this it reflects Wuthnow's (1996) finding that religious divides in America are typically now greater between conservatives and progressives within most denominations than among different denominations.

Agliardo identifies two different discursive frames of climate change—stewardship and eco-justice. Different actors and institutions appropriate different frames, although presumably both orientations should lead to action. Catholic climate change educational material is usually deliberately tied to papal statements and Catholic Social Teaching, which means that there should be some portability to other nations. Indeed, Agliardo does mention, at times, transnational networks within the global Catholic community, such as the environmental advocacy network set up by the Jesuits or the international Catholic health community. This could be useful, for Catholics are active in many social sectors. For example, they have extensive health care networks and social service agencies and run numerous educational institutions. It would be a useful empirical exercise to examine the Catholic curriculum for climate change-related material, especially since both Pope Benedict XVI and Pope John Paul II actively promoted Catholic engagement with climate change and environmental issues in general. On the other hand, Agliardo explains that the Roman Catholic Church is not as hierarchical as it is generally perceived to be, with considerable autonomy among its various sections, which means it will not operate as a single unit on an issue like climate change.

In the United States, there is an active Catholic climate coalition supported by the U.S. bishops conference. In contrast, Lysack (Chapter 11) reports that Canadian Catholics have been nearly silent on climate change recently. While many Catholic organizations endorsed the *Canadian Interfaith Call for Leadership and Action on Climate Change*, no bishop from the Canadian Conference of Catholic Bishops (CCCB) agreed to sign. In other interfaith actions on the issue, Catholics were also absent. Of note is that, while both the U.S. and the Canadian bishops' conferences have issued statements about the environment, only the U.S. bishops have issued a statement on climate change. However, the U.S. document was produced in 2001, so how this may relate to the increasingly contested and politicized nature of climate change in North America is unclear.

In Latin America, Roman Catholicism exists alongside indigenous religions and new and burgeoning evangelicalism. In Chapter 8 by Hermesse we learn that a sort of syncretism has formed between the Mayan and Catholic faiths (although not with official church approval). Hermesse argues that indigenous spirituality is more respectful of the natural world, in contrast to the Pentecostal Christians in the region, with Catholics placed somewhere in between. Still, in Guatemala, Catholics pay relatively little attention to climate change and they often connect it to divine action (although not divine retribution or end-times theology as do the Pentecostals).

Worldwide, Caritas Internationalis, the Vatican's international development agency, places climate change as one of its top issues[4]—Reale (Chapter 7) briefly notes Caritas's involvement in the Solomon Islands, and it has an active climate refugee project. This review makes clear that many more social scientific assessments, over a broader range of contexts, are needed to understand the Roman Catholic engagement with climate change.

Islam

With about one and a half billion adherents, Islam is the second-largest religion on earth, and it is growing rapidly. It is currently concentrated mostly in South Asia,

Southeast Asia, the Middle East, North Africa, and sub-Saharan Africa, with considerable populations in Europe and North America due to immigration. Regrettably, we are missing research from many nations where Islam is the major religion, and only one of the chapters (Chapter 6 by Amri) explores Islam's role in great detail. Several others mention Islamic responses in the context of climate activism (Chapter 15 by Nita), interfaith engagement (Chapter 11 by Lysack; Chapter 18 by Kerber), and comparative research (Chapter 9 by Chérif and Greenberg; Chapter 10 by Sarfo-Mensah and Awuah-Nyamekye). What we can conclude from these few chapters is that Islam is certainly capable of promoting activism, but can also work against concern in certain contexts.

Chapter 6 by Amri provides excellent detail on the discursive work that Indonesian Muslims have done to find a green message in the Qur'an and Hadiths, in the nation that has the largest Muslim population in the world. Amri also provides a good description of the way that self-described "eco-jihadists" there have sought to move beyond words to deeds. It would be of great interest to find out if similar processes are playing out in other places where Islam is dominant or, if not, why not.

Islam also appears in both of the chapters on Africa. In both Ghana and the Ivory Coast a substantial part of the population are followers of Islam—16 percent in Ghana and 37 percent in the Ivory Coast. However, Islam is not a major focus in either of the chapters, both of which focus more on the role played by indigenous faiths and Christianity. To the degree that the narratives do address the role of Islam, the chapters show some interesting differences. According to Chérif and Greenberg (Chapter 9), Islam in the Ivory Coast has certain similarities to indigenous beliefs and practices, much more so than Christianity does, with the greater compatibility making it more possible to forge a certain syncretism between Islam and traditional beliefs. In Chapter 10 by Sarfo-Mensah and Awuah-Nyamekye, Islam in Ghana is depicted as critical of and hostile toward the older, local faith's traditional beliefs and taboos. Similarly to the role other chapters described for Christianity, Islam is depicted here as undermining the traditional knowledge practices embedded in indigenous beliefs.

Finally, combing other chapters for potentially relevant information, we can observe that in Canada Muslims did not have a public statement about climate change, but did participate in an ecumenical climate coalition (Chapter 11 by Lysack). They also participated in some climate-related events organized primarily by the World Council of Churches at the annual UNFCCC Conferences of the Parties (Chapter 18 by Kerber). That Islam does not have the same sort of institutional forms as does Christianity with its denominations and ecumenical forums likely affects the way this religion can respond on a large-scale or institutional level. Amri (Chapter 6) describes the important role of Muslim civil society organizations, which would be more comparable to that of aid organizations like World Vision that Brown, Brown and Shore (Chapter 17) describe. In Great Britain Muslims were present in climate action only in tiny numbers (Chapter 15 by Nita), which Nita attributes in part to their faith identity in British society, which was difficult to merge with an environmental activist identity.

Studies from Bangladesh and Morocco were among the research that was proposed but not able to be completed in time for this collection, so we may well see more social science on Muslim responses to climate change forthcoming. Another reason scholarship in this area may be sparse is because Islam does not conceptualize itself as a separate *religious* institution in the same way as religions in Western modernity are understood to be. This means that some research may not have noted the specifically Muslim aspects of responses to climate change but subsumed them as ethnic or national responses. However, in the broader religion and environment literature, there is growing scholarship on Islamic environmental ethics, which may mean that more attention will be devoted to the social science of Islam and the environment or climate change (Finnegan 2011). This would be a positive development, since many Muslim or predominantly Muslim countries will be experiencing greater climatic changes than other parts of the globe.

Indigenous religions

While indigenous religions do share some characteristics in common, we acknowledge the glaring ethnocentricity of lumping them together, and repeat our caveat that generalizations based on a small number of studies are bound to be premature.[5]

The chapters by Manandhar, Schmidt-Vogt, Pandey and Kazama (Chapter 3), Hermesse (Chapter 8), Chérif and Greenberg (Chapter 9), Sarfo-Mensah and Awuah-Nyamekye (Chapter 10), and Johnson (Chapter 16) directly discuss the role of indigenous religion. In all five, factors beyond religion play an important role in determining local dynamics. Specifically, they show that the older, indigenous faith must be understood in the broader, post-colonial situation in these societies. This includes the massive recent and ongoing disruptions, including but not limited to population outpacing economic development, in-migration from other nations, mass migration from countryside to metastasizing urban megaslums, increasing replacement of older economic relations by the market-driven logic of modern economics, the ever-growing influence of the culture of the market, and/or a generalized alienation from "old ways." In addition, these studies (except Chapter 3 by Manandhar, Schmidt-Vogt, Pandey and Kazama) all describe aggressive proselytizing by Islam, conservative Christianity, or both. In short, these chapters describe a society–nature relationship under considerable stress from the combined impacts of all these facets of "modernization."[6] Davidson (2012) found similar stressors operating in Guinea-Bissau.

Within this context, these chapters describe indigenous religions as having encoded within their beliefs, rituals, and approved social and economic practices ways of being that were, at an earlier time, something we could reasonably describe as "sustainable." Although there is of course a wide variety of indigenous perspectives on the relationship of humans and nature, and on the source, role, and efficacy of locally based knowledge, numerous anthropological studies have found similar relationships between indigenous religions and the environments in which people lived.[7] Yet, while as social scientists we caution against idealized notions of indigenous peoples' environmental behavior, we also acknowledge the considerable body of literature showing that indigenous religious and cultural beliefs support dynamic but sustainable relationships with the land. The research in this book echoes these findings, while also noting that indigenous religions' ecological efficacy has been waning as a result of the combined influence of (a) the general "modernization" of social conditions and (b) rising adherence to faiths which are actively hostile to older, indigenous faiths, including their ecologically relevant beliefs. It is in that distressed context that climate change then arrives as a new stressor, causing changes in temperature and changes in rainfall patterns severe enough to threaten rural livelihoods, which in primarily agricultural societies inevitably threaten social life in general.

How have the various indigenous religions responded? The answers we find in these chapters are sobering, though there are important differences between them. In the African chapters, adherents of the indigenous faiths interpreted the signs of climate change as evidence of moral failure—of the new Christians, the urbanites who have abandoned old ways, or the young and faithless. Thus it appeared that these people did not perceive the threat of climate change in ways that would increase their capacity to adapt. Instead, their representation of causes seems most likely to disable any appropriate response. As Roncoli, Crane and Orlove note, similar attributions of weather and climate changes to "violation of religious, moral and social norms" (2009: 97) have been found in many other communities. Johnson's Inuit informants (Chapter 16) also interpreted the environmental changes they were seeing as a response to human moral behavior, but in this case they viewed it positively, as an indication of divine blessing. They were encouraged to do so by a new, evangelical movement that accorded well with existing Inuit interpretations of the interrelationships of land, human, and spirit (see also Leduc 2011). These views did not, however, accord with climate science or promote a preventative response.

On a slightly more hopeful note, Manandhar, Schmidt-Vogt, Pandey and Kazama (Chapter 3) refer to indigenous knowledge in terms of both Buddhism and the Bon religion that predated Buddhism in the Thini region of Nepal. They conclude that traditional beliefs can mislead people as they try to make sense and act in the face of climate-related changes, but also that these belief systems have built-in flexibility that may still serve. In Hermesse's case study (Chapter 8), the shamans she interviewed seemed well aware of the anthropogenic cause of climate change, and they advocated a reasonable response—taking better care of nature—while still expressing the view that extreme weather events were due to human moral decline. Unfortunately, she cautions that given endemic poverty in the region they were unable to promote better ecological practices.

While it seems that indigenous religious beliefs encourage a variety of interpretations of climate change, we do not know for sure what forces lead to the divergent interpretations. One possibility is that access to international media coverage of climate change plays a role. As Salick and Byg suggest in their study of indigenous peoples' engagement with climate change, "In some places, media and their coverage of climate change issues dominate [indigenous] people's understandings of climate changes. . . . In contrast, where media play a limited role, interpretations are more closely dependent on people's own observations and local cultural framework" (2007: 17–18)—leading especially to interpreting adverse or extreme weather as punishment for moral failure, as in the Africa case studies. It would seem, then, that the media can play a beneficial role when it comes to encouraging a response to climate change.

As Manandhar, Schmidt-Vogt, Pandey and Kazama (Chapter 3) point out, it may be that "religion-based indigenous knowledge and practices threaten to become inadequate in the face of current climatic variability." On the other hand, it may be that something more helpful may yet emerge—several researchers argue, based on historical evidence, that these local, indigenous faiths possess a certain flexibility that may yet help them lead their people to a more appropriate and adaptive understanding of and response to climate change. A delicate combination of education about the science of climate change and sensitivity toward local cultural mores—coupled with material resources and enlisting the social capital that Reale (Chapter 7) observes in the Pacific Islands—seems essential if indigenous religious views are to contribute to a proactive response to climate change. Yet, given how little these groups contribute to the problem, we must remember that action in the international context is equally—indeed arguably more—important.

Buddhism

Two chapters offer observations about coping with climate change in societies where Buddhism is the main faith: Manandhar, Schmidt-Vogt, Pandey and Kazama (Chapter 3) on the situation in one part of Nepal, and Branch (Chapter 4) on Bhutan. In the Nepalese case, climate change is, over time, compromising the value of the traditional knowledge and traditional ritual practices encoded in Buddhist beliefs and interpreted by Buddhist lamas. On the other hand, these beliefs are depicted as flexible enough in actual application for the villagers to be able still to believe in old ways while, in actual behavior, they are, in fact, adapting to new conditions. In Bhutan Buddhism has been reinterpreted to the point that it has become the basis for the governing philosophy known as gross national happiness (GNH), and GNH is currently in the process of being discursively expanded to address some of the impacts of climate change (such as glacial lake outburst flooding).

These studies correspond to other research on Buddhism and climate change. Like Branch, Daniels (2010) positions happiness as a key Buddhist principle that can improve ecological economics, and a team led by Jan Salick has conducted long-term research on Tibetan Buddhist practices and cosmology vis-à-vis climate change in Tibet and Yunnan, China (Salick and Byg 2007; Salick, Byg and Bauer 2012). Buddhism does not have central authorities, although respected figures like the Dalai Lama have actively expressed views on climate change. Along with the studies in this volume, such research suggests that Buddhism is characterized by a certain flexibility, which could make it useful as societies where it is predominant adapt to a changing climate. However, with such a small sample it is too early to make any definitive conclusions.

Hinduism and other religions

If one thinks it hazardous to generalize on the basis of two cases, generalizing on the basis of just one is simply impossible. And, unfortunately, we are down to a single study of one of the world's great religions, Hinduism (Chapter 2 by Drew). That study doesn't tell an optimistic story. The Ganga River's flow has decreased; everyone sees and acknowledges that. But even the best-educated segments of the population—those who acknowledge having learned about scientists' explanations of climate change—cite Hindu texts and mythologies when asked to explain why the river is changing, beliefs that lead the author to observe "apparent lack of concern for the Ganga's future condition." Further studies will either confirm these somewhat dispiriting findings or, hopefully, offer the opportunity to come to more sanguine conclusions.

Research on the planet's other religions is very incomplete. We do not have any research on Jewish, Sikh, or Bahai'i religions. We also wish we had information on value construction and religion in China, now the largest carbon emitter. Over 100 million people belong to faith groups in China, although they represent only about 7 percent of the population. Chinese indigenous religions account for a large number, and other Chinese approaches, like Confucianism or the general "harmony" orientation in Chinese philosophy (Cao 2012), may be considered religion-like. Importantly, any and all of these faiths will likely have different effects in contexts where they are majority or minority, culturally homogeneous or not, diasporic or historically rooted, and myriad other contextualizations.

What factors help religions in general engage with climate change?

Another mode of analysis looks *across* religious traditions, examining what aspects of religions (worldviews; reach and moral authority; institutional and economic resources; and social capital) have influenced the response to climate change. In Chapter 1 (in the section "Religions' potential to motivate concern and activism") we listed four reasons that religions are thought to have significant potential to aid in the global response to climate change. It is now time to consider these reasons in light of the chapters in this volume. We do so tentatively, aware that the limited number of case studies allows only preliminary conclusions to be drawn.

The first argument is that religions may be able to motivate a response to climate change via their influence on believers' worldviews or cosmologies, which in turn could motivate appropriate, climate-responsive behavior. Much of the effort to green various religious traditions has evolved from this rationale, and we can see in Chapter 5 by Pender, Chapter 6 by Amri, Chapter 11 by Lysack, Chapter 12 by Agliardo, Chapter 13 by Townsend, and Chapter 18 by Kerber that this has frequently been a first step on the way to climate activism, some of it with significant on-the-ground impacts. However, as noted above, in some places it has been a challenge to change the frameworks of those not already convinced climate change is an important issue, reminding us that religious worldviews sometimes struggle to overcome other cultural and political influences. In Chapter 15 by Nita we see real passion among religious climate activists as they work to merge their religious views with concern for the planet, yet she notes how their religious backgrounds also sometimes hinder their participation, especially if they wish to continue practicing their tradition. So we are reminded that in practice it is not necessarily easy to knit together religious worldviews and concern over climate change. Worldviews may also have been a significant impediment to other religious participation in climate activism. We have noted the eschatological impacts; Curry (2008) identifies understandings of human capacity and God's sovereignty (for example, that humans are too puny to impact the entire planet, or that only the Creator really controls the earth and its environment) as significant influences on acceptance of climate change science for American evangelicals. In the case of indigenous religions, cosmologies have a particularly ambiguous influence. The tendency to interpret environmental disasters and anomalies as resulting from human moral failure can promote environmentally beneficial behavior under the right circumstances, but in the case of climate change we have seen that it can also lead adherents astray. Similarly, traditional knowledges and their cosmological properties may be at odds with scientific knowledge and its foundations; it is sometimes difficult to accept *both* indigenous knowledge and science.

The second argument concerns the many people that the world's religions are able to reach, and the moral authority they wield with those audiences. Religious emphasis on climate justice, for example, appears to be contributing to a growing movement that facilitates interfaith collaboration (Chapter 11 by Lysack; Chapter 12 by Agliardo; Chapter 13 by Townsend; Chapter 18 by Kerber).[8] Religious orientations provide a counterpoint to economistic, laissez-faire ways of thinking (Chapter 4 by Branch; Chapter 12 by Agliardo), although a specific religious critique of the growth imperative of global capital or industrialization as among the root causes of climate change was mostly implicit rather than explicit. An important caveat these chapters suggest is that, while most of the world's people are indeed religious, they do not all belong to religions that encourage concern or action, and some of them belong to groups that discourage action. For example, Pentecostal Christianity, which several chapters depicted as discouraging concern, is among the fastest-growing religious groups in the world, especially in regions where people are experiencing some of the early effects of climate change (they currently constitute 12.8 percent of the world's Christian population, or 4 percent of the world population[9]). Indeed, these early effects may actually be aiding their evangelism. In addition, numerous chapters indicate that religions' moral authority is weaker than is sometimes supposed, especially in religions with a diffuse institutional and teaching structure. It is also not clear whether religion actually engenders concern or is merely able to capitalize on concern deriving mainly from other (secular) sources. In chapters such as Chapter 4 by Branch and Chapter 6 by Amri we seem to see evidence of the former, while in Chapter 13 by Townsend and perhaps Chapter 11 by Lysack we may be seeing evidence of the latter. Establishing whether there is anything unique and/or uniquely powerful about religions' participation—or whether their efficacy relies on the same strategies we see in the secular world—is an important theoretical point that will be of particular interest to scholars of religion (e.g., McCutcheon 1997). Finally, we have very little information about several of the world's largest religions—Islam, Hinduism, and Buddhism, which together account for over 3 billion people. Therefore, the extent to which religions can, do, or will use this aspect of their potential positively to address climate change remains very much an open question.

The third argument, concerning religions' institutional and economic resources, may be the most significant—or at least it is the one for which we currently have the greatest evidence. Chapter 5 by Pender, Chapter 6 by Amri, Chapter 7 by Reale, Chapter 11 by Lysack, Chapter 12 by Agliardo, Chapter 13 by Townsend, Chapter 18 by Kerber, and to some extent Chapter 4 by Branch and Chapter 17 by Brown, Brown and Shore all show the difference these resources can make when applied to climate change. Some of the resources have effects in terms of providing adaptation strategies, education about climate change, access to international advocacy networks, and leadership. Better access to transnational networks is another resource that was useful in some studies, especially for people in under-resourced societies (e.g., Chapter 5 by Pender; Chapter 7 by Reale). For at least some traditions that emphasize service (especially to the poor and vulnerable), climate change adaptation or advocacy thus seems to be a natural outlet. It would appear that more institutionalized faiths (such as Christian denominations or Muslim organizations) are better positioned to be useful than faiths with lower degrees of organization or centralization (such as Hinduism and most indigenous faiths). However, the same resources can be utilized contra climate change, as Hermesse (Chapter 8) found in Guatemala, and Zaleha and Szasz (Chapter 14) discerned among evangelicals in the United States. Similarly, Agliardo (Chapter 12) found that the structure of the American Roman Catholic church shaped the discourses that institutional actors used to discuss and act on climate change. Institutions are generally slow to change. In addition, religious organizations have more interests or concerns than just climate change, which both Amri (Chapter 6) and Agliardo (Chapter 12) point out from Muslim and Roman Catholic perspectives. Their resources will be used for many needs which could make them appear less concerned about climate change. None of the studies assessed actual political, economic, or financial data for religious groups.

The fourth argument highlights one of religion's most significant resources—its potential to provide social connectivity and collective action. Scholars of social movements are among those who have identified religious resources as often being significant in social change (e.g., C. Smith 1996; Young 2002). Religious groups not only are a significant portion of civil society, but religion generates—at least internally and in some cases external to the faith community—norms of cooperation and durable networks of meaningful affiliation. This social capital can facilitate the deployment of institutional capital. Moreover, especially in homogeneous groups, a common faith can be an important form of social cohesion. This is a significant point made by Reale (Chapter 7) in her study of Solomon Islands, which found that church networks facilitated disaster recovery. Reale notes that churches in Solomon Islands were well integrated culturally, being well respected and closely involved with the day-to-day running of villages. But it is worth recalling a caveat she mentions: that communities with a single church fared better with disaster recovery than those with numerous competing churches. This suggests that competition between denominations or faiths could undermine their potential to facilitate collective action on climate change.

Several other chapters also identified how institutional resources and discourses of climate flowed along with social capital, such as the analysis by Amri (Chapter 6) of the relatively informal system of *pengajians*. The research on this potential for religion's contribution to addressing climate change was clearest in the case studies of localized situations (e.g., Chapter 3 by Manandhar, Schmidt-Vogt, Pandey and Kazama; Chapter 9 by Chérif and Greenberg; Chapter 16 by Johnson). Nevertheless, we have to admit that relatively little attention specified how social capital influenced religions' engagement with climate change. Though we can guess that it would be especially significant in disaster response and climate change adaptation efforts, more research needs to be done in this area to assess its impact. Similarly, we do not understand the conditions by which religions may facilitate collective action on this issue.

As for impediments to engagement on climate change, we confront four situations in these chapters. We have religions that are not doing or saying much at all, or not yet—Buddhism in Nepal; Roman Catholicism, Judaism, and Islam in Canada; Hinduism in India; and indigenous religions in the Ivory Coast and Ghana. Then we have religions or religious groups that are engaging unevenly or just beginning to engage—Buddhism in Bhutan; Roman Catholicism in the U.S.; and World Vision, internationally. Third, we have religions that are tacking in the opposite direction—chiefly Pentecostalism in the Ivory Coast, Ghana, and Guatemala, and evangelical Christianity in the U.S. Finally, we have religions that are trying hard to make something happen, to varying degrees of success—the Church of Bangladesh, the Muslim organizations Nadhlatul Ulama and Muhammadiyah in Indonesia, the World Council of Churches, the Presbyterian Church, U.S.A., Christians and Muslims in Britain, and Christian, Muslim, Hindu, Baha'i, and Buddhist signatories of the *Canadian Interfaith Call for Leadership and Action on Climate Change*. Lack of information is almost certainly a factor preventing action in some locales, especially in the Global South. In the other cases the reason for lack of action is harder to diagnose, much less attempt to remedy. More studies like those by Lysack (Chapter 11) and Kerber (Chapter 18), which identify barriers specific to the context in which the religion is working, and identify possible solutions, are needed.

What does all this mean? Certainly there is tremendous diversity in "religion" and "religious responses" to climate change. Even Roman Catholics in one country had different approaches to the issue of climate change, much less all over the world. This volume should eliminate the notion that religion offers a unified category for analysis. Instead, as noted at the beginning of this chapter, context matters. So does scale, as what happens in local circumstances may not reflect the stance at a larger, say denominational, scale. This is not the same as saying that context is all-important, for there may be emergent properties as one transcends along a continuum of scale from individual to faith community to organization to transnational tradition. In fact, just as climate change is a global phenomenon with numerous local manifestations, so too are many of the world's religious traditions, which are actively global while also having a tendency to hybridize with local traditions. This global–local interaction—or what has come to be known as "glocalization" (Robertson 1992)—means that one must attend to multiple scales simultaneously when dealing with both phenomena. Perhaps most importantly, we still need more research into the grounding conditions for acceptance or denial of climate change, and the resulting willingness or unwillingness to devote religious resources to addressing it.

Final thoughts

As we have repeatedly cautioned, any conclusions based on the kind of collection of studies contained in this volume have to be considered, at best, preliminary and provisional. But that is generally the situation at the fledgling stages of any study of complex phenomena. We are, necessarily, at those early stages. Only relatively recently have the world's religions made the effort to seriously engage with environmental issues, and social scientists have paid even scanter attention to this engagement. On the one hand, this may be surprising, since religious worldviews and institutions are such significant parts of life and human societies all around the globe. On the other hand, this relatively scant attention may be understandable, since modern environmentalism is scarcely five decades old. And it has been more recently still that climate change was identified, first as one more environmental issue, and then as not just another issue but the largest and most dire of environmental threats. So first came the science; then came the political organizing that turned scientific findings into an *issue*; then various faiths become interested enough to engage, to do the discursive work to articulate what believers should think and what they should do. Only then did the conditions—the raw materials—come into existence for social science scholarship.

In the end, while the works collected herein are substantial enough to warrant some initial conclusions—as we have attempted to offer in this final chapter—the most obvious conclusion is that much remains to be done. Fortunately (for scholarship), but certainly unfortunately (for humanity and for the world's ecosystems), the future holds plenty of opportunity for more good studies. As we write, early in 2013, the world's nations, collectively and individually, are very far from having agreed on and implemented the kinds of reforms that are needed to deal with this impending crisis. It does appear, then, that conditions will continue to deteriorate. As they do so, there will be much greater human distress and suffering, and religions will increasingly feel the need to respond. Hopefully that will be in helpful ways, but, as we have seen here, we cannot expect this to be always the case. In either case, paying close attention to the evolving relationship between religions and climate change will become increasingly vital.

Notes

1 It is particularly important to note that different faiths do not serve as functional alternatives, that they occupy different societal positions in most societies—especially in complex multi-faith, multi-ethnic, and multi-cultural societies—and that they may operate very differently. For example, many religions do not place as much emphasis on propositional exposition or doctrinal fidelity as does Christianity.
2 http://www.pewforum.org/Christian/Global-Christianity-orthodox.aspx.
3 For example, Christianity in Oceania has rethought the Hebrew attachment to the land in terms more appropriate for oceanic spirituality, with implications for the faith in the context of climate change (Rubow 2009).
4 Caritas's main page on the issue is http://www.caritas.org/activities/climate_change/its_the_climate_stupid.html.
5 What counts as an "indigenous" religion is often conceptually unclear, especially in terms of place-based ecological knowledge. For example, Christianity is rarely so identified, yet Gómez-Baggethun and colleagues (2012) consider Catholic ritual practices as traditional ecological knowledge in a study located in the southern part of Spain. Falk (2010) makes no distinction between Buddhist and local knowledge in a study of responses to tsunami, nor do Salick, Byg and Bauer (2012).
6 We acknowledge that there are "multiple modernities" rather than any single narrative (Asad 2003; Hefner 1998; Schmidt 2006).
7 The same outcomes for environmental management can have widely different foundations, as Milton (1996) explains. Berkes (2012) is a good place to start on the literature on indigenous knowledge and cosmology, religion, or spirituality.
8 Another long-running partnership is the Alliance on Religions and Conservation (www.arcworld.org) in which 11 faiths have worked with the World Wildlife Federation and United Nations. There are also numerous local, regional, and national examples, such as Southern African Faith Communities' Environment Institute (www.safcei.org), whose website includes pages for Buddhist, Christian, Muslim, Hindu, and Bahai'i and lists the Indigenous Peoples of Africa Coordinating Committee as a partner.
9 http://www.pewforum.org/Christian/Global-Christianity-movements-and-denominations.aspx (accessed February 10, 2013).

References

Asad, Talal. 2003. *Formations of the Secular: Christianity, Islam, Modernity*. Stanford, CA: Stanford University Press.

Berkes, Fikret. 2012. *Sacred Ecology*, 3rd edn. Philadelphia, PA: Routledge.

Braun, Gerald W., Monika K. Hellwig and W. Malcolm Byrnes. 2007. "Global Climate Change and Catholic Responsibility: Facts and Faith Response." *Journal of Catholic Social Thought* 4, 373–401.

Cao, Shixiong. 2012. "Socioeconomic Value of Religion and the Impacts of Ideological Change in China." *Economic Modeling* 29, 2621–2626.

Carr, Wylie, Michael Patterson, Laurie Yung and Daniel Spencer. 2012. "The Faithful Skeptics: Evangelical Religious Beliefs and Perceptions of Climate Change." *Journal for the Study of Religion, Nature and Culture* 6.3, 276–299.

Curry, Janel M. 2008. "Christians and Climate Change: A Social Framework of Analysis." *Perspectives on Science and Christian Faith* 60.3, 156–164.

Daniels, Peter L. 2010. "Climate Change, Economics and Buddhism—Part 2: New Views and Practices for Sustainable World Economies." *Ecological Economics* 69, 962–972.

Davidson, Joanna. 2012. "Of Rice and Men: Climate Change, Religion, and Personhood among the Diola of Guinea-Bissau." *Journal for the Study of Religion, Nature and Culture* 6.3, 363–381.

Falk, Monica Lindberg. 2010. "Recovery and Buddhist Practices in the Aftermath of the Tsunami in Southern Thailand." *Religion* 40.2, 96–103.

Finnegan, Eleanor Daly. 2011. "What Traditions Are Represented in Religion and Ecology?" In *Inherited Land: The Changing Grounds of Religion and Ecology*, ed. Whitney A. Bauman, Richard R. Bohannon II and Kevin J. O'Brien, pp. 64–79. Eugene, OR: Pickwick Publications.

Gómez-Baggethun, Erik, Victoria Reyes-García, Per Olsson and Carlos Montes. 2012. "Traditional Ecological Knowledge and Community Resilience to Environmental Extremes: A Case Study in Doñana, SW Spain." *Global Environmental Change* 22.3, 640–650.

Hefner, Robert W. 1998. "Multiple Modernities: Christianity, Islam, and Hinduism in a Globalizing Age." *Annual Review of Anthropology* 27.1, 83–104.

Jacka, Jerry. 2009. "Global Averages, Local Extremes: The Subtleties and Complexities of Climate Change in Papua New Guinea." In *Anthropology and Climate Change: From Encounters to Actions*, ed. Susan A. Crate and Mark Nuttall, pp. 197–208. Walnut Creek, CA: Left Coast Press.

Kearns, Laurel. 2011. "The Role of Religions in Activism." In *The Oxford Handbook of Climate Change and Society*, ed. John S. Dryzek, Richard B. Norgaard and David Schlosberg, pp. 414–428. Oxford, U.K.: Oxford University Press.

Leduc, Timothy. 2011. *Climate, Culture, Change: Inuit and Western Dialogues with a Warming North*. Ottawa, ON: Ottawa University Press.

McCutcheon, Russell T. 1997. *Manufacturing Religion: The Discourse on Sui Generis Religion and the Politics of Nostalgia*. Oxford, U.K.: Oxford University Press.

Milton, Kay. 1996. *Environmentalism and Cultural Theory*. London: Routledge.

Robertson, Roland. 1992. *Globalization: Social Theory and Global Culture*. London: Sage.

Roncoli, Carla, Todd Crane and Ben Orlove. 2009. "Fielding Climate Change in Cultural Anthropology." In *Anthropology and Climate Change: From Encounters to Actions*, ed. Susan A. Crate and Mark Nuttall, pp. 87–115. Walnut Creek, CA: Left Coast Press.

Rubow, Cecilie. 2009. "Metaphysical Aspects of Resilience: South Pacific Responses to Climate Change." In *The Question of Resilience: Social Responses to Climate Change*, ed. Kirsten Hastrup. Copenhagen: Royal Academy of Sciences and Letters.

Salick, Jan and Anya Byg. 2007. *Indigenous Peoples and Climate Change*. Oxford, U.K.: Tyndall Centre for Climate Change Research.

Salick, Jan, Anya Byg and Kenneth Bauer. 2012. "Contemporary Tibetan Cosmology of Climate Change." *Journal for the Study of Religion, Nature and Culture* 6.4, 447–476.

Schmidt, Volker H. 2006. "Multiple Modernities or Varieties of Modernity?" *Current Sociology* 54.1, 77–97.

Smith, Christian, ed. 1996. *Disruptive Religion: The Force of Faith in Social-Movement Activism*. New York: Routledge.

Smith, Jonathan Z. 2004. *Relating Religion: Essays in the Study of Religion*. Chicago, IL: University of Chicago Press.

Spalek, Basia and Alia Imtoual, eds. 2008. *Religion, Spirituality and the Social Sciences: Challenging Marginalisation*. Bristol, U.K.: Policy Press.

Wuthnow, Robert. 1996. "Restructuring of American Religion: Further Evidence." *Sociological Inquiry* 66.3, 303–329.

Young, Michael, P. 2002. "Confessional Births: The Religious Roots of U.S. Social Movements." *American Sociological Review* 67.5, 660–688.

Index